D0948622

THE RELATIONSHIP BETWEEN
NUMERICAL COMPUTATION AND
PROGRAMMING LANGUAGES

IFIP TC 2 Working Conference on
The Relationship between Numerical Computation
and Programming Languages
Boulder, Colorado, U.S.A., 3-7 August, 1981

organized by
IFIP Working Group 2.5 *(Mathematical Software)*
on behalf of
IFIP Technical Committee 2 *(Programming)*
International Federation for Information Processing

programme committee
C. L. Lawson and J. K. Reid *(Co-Chairmen)*
Th. J. Dekker, B. Ford, C. W. Gear, C. A. R. Hoare,
T. E. Hull, B. T. Smith, S. G. van der Meulen, W. Waite

NORTH-HOLLAND PUBLISHING COMPANY
AMSTERDAM • NEW YORK • OXFORD

THE RELATIONSHIP BETWEEN NUMERICAL COMPUTATION AND PROGRAMMING LANGUAGES

Proceedings of the IFIP TC 2 Working Conference on
The Relationship between Numerical Computation
and Programming Languages
Boulder, Colorado, U.S.A., 3-7 August, 1981

edited by

John K. REID

Computer Science and Systems Division
Atomic Energy Research Establishment, Harwell
Didcot, Oxfordshire
U.K.

1982

NORTH-HOLLAND PUBLISHING COMPANY
AMSTERDAM ● NEW YORK ● OXFORD

ISBN: 0 444 86377 X

Publishing by:
NORTH-HOLLAND PUBLISHING COMPANY-AMSTERDAM ● NEW YORK ● OXFORD

Sole distributors for the U.S.A. and Canada:
ELSEVIER SCIENCE PUBLISHING COMPANY, INC.
52 Vanderbilt Avenue
New York, N.Y. 10017

Library of Congress Cataloging in Publication Data

IFIP TC2 Working Conference on the Relationship
    between Numerical Computation and Programming
    Languages (1981 : Boulder, Colo.)
    The relationship between numerical computation
and programming languages.

    "Organized by IFIP Working Group 2.5 (Mathe-
matical Software) on behalf of IFIP Technical
Committee 2 (Programming), International Federation
for Information Processing"--P.
    1. Numerical analysis--Data processing. 2. Pro-
gramming languages (Electronic computers) I. Reid,
John Ker. II. IFIP Working Group 2.5--Mathematical

Software. III. IFIP Technical Committee 2--
Programming. IV. Title.
QA297.I34 1981   519.7   82-2203
ISBN 0-444-86377-X (Elsevier)   AACR2

PRINTED IN THE NETHERLANDS

PREFACE

Scientific computing represents a large share, perhaps 50%, of today's computing expenditure and there is growing concern about the total cost of software, which includes not only its initial implementation but also its maintenance and modification over its life history. Numerous studies have cited the high costs associated with conversion or reprogramming of software due to changes in computer systems or programming languages. Rapidly advancing computer hardware technology is further complicating the task of constructing reliable and efficient software.

The purpose of this working conference (the second organized by IFIP Working Group 2.5 on behalf of IFIP Technical Committee 2) was to bring together specialists in numerical software and in programming languages to discuss needs in the programming language area as perceived by developers of numerical software, and approaches to satisfying these needs proposed by developers of programming languages.

Plenty of time was allocated for discussion, which was particularly lively. Edited versions of these discussions were prepared by "discussants" for each half-day session, and are attached to the ends of the papers. Each includes questions and comments made immediately following the presentation of the paper or in the general discussion held at each session end. I would like to thank the discussants (B. Ford, R.P. Brent, W.S. Brown, P. Kemp, G. Paul, E.L. Battiste, F.N. Fritsch and Th.J. Dekker) for their help and commend their work to the reader as well worthy of attention.

Financial assistance for the travelling expenses of some of the speakers was provided by the U.S. Army Research Office. The committee gratefully acknowledges this and the administrative assistance of AFIPS (American Federation of Information Processing Societies) in distributing these funds.

The conference chairman was C.L. Lawson. The programme committee consisted of C.L. Lawson and J.K. Reid (Co-chairman, WG2.5), Th.J. Dekker, B. Ford, C.W. Gear, T.E. Hull, B.T. Smith (all of WG2.5), C.A.R. Hoare (WG2.3), S.G. van der Meulen (WG2.1) and W. Waite (WG2.4).

Local arrangements were organized by L.D. Fosdick with the assistance of Gloria Farler. Secretarial help to the editor was given by Rosemary Rosier. A gift towards local travel expenses from Cray Research Inc. is gratefully acknowledged.

J.K. REID
Harwell

LIST OF CONTENTS

SESSION 1
INTRODUCTION

Chair
C.L. Lawson

Discussant
B. Ford

THE RELATIONSHIP BETWEEN NUMERICAL COMPUTATION
AND PROGRAMMING LANGUAGES, J.K. Reid (editor)
North-Holland Publishing Company
© *IFIP, 1982*

PROGRAMMING LANGUAGES: POWER, TRENDS AND FACILITIES FOR
NUMERICAL COMPUTATION

John R. Rice

Department of Computer Science
Purdue University
West Lafayette, IN 47907
U.S.A.

A general discussion of the objectives, costs and power of
programming languages is given. Power is defined in terms of
the time to program a given task. This is then related to
the richness of operators, operands and control structures
in the language plus the effectiveness of the support
facilities. The future trend will be to increase the power
of programming languages by increasing the richness of
operators and operands. This will be accomplished by the
extensive use of software parts and lead to very high level
languages and problem solving environments. A scenario for
software production in this environment is described. The
final part examines in more detail the facilities (operands,
operators and control) that are desired, needed and feasible
for numerical computation languages.

PART 1: THE GOALS, COSTS AND POWER OF PROGRAMMING LANGUAGES

THE GOALS AND COSTS OF PROGRAMMING LANGUAGES

The fundamental purpose of a programming language is to allow us to exploit the
computational power of computers. The cost of a computation is roughly divided
into three parts: the cost of writing and maintaining the program, the cost of
processing the program (compilation, assembly, library support, etc.) and the cost
of actually doing the computation (production or execution). We call the
accumulated costs of writing, maintaining and processing the program the
programming cost. The programming cost may be spread over a long period and involve many computer runs, the computation cost may be a one time or a recurring
cost. There have been many discussions of the trade-offs between programming and
computation costs, these are not repeated here. We simply note that the programming cost is very significant for most computations and the dominant cost for many
of them.

Thus the primary goal of programming languages is to reduce the programming cost.
This cost, in turn, consists of the computer cost of processing and the human cost
of devising, debugging and documenting the program. Again there are trade-offs
between the computer costs and human costs which are not discussed here. We simply
note that the human cost (including such intangible costs as not having the program
available) is a very significant (usually dominant) part of the programming cost.
Thus, we conclude that the goal of programming languages is to reduce the human
costs of programming.

This conclusion is not novel and this goal has been used implicitly or explicitly
in the design of many languages. There have been heated debates about what affects
the human costs (e.g. iteractive vs. batch, structured vs. unstructured, APL vs.
Cobol, Fortran vs. assembly language) but there is general agreement on this goal.

THE POWER OF A PROGRAMMING LANGUAGE

We conclude that the principal part of the programming cost can be measured by the
number of hours required for someone to write the program, to get it to run properly and to keep it running. The power of a programming language (for a particular
computation) is the reciprocal of the human programming cost. Power can be
measured in reciprocal hours and one must count all the time. For a one shot
computation, the hours for writing and debugging the program are all that are
counted. For programs of more permanent value and use, one must also include the
hours required for documentation, maintenance and similar activities.

There have been several approaches to estimating the programming effort for various
computations. Approaches that estimate the effort in advance have a high level of
uncertainty due to the possibility that unforeseen problems may arise. For further
information see [Putnam, 1980], [Boehm, 1979]. There are also methods to estimate
the programming effort (original writing, primarily), of an existing program, see
[Fitzsimmons and Love, 1978], and [Halstead, 1977]. All of these approaches involve the length of the program, which can be measured in various ways and merely
counting lines is obviously a crude approach. It is clear that programming cost is
very strongly related to program length, however it is measured. Thus, the power
of a programming language is directly related to the length of the program required
for a particular computation. It is intuitively plausible that if the program is
shorter, then it requires less time to write it, less time to debug it, less time
to document it and less time to make maintenance modifications. Thus all aspects
of programming language power benefit from shorter programs. There are some
variations due to the logical complexity of programs although some aspects of complexity are included in some measures of length.

At the risk of oversimplification, we say that the way to increase the power of a
programming language is to allow shorter programs to be written.

We identify three general components of a programming language: its operators,
operands and control facilities. There has been intense study of control
facilities over the past 15 years. See [Dahl et al, 1972], and [Mills et al, 1980]
for further information. I believe that we are at the point where significant
improvements from better control facilities are unlikely to be forthcoming.
Modern programming languages already have about as good control facilities as one
can expect; at least we are at the point where new control structures are unlikely
to reduce significantly the length of most programs.

More recently, the operands in programming languages have received serious study in
the computer science community, usually under names such as data structures,
abstract data types, etc. There is a well established move to introduce new and
somewhat general operands into programming languages. It is intuitively clear that
the natural availability of stacks, trees, graphs, matrices, files and such things
can greatly simplify some programs and thereby reduce their length. The further
introduction of operand facilities is a promising approach to increasing the power
of languages and its potential probably far exceeds that of perfecting the language
control facilities.

Unfortunately, little attention has been given to the role of operators. A few new
ones for files and strings have been introduced in some languages but, by and large,
we are making do with the same operators that were available in 1960 with Fortran
and Algol. APL provides a fair number of additional operators and much of the
popularity of APL is probably due to this power in a compact notation. Snobol and
descendents have pattern matching operators which dramatically reduce the effort of
certain types of computations. The lead of APL and Snobol has not been followed up
and most discussions of programming languages within computer science circles
ignore operators.

It is obvious that the full potential for benefits from a richer set of operands requires some associated useful operators. It is apparently not so widely recognized that the more difficult task is usually to implement the operators. In my own experience, I have written a few hundred or a few thousand lines of code to implement certain operators and then the few dozen (at most) lines of code I might save by having the operands as part of the language are simply irrelevant.

Consider, as a realistic example, introducing systems of ordinary differential equations into a language. The required operands are vectors of functions. The required operators are differential operators and their inverses. While it is a considerable chore to include vectors of functions in a language, it is one or two orders of magnitude more effort to include differential operators and their inverses if the effort must be made from scratch. One can and should implement these operators using well known, high quality and publically available routines; this would make their implementation quite feasible.

In summary, I believe that we have made a lot of progress in the past 20 years in control facilities and that there is little more power to be gained here. We have made only modest progress in operands (files and character strings for Fortran, some more general data structures in Pascal and Ada) and practically no progress in operators. Most programmers are using languages with about the same power that we had 20 years ago, power has increased by a factor of 2 or 3 perhaps. This is equivalent to programming productivity gains of only about 3-6 percent a year.

The preceding discussion is in the context of mainstream computer science and reflects a discouraging lack of progress. The real excitement in programming languages lies in the application areas. There are programming languages being introduced and widely used which have power far beyond the mainstream computer science languages. The people behind these languages often have only a modest knowledge of programming language theory and practice and the resulting languages are sometimes awkward in their design and implementation. Yet these people have had their eye on the main objective: power. Many of these languages have managed, in one way or another, to provide impressive power. One can actually write, in a fairly direct way, things similar to "Analyze the following data", "Give me the following information". "Solve the following partial differential equation" or "Optimize the following model". Such power requires operands and operators well beyond things currently considered in programming language circles.

PROGRAMMING LANGUAGE SUPPORT.

There is an aspect of programming language power that has been ignored in the above discussion and which must be mentioned. This is the support facilities of the language. These consist of the

Translator: Must be accurate, should be efficient in using computer resources both in translation and execution.

Diagnostics: Meaningful messages for incorrect programs.

Static Analyzer: Provides simple things like cross reference tables and language standards checking or complex things like flow and control checking or test data generation.

Dynamic Analyzer: Provides things like subscript checking, execution profiles or symbolic execution.

Editorial Assists: Editors, formatters, version controllers, program concatenators, etc.

High quality support facilities can increase the effective power of a programming language significantly, perhaps by a factor of 2 or 3.

There have been numerous good ideas in programming languages which have suffered,
even failed, for the lack of adequate support. There is nothing quite as annoying
as unreliable translators or misleading and inadequate diagnostics. The translators and diagnostics are directly "connected" to the programming language but
much of the other support is ad hoc and somewhat divorced from the language itself.
We are just beginning to see the emergence of programming language environments
where all the support facilities are integrated with the language processing. I
believe this is a very useful direction of development which will give a
significant, even if not dramatic, increase in the power of programming languages.

                        PART 2:   FUTURE TRENDS

THE HIERARCHY OF PROGRAMMING LANGUAGES

There will be four distinct levels of programming languages:

    machine language (an absolute numeric code to operate computers)

    assembly language (a symbolic form of machine language which provides
       access to all computer facilities)

    algorithmic language (Fortran, Algol, Pascal, Ada, Cobol, Lisp etc.)

    problem solving environments   = very high level languages

                       = problem statement language

The first three levels are already recognized and examples of the fourth level are
appearing with increasing frequency. A problem solving environment exists in a
context of a set or class of problems to be solved, its language interprets
everything within this context and it contains all (or, at least, many) of the
standard operands and operators relevant to the class of problems. There will be
problem solving environments for solving partial differential equations, carrying
on correspondence, designing highway bridges, making travel arrangements, building
new problem solving environments, etc. Most end use of computing will be through
problem solving environments, but I do note that they require good human engineering and substantial (but readily available) computing resources. They will
provide a level of power to the programmer that is unapproachable with conventional algorithmic programming languages. Increases in power by factors of 10 to
1000 can already be demonstrated [Rice, 1979], [Tutsch, 1981].

The conventional programming languages will increase modestly in power, but it
does not seem likely or even desirable that they evolve into very high level
languages. There is an important role for each level of language even though the
machine and assembly languages may well become invisible to the ordinary programmer as the translation of algorithmic language to assembly language to machine
language is incorporated into the hardware.

PROGRAMMING AND SOFTWARE PRODUCTION

I believe that the lack of increased productivity in programming and software production is directly tied to the lack of increased power in the programming
languages themselves. There are powerful languages in widespread use, (e.g.
the various statistical systems) but "computer scientists" view them as something
only for naïve, uninformed or low level programmers. "Real" programmers with
"real" projects use algorithmic languages. Until this view of programming is
abandoned, we cannot expect much improvement in programming productivity.

The key to increasing programming productivity is to bring the problem solving
environments into the mainstream of programming. This can be accomplished best by
simultaneously developing a software parts technology. A <u>software part</u> is the
logical extension of a library routine, it consists of a prologue, an epilogue and
a nucleus. The nucleus is an algorithm that performs some standard computational
task. The input, output and characteristics of the task are well understood and
one can put these parts in a catalog with brief descriptions similar to the way
electronic and mechanical parts are currently catalogued. The prologue and
epilogue perform three functions: they translate to and from a standard data representation, the prologue checks the input data to verify that it is admissable
for the part and the epilogue marks the output data so it can be checked later.

Problem solving environments and larger parts are <u>composed</u> from software parts.
Basic software parts themselves are built by the current programming methodology
using algorithmic or lower level languages. The prologue and epilogue provide
great flexibility and high reliability at the cost of efficiency. When a new part
or problem solving environment is being developed, the overhead of these interfaces
is ignored at first. As the design becomes firm the interconnections are optimized by removing as much of the representation translation and checking as
appropriate.

Problem solving environments have their own context, their own limited world
knowledge. A programmer "steps into" a problem solving environment in order to
carry out tasks within its domain. Certain basic information and facilities (user
variables, the parts catalog, files, the editor) move along with the programmer as
he moves from environment to environment.

It is beyond the scope of this paper to develop this scenario in more detail, see
[Comer et al, 1981] for more information. However, many people in mathematical
software will, I believe, see this scenario as a logical evolution of current
trends. Figures 1 and 2 schematically display the software production methodology
and the high level computing environment envisaged by this scenario.

There are many aspects of this scenario which require further specification, but
that goes beyond the scope of this paper. The key points relative to the interface between programming languages and numerical computation are:

   A.   <u>Software Parts</u>: These parts have standardized interfaces and
        <u>functionality</u>. The syntax of the interfaces is handled
        automatically by the prologue and epilogue while the semantics
        is handled through universal definitions of objects (i.e.,
        matrices, character strings, code segments, vector length,
        integrate from A to B, print, etc. have an agreed upon
        definition in each context where they are relevant). The parts
        can be cataloged along with their functionality and performance
        characteristics.

   B.   <u>Problem Solving Environments (PSEs)</u>: Each problem solving
        environment has a semi-natural language appropriate for its
        context. These languages have a consistent syntatic pattern
        that comes from automatic generation of language processors.
        Each PSE is supported by a prompting, language intelligent
        editor to ease the burden of recalling the details of the
        PSE syntax and semantics.

Computing Environment   Standard Support

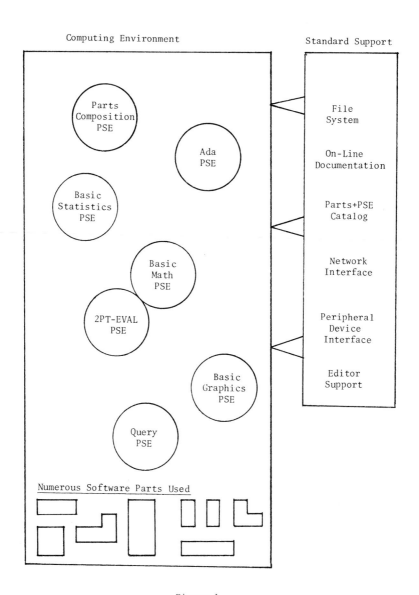

Figure 1

Schematic of the computing environment with very high level languages implemented
within Problem Solving Environments (PSEs). A snapshot is shown from an example
where the new PSE 2PT-EVAL is being created and tested. The other PSEs shown are
standard ones used during the construction process, the programmer moves freely
between PSEs as he progresses in building 2PT-EVAL. This example is given in full
detail in [Comer et al, 1981], pages A-4 to A-10.

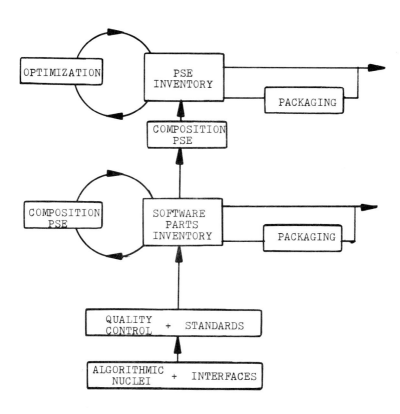

Figure 2
Organization of a software factory producing software parts and problem solving
environments.

PART 3:   FACILITIES FOR NUMERICAL COMPUTATION

SOFTWARE PARTS AND PROBLEM SOLVING ENVIRONMENTS FOR NUMERICAL COMPUTATION

I believe that the scenario described above has the potential to make a quantum
jump in programming productivity. This scenario builds naturally on the trends
and current activities in mathematical software. We are already deeply involved
in standardized, reliable, versatile, interchangeable and widely available software. The IMSL and NAG libraries, the LINPACK, FUNPACK, and EISPACK projects, the
BLAs and the ACM-TOMS algorithms are all examples of this type of software. While
we all know there is room for improvement in our efforts we also know that we have
made a great deal of progress in the past 15-20 years and, I believe, the mathematical software community is well ahead of other computing areas in establishing
the foundation for a software parts technology. The key lesson that we have
learned is that quality control is critical to the success of any attempt to distribute software parts. The key to quality control is a central organization to
set and enforce standards (the catalog keepers) and professionals to produce
software parts. By professionals, I mean professionals in the software parts
business, people who do not have an emotional attachment to the software, who can
objectively balance ease of use and flexibility, who can prune superfluous bells
and whistles and who will invest talent in improving performance, usability and
documentation.

The mathematical software community is also well into stand alone problem solving
environments. The statistical software community is even further advanced in this
direction and it behoves us to study their methods and products. Examples of
PSEs include MATLAB, ELLPACK and TOOLPACK; all of which create powerful environment for certain classes of tasks. The door is open for many more experiments and
approaches; I foresee a lot of exciting and very useful research and software
coming from these efforts.

FACILITIES DESIRED, NEEDED AND FEASIBLE FOR NUMERICAL COMPUTATION

It is clear that a large number of problem solving environments are appropriate
in numerical computation areas. It is also plausible that there should be a
"basic math" environment which overlaps part of a number of other PSEs. Thus the
basic math PSE might provide a minimal capability to solve ordinary differential
equations, suitable for routine problems. More difficult and complex differential
equations would require the capabilities of the ODE PSE or perhaps a tailor made
part.

It is less clear what facilities the basic math PSE should contain or whether it
should be a general purpose algorithmic language for numerical computation. There
are several alternatives that suggest themselves:

    (A)   Enhance Fortran. The facilities for a basic math PSE could be added
          to Fortran or, almost equivalently, a completely new language could
          be designed which repackages these facilities. Such languages were
          experimentally implemented in the 1960's and were successful in
          principle, in practice they failed because the lack of sufficient
          computing resources (see [Rice and Rosen, 1966], [Rice, 1973]). The
          main liability of this approach is the risk of developing an overly
          complex language which is cumbersome to learn, use and process.

    (B)   Shrink Fortran, then add a basic math PSE. The idea is to define a
          basic algorithmic language (core Fortran?) and then access the
          math facilities through a PSE. This is similar in spirit to the
          idea of a core Fortran plus "application" modules being considered by
          the Fortran standards committee, X3J3. The main difference is that
          the PSE approach does not require compatible syntax in the core and
          the basic math PSE. The two languages would have independent

processors so no conflicts arise. Note that the computing environment in the above scenario allows one to literally "bring" the basic math PSE inside a program in the core language.

(C) Keep Fortran as it is, add a basic math PSE to it. This is a compromise between the preceeding two approaches with the significant advantage that one does not have to wait for a substantial reorganization of the computing environment in order to try· it out.

However one implements the basic math PSE, there is the key question of "what should it contain?" The danger is that one throws so much into it that it becomes too cluttered and complex. The Fortran X3J3 committee is now facing the very difficult problem of trying to include many nice new things and still retain a reasonable simplicity in the language. One of the attractions of the PSE concept is that each PSE is kept clean, natural and fairly narrowly focused so that the learning, use and language processing of any one PSE is efficient. The PSE designer will make a thoughtful (or random?) selection of facilities and mold them into an attractive environment.

I have initiated a small project to categorize and evaluate the facilities most relevant to numerical computation. This project might be adopted as a WG 2.5 project and, in any case, I am grateful for many suggestions and much input from my colleagues of WG 2.5. This project has led a survey questionaire in the September, 1981 issue of the SIGNUM Newsletter. This questionaire provides a very condensed form of the facility categories, a more detailed but still incomplete presentation is given in [Rice, 1980]. It is hoped that this survey will produce a useful picture of the programming language needs as perceived by a large cross section of the numerical software community. The 16 categories of facilities included in the questionaire are:

1. Array construction and manipulation
2. Vector-Matrix operations
3. Sum, max, min plus calculus operators
4. Function definition and manipulations
5. Problem solving statements
6. Data structures
7. Input/Output for arrays and functions, to system storage, etc.
8. Elementary and higher mathematical functions
9. Variable types (real, integer, etc.)
10. Precision control
11. Declarations, names, character set, etc.
12. Storage management
13. Library support
14. Communication between programs
15. Exceptions and error handling
16. Environment inquiry (machine and system parameters)

The ultimate goal of this project is to produce a document to guide the programming language and PSE designer. It would catalog the facilities pertinent to numerical computation and provide guidance as to their perceived usefulness. There might also be a brief discussion of the expected difficulty in implementing certain facilities. The project is not intended to design any new language and it is specifically intended to avoid any discussion or proposals about syntax or details of semantics. It is the task of the language designer to study his objectives, select the most appropriate combination of facilities and then "package" them in an elegant syntax with clean semantics.

Some of my colleagues on WG 2.5 have already completed the questionaire and, while they are perhaps an atypical group, I made a few comments on their responses. This group gives strong support for one type of facility that they are particularly concerned with (library facilities) plus several general ones: arrays, elementary functions, variable types (real, integer, high precision, complex and character), two levels of precision and the identification and location of errors. Only three facilities received consistently low support: calculus operators, mathematical typing of functions and mandatory declarations. Opinion was widely varying on the value of many facilities, most significant of these were: I/O (all types), reserved words, significant blanks, inter-program communication (all types), automatic or block storage allocation. Other facilities with very inconsistent ratings were: range variables for arrays, function definitions, and extended or dynamic control of precision. The remaining facilities in the survey **received** a medium level of support (even though there were significant differences).

Perhaps the main conclusion one reaches from these responses is that there are a few obvious facilities that everyone wants, there are a few that are not perceived as needed, there are considerably more that are **inconsistently** evaluated, (even controversial in nature) and finally the largest number are perceived as nice, but not absolutely essential. The task of the language designer is to pick and choose from the latter two groups and put together an attractive, useful and efficient language.

In conclusion, I believe that superficially, the landscape of numerical computation has not changed much in the past 20 years. We are still using Fortran. We had subroutine libraries and still use them. Our keypunch machines now have screens and are called terminals. We still worry about getting the program into memory (we have more now) and about the time it takes to get it to run (we have less now). It still takes several weeks to write and debug a moderate sized program which does about the same thing a moderate sized program did 20 years ago. Underneath, some things have changed profoundly. The libraries and standard software of today are orders of magnitude better now than then. Worrying about fitting into 200,000 words of memory with a fast disk is a whole new ball game compared to using 4 or 8,000 words of memory with slow tapes. Some terminals give fast, interactive access to computing power (others give fast access to operating system queues). Fortran 77 is a significant improvement over "Fortran 61" even if not a quantum leap. But, programming power has improved only by a small factor and programming has become the bottleneck in getting numerical computation done.

I believe we are entering a transition period that will radically change the landscape. Even Fortran will look different. More importantly, we will be freed from Fortran (and algorithmic programming) by building on our foundation of software parts. The typical user will stop the absolute nonsense of writing code like

```
          SUM = 0.0
          DO 10 I = 1,KTERMS
   10         SUM = SUM + TERM(I)
```

One of the original goals for Fortran stated by John Backus in 1954, was to eliminate such code for sums. Programming in algorithmic languages will be done only when special circumstances demand that level of detail in the control of the computation. Most routine things will be done by saying things like "sum these terms", "solve these equations", "plot these functions", "analyze this data". We have created the foundation upon which to accomplish this. It remains to apply our ingenuity to see the best ways to exploit the potential that is present.

REFERENCES

[1]   Boehm, B.W., Software Engineering: R&D Trends and Defense Needs, in
       [P. Wegner, (ed.)], Research Directions in Software Technology (MIT
       Press, 1979, 44-86).

[2]   Comer, D.E., Rice, J.R., Schwetman, H.D. and Snyder, L., Project Quanta,
       CSD-TR 366 Computer Science Dept., Purdue University, 1981.

[3]   Dahl, O.J., Dijkstra, E.W. and Hoare, C.A.R., Structured Programming
       (Academic Press, 1972).

[4]   Fitzsimmons, A. and Love, T., A review and evaluation of software science,
       ACM Comput. Surveys, 10 (1978) 3-18.

[5]   Halstead, M.H., Elements of Software Science, (Elsevier North-Holland,
       1977).

[6]   Mills, H.D., Basili, V.R., Goldberg, J.S., Gries, D.J., Hansen, P.B.,
       Parnas, D.L., Ramamoorthy, C.V., Vyssotsky, V. and Yeh, R.T., Software
       Methodology, Chapter II in [B. Arden, (ed.)], What can be Automated
       (COSERS), (MIT Press, 1980, 791-820).

[7]   Putnam, L.H., Software Cost Estimating and Life-Cycle Control: Getting
       the Software Numbers,No. EH0165-1, IEEE, 1980 (a collection of articles).

[8]   Rice, J.R., Programming effort analysis of the ELLPACK language, SIGNUM
       Newsletter, 14 (1979) 109-111.

[9]   Rice, J.R., NAPSS-like system: Problems and prospects. Proc. Natl. Comp.
       Conf., (1973) 43-47.

[10]   Rice, J.R. and Rosen, S., NAPSS - A numerical analysis problem solving
       system. Proc. ACM Natl. Conf. (1966) 51-56.

[11]   Rice, J.R., Programming language facilities for numerical computation,
       Tech. Rpt. #2033, Mathematics Research Center, University of Wisconsin
       (January, 1980).

[12]   Tutsch, J.H., The use of very high level languages for numerical applica-
       tions, Univ. of Wisconsin Computing Center Report (January, 1981).

DISCUSSION

Kahan The cost of programming can, for the purpose of comparing languages and
their implementations, be separated from the cost of analysis. I have come to
doubt that the cost of numerical programming depends merely upon the number of
lines written. Rather certain kinds of program structures cost several times as
much as others. The costliest structures seem to be:
          Subprogram calls, especially with many arguments;
          Tests, especially against non-trivial thresholds;
          Branches and the labels they branch to;
          Repeats (DO, FOR, WHILE, ...) and their limits.
      Subprogram calls cost human time to discern dependency relations among
constituent subtasks into which a program must be organised, but the time is well
spent. Repeats often substitute for a lack of adequately strong operators like
SUM, MATRIXMULTIPLY, POLYNOMIAL .. for which subprogram calls might be deemed
either inefficient or inscrutable, sometimes both; but the unavoidable repeats
cost human time mostly to determine correct initial and terminal conditions.
      The other costly structures are associated with the classification of
cases, and coping with exceptions. Here the issues are not so much linguistic
(syntactive) as semantic; in computing environments where exceptions and special
cases abound, programming must in consequence be very costly. Economies in this
area come less from language development than from carefully thought out
implementations of the various arithmetic and other operations, plus a non-
punitive attitude towards execution time errors, error-indicators, and
diagnostics. Put simply, robust procedures cost less than others for programmers
to use.

When one seeks to compare languages versus their implementations it is probably impossible for a programmer to measure the quality of a language separately from the quality of its implementation. Some quantitative measure is needed. I suggest <u>number of successful runs</u>.

The software being run are subprograms, 20 to 100 records long. A successful run is one which returns a result that the program, as originally intended deserves to get; hence a result that causes a change in the program's mathematical foundation is a successful run. An attempted run is a failure when it encounters a diagnostic message at compile time, or when it exposes my mistaken understanding of how the language processor interprets strings of characters.

<u>Rice</u> There can be several measures of cost, a number of them of increasing sophistication. However many programmers find a high correlation between the cost of preparation and the number of lines in a program. Hence one ignores this, the simplest measure, at one's peril. But obviously there can be other useful measures too.

<u>Gentleman</u> Granting that the human cost of programming is a function of program length, why specify that the cost is proportional? Regardless of how program length is measured, the linearity of the proportionality assumptions implies that the cost of working with (writing, reading or modifying) two programs is the sum of the costs of working with each alone. Experience seems to indicate that the cost rises faster than this.

In your talk you implied that control structure design questions are all solved, and I would agree that little advance can now be expected for the standard context.

However why should we assume that parallel computations cannot benefit from different control structures than those that have been studied so far? Similarly, why should we assume that interactive execution with manual intervention cannot benefit from different control structures too?

<u>Rice</u> Some of the measures of program length I refer to are not linear in the ordinary sense, as they are aimed at estimating programming effort. Thus, they include terms like xlogx where x is the number of distinct operands or total number of operators.

I agree that special control structures for parallel computation should be investigated. The people in operating systems have made a start (e.g. concurrent Pascal), but their viewpoint is somewhat different from that of the program writer and they do not provide adequate control for asynchronous parallel numerical computation.

<u>Fritsch</u> You referred to gross inefficiencies due to use of universal data representations. I am also concerned about potential loss of <u>accuracy</u>. For example requiring splines to be treated only by the piecewise polynomial representation.

On a separate issue I seriously doubt that your survey will reach the "right" audience. In my experience, very few serious <u>users</u> of programming languages are members of SIGNUM.

<u>Rice</u> In answer to your first point regarding data representations you are being much more sophisticated than I intended. I was thinking of representations such as character sets.

With the survey I shall sample members of SIGNUM and then consider the matter again.

<u>Adams</u> X3J3 have undertaken a number of surveys. The questionnaire is difficult to prepare and the results are difficult to interpret. Do you have any results yet?

<u>Rice</u> A trial sample from members of IFIP Working Group 2.5 has been taken to date. Whilst surveys rarely provide the complete information that one would like you nevertheless do learn things from them. For example from the trial sample there are areas of enquiry where the answers are in general agreement.   In other areas there is apparently no concensus of view at all.

<u>Adams</u> Please let X3J3 have a copy of your results.

<u>Meissner</u> A similar survey was recently m'ade by Metcalf at CERN.   57 replies were received, mainly from European physicists. The results were reported in For-word Volume 7, No.3, June 1981.

<u>Ford</u> In designing programming languages for numerical computation should we not show greater concern for issues of reliability, through the choice of constructs in the language, and in the preparation of the programming support environment? The refusal by our community to accept responsibility for results computed by numerical software is in marked contrast to the technical standards and warranties in the vast majority of technical and scientific fields and in the professions.   Can we claim to be professional when we have so little apparent confidence in the reliability of our numerical computations?

<u>Rice</u> Issues of reliability are covered by the quality control phases of software preparation.   Many preparers of numerical software parts (such as the NAG Library) do take some responsibility for their results in that they return error flags when numerical procedures fail.   I believe this practice will be developed into really sophisticated reliability analysis within software parts and future programming environments will handle numerical failures in a general way.   That is, not only will SQRT(-1.2) or EXP(123456.7) be signalled and handled as an exception, but so will events like "LINEAR SYSTEM WITH SINGULAR MATRIX" or "NUMBER OF INTERPOLATION POINTS MOST EQUAL NUMBER OF BASIS FUNCTIONS".

<u>Hehner</u> Your equation

Human Programming Cost = k* program length

suggests that effort is expended only to increase program length. Much of my effort is directed to simplifying a program, and that often means reducing its length. Of course, I can still agree with your conclusion that program length is a measure of language power, but disagree with your argument for that conclusion. I also want to point out that APL has taught us not to pursue brevity alone; brevity is not synonymous with readability (understandability).
    You said that our increase in language power has not been as great as we, in the computing community, have come to expect. First, I would like to say that language power and programming power are not identical, and there have_been significant advances in programming methodology. Second I ask where this expectation comes from. There have been wonderful increases in power per cost of hardware, but that should no more raise expectations for software than an advance in the production of paper should raise expectations of improved productivity of novelists.
    Language designers have mostly decided not to try to invent more powerful primitives for particular applications. Instead, the procedure definition is a way for the programmer to invent more powerful "language features" on the spot, tailored to the application. More recently, the invention of packages serves the same purpose.

Rice   A tacit assumption in my, and others, use of this equation is that there is
a constant level of effort spent in "polishing" a program.   The usual   measures
are oriented toward "routine" programming, where the basic methods and algorithms
are   understood;   there   are   no   special subtleties in the implementation and
exceptional   requirements for clarity, etc. (such as for an ACM Algorithm or a
text book example).   I agree that extreme brevity is not always beneficial.

     There have been significant advances in programming methodology; I have
included this effect in my estimate of the increase of programming productivity.

     While I do not propose that I can or should increase the productivity of
novelists, I do not find programming that closely related to writing a novel.
The objectives of the novelist and programmer are distinct, perhaps a more
realistic analogy is between programming and cooking.   In the beginning, there
were long apprenticeships, followed by various levels of artistic ability.   Then
there were recipes that could be followed with various degrees of success.   Now,
we have frozen TV dinners and library routines.   We hope someday to have an
automatic gourmet kitchen (just push STEAK DIANE - MEDIUM RARE) and automated
problem solving (just push SOLVE ODE #3 - ERROR = .0004).

     I disagree with your statement that language designers have not tried to
invent more powerful primitives.   It is true that the pure linguists in computer
science have used the existence of procedures as an excuse to avoid this task.
This is a mistaken approach, programmers do not want to invent language features
on the spot and most of them are incapable of doing it well.   Language designers
outside the "core" of computer science are heavily engaged in providing more
powerful primitives and this approach has received enthusiastic support from the
user community.

THE RELATIONSHIP BETWEEN NUMERICAL COMPUTATION
AND PROGRAMMING LANGUAGES, J.K. Reid (editor)
North-Holland Publishing Company
© *IFIP, 1982*

PROGRAMMING LANGUAGES FOR NUMERICAL
SUBROUTINE LIBRARIES

J.J. Du Croz
(Presented by B. Ford)

NAG Central Office
7 Banbury Road
Oxford
U.K.

This paper examines and compares some ways in which the design
and provision of a numerical subroutine library is affected by
features of different programming languages. Much of the discussion is based on the experience of the NAG Project in
developing libraries in FORTRAN (66), ALGOL 60 and ALGOL 68,
but the paper also considers the design of other libraries, and
notes the potential impact of other languages such as FORTRAN
77, ADA and FORTRAN 8X. The main topics discussed are: calling
sequences; numerical precision; the computational environment;
and library mechanisms.

1. INTRODUCTION

In this talk we aim to examine and compare some ways in which the design and provision of a numerical subroutine library is affected by features of different programming languages. Much of the best numerical software is nowadays made available
in libraries (or packages or collections) of subroutines, and the use of such
libraries has steadily increased over the past decade. We assume throughout this
paper that libraries are intended for use on many different types of computer, and
hence the software must be transportable (see, for example, Hague and Ford (1976)).

Those who contribute to the development of subroutine libraries may constitute a
rather small proportion of all those engaged in numerical computation; and what
library developers require of a programming language may show a different emphasis
to the requirements of average programmers. But the proportion of programmers who
use numerical subroutine libraries is certainly much larger. Just as ordinary languages should be responsive to the needs of lawyers and technologists, as well as
to those of everyday speech, so programming languages should meet the needs of
library developers as well as those of average programmers. In some areas it may
be true that what library developers need to-day, an average programmer will need
to-morrow; for example, the importance of adaptability of algorithms and transportability of software has long been apparent to library developers, but is increasingly important to many programmers who have access to different computers via a
network.

Much of the content of this paper is based on the experience of the NAG project,
which has been developing subroutine libraries for numerical computation, in
FORTRAN 66 and Algol 60 since 1970, and in Algol 68 since 1973. (We shall use the
designation "FORTRAN 66" to distinguish the 1966 ANSI standard from FORTRAN 77, and
use "FORTRAN" without qualification to cover both standards.) We also consider
aspects of the design of other libraries and packages written in FORTRAN 66, and
note the potential impact of other languages, namely FORTRAN 77, Pascal, ADA and
FORTRAN 8X. (NAG has recently begun to develop a subset library in Pascal,
specifically for use on microcomputers, but has not yet developed any library
routines in any of the other languages.)

Many features of programming languages could have a considerable impact on the
code of library routines, as on any numerical software - for example, array-
handling features or control structures. However in this talk we concentrate on a
limited number of linguistic features which we believe to be particularly relevant
to libraries. These are:

  Calling sequences

  Numerical precision

  The computational environment

  Library mechanisms.

2. CALLING SEQUENCES

The calling sequences of routines within a library are normally the only linguistic feature of the library that is visible to an average user. They are the subject of frequent comment from users who, quite reasonably, desire both simplicity
and flexibility. No existing language appears to offer all the facilities that
could be wished for in order to meet both these requirements. Improvements in
this area would, we believe, have a bigger impact on the average user of a library
than any of the other topics that we shall discuss.

2.1. Array Parameters

Array parameters, with dimensions dependent on the dimensions of the problem to be
solved, are required by the great majority of numerical subroutines - by 77% of
the routines in the NAG FORTRAN Library. (The lack of this feature in Pascal,
prior to the current proposed standard, has been a serious limitation on the use
of this language for numerical subroutine libraries.) FORTRAN 66 in particular
makes the passing of array parameters to subroutines unnecessarily cumbersome.

We begin by comparing calls to NAG Library routines in different languages for the
solution of a set of simultaneous linear equations with multiple right hand sides
(a comparison also used by Delves (1977)); that is, the mathematical problem is:

  compute $X = A^{-1}B$

where A is an N-by-N matrix, and B and X are N-by-M. In each language the routine
requires a failure indicator, FAIL or IFAIL.

Algol 68:

  F04AAB(A,B,X,FAIL)

Algol 60:

  F04AAA(A,B,N,M,X,IFAIL)

FORTRAN:

  CALL F04AAF(A,IA,B,IB,N,M,X,IX,W,IFAIL)

Here in the FORTRAN routine the array parameters have the specification

  REAL A(IA,N), B(IB,M), X(IX,M), W(N)

The parameters IA, IB, IX are included solely to specify the leading dimensions of
the 2-dimensional arrays. No such parameters are required in Algol 60 or Algol 68
because the information is passed implicitly. For both the FORTRAN and Algol 60
routines, the dimensions of the arrays A, B and X can be larger than the
dimensions of the corresponding mathematical entities, the matrices A, B and X.

The Algol 68 routine requires that the arrays should have the same dimensions as
the matrices (this can always be achieved by slicing if necessary), and can then
dispense with the parameters N and M, since these can be derived from the array
parameters by the operators lwb and upb. All routines require an array of workspace of length N: in Algol 60 and Algol 68 this can be declared dynamically within the routine, in FORTRAN it must be passed as a parameter.

To illustrate a further point, we quote the specification of a NAG FORTRAN routine
for the Singular Value Decomposition:

    SUBROUTINE F02WCF(M,N,MN,A,IA,U,IU,S,V,IV,W,LW,IFAIL)
    REAL A(IA,N), U(IU,MN), S(MN), V(IV,N), W(LW)

Here, not only do we have the parameters IA, IU and IV to specify the leading
dimensions of 2-dimensional arrays. We also have parameters MN and LW included
solely to specify the second dimension of U and the dimensions of S and W. They
must satisfy:

    $MN \geq \min(M,N)$
    $LW \geq 3*\min(M,N)$

(We are ignoring an additional use of the parameter LW in the actual routine, in
order to illustrate a point.) FORTRAN 77 makes some such parameters redundant,
since it allows integer expressions in array declarations, e.g. REAL Z(N+1). But
that would not apply in this example, because function references are prohibited;
writing REAL S(MIN(M,N)) is not allowed.

The similar LINPACK routine has the specification

    SUBROUTINE SSVDC(A,IA,M,N,S,E,U,IU,V,IV,W,JOB,INFO)
    REAL A(IA,1), S(1), E(1), U(IU,1), V(IV,1), W(1)

This uses the non-standard convention of specifying the final dimension of an
array parameter as 1 to indicate that the dimension can be as large as that of the
corresponding actual array, as declared in the calling program. The convention
has been standardized by FORTRAN 77 as the assumed size array declaration, e.g.

    REAL A(IA,*), S(*), E(*), U(IU,*), V(IV,*), W(*)

Note however that this feature does not permit the full power of array-bound
checking. It only allows an error to be detected if the routine attempts to
access an array element outside the bounds of the actual array as declared in the
calling program. But, if the dimension of the array S is specified as MIN(M,N),
then an error can be detected on entry to the routine if it is called with the
actual dimension of S smaller than M or N; and also within the routine if it
attempts to access the element S(M+1), say. It would therefore be preferable if
adjustable dimensions of array parameters could be specified by expressions of
full generality (including conditional expression and function references) - as
are allowed by Algol 60 and Algol 68 in array declarations. It is true, though,
that in some instances the required minimum dimension may not be known before
entry to the routine (e.g. it may depend on the number of eigenvalues in a specified range, or on the number of non-zero elements in the decomposition of a sparse
matrix).

We now summarize the implications of FORTRAN 66 on passing array parameters, and
quantify their impact on the calling sequences of NAG FORTRAN Library routines.
The figures relate to the Mark 8 Library which contains 466 routines with a total
of 4,380 parameters.

a.) Parameters are required solely to specify leading dimensions of 2- or 3-
dimensional arrays: 376 parameters (8.6%), affecting 183 routines. The number of

such parameters could be reduced to at most one per affected routine (unless 3-dimensional arrays are involved) by requiring that all array parameters for a routine should have the same leading dimension. This has been done in EISPACK - but is less flexible.

b.) Parameters are required solely to specify the final dimensions of arrays: 237 parameters (5.4%), of which 139 could be avoided in FORTRAN 77 by using integer expressions as array dimensions.

c.) Workspace arrays, if of variable dimension, must be passed as parameters: 370 parameters (8.4%) are workspace arrays, and 163 parameters are required solely to specify their dimensions; 189 routines (41%) are affected. Note that the designers of the PORT Library thought it worthwhile to devise their own mechanism to make workspace dynamically available to library routines (Fox, Hall and Schryer (1978)).

d.) "Empty" array parameters are not allowed: i.e. if a routine has an array parameter of dimension N, the routine cannot be called with N equal to 0, even if no reference is made to the array in this case. (The FORTRAN 77 standard definitely forbids this; the FORTRAN 66 standard is vague, but some FORTRAN 66 compilers forbid it.) Hence either the routine cannot be called with N equal to 0, or an additional parameter must be included to specify the actual dimension. 19 routines (4.1%) are affected by this restriction; examples are: an array of user-specified break-points in a quadrature or O.D.E. routine; an array of constraint values in a curve-fitting routine; or an array containing a strictly lower triangular matrix (which is "empty" when N = 1).

A limitation on arrays, shared by both FORTRAN and Algol 60, is that only rectangular arrays are permitted. But numerical computation often uses non-rectangular structures such as:

   triangular matrices (which may stand for half of a symmetric matrix);

   banded matrices of constant bandwidth;

   banded matrices of variable bandwidth;

   vectors of vectors (of unequal length) (for example, function and derivative values of varying order to be interpolated at different points; or unequally sized groups of statistical observations).

In order not to set aside storage for large numbers of unreferenced zero elements, non-rectangular structures must be mapped compactly onto a linear or rectangular array. The actual layout of elements in the linear or rectangular array may require considerable care in documentation (for example, a page of the LINPACK manual is used to describe how to store a general banded matrix) or else is a source of confusion to users; in the case of a vector of vectors, additional parameters are required to define the structure. 34 NAG FORTRAN routines (7.3%) are obviously affected by this difficulty. Algol 68 allows a non-rectangular structure to be declared more naturally as a "row of reference to row" so that matrix element a_{ij} can be referred to as a[i][j]. The penalty for this is of course that we then have a "row" of descriptors. It would be preferable if certain special array shapes could be set up without these descriptors.

Thus improved facilities for handling array parameters would allow considerable simplification of calling sequences.

2.2. Omitting parameters from routine-calls

We now shift the emphasis to ways of achieving greater flexibility in calling sequences. We return, first of all, to the example of routines to solve simultaneous linear equations, and suppose that we require the solutions to overwrite the right hand sides. In FORTRAN we write:

    CALL F04AFF(A,IA,B,IB,N,M,B,IB,W,IFAIL)

supplying the same array as the actual parameter corresponding to two distinct
dummy parameters. The routine does not 'know' whether or not these two parameters
are identified in the call, but is written so that it performs correctly in either
case. In fact assigning a value to either of the identified parameters is non-
standard in both FORTRAN 66 and FORTRAN 77 - it could lead to incorrect results if
array parameters were passed by a "copy-in, copy-out" mechanism - though we have
encountered no FORTRAN system in which this particular use of the routine does not
work. However the corresponding Algol 68 routine is written so that the desired
result can be achieved by:

    F04AAB(A,B,nil,FAIL)

The routine tests for the special symbol nil, so it does "know" that the solutions
are to overwrite the right hand sides. This usage seems more elegant and potentially more powerful than the FORTRAN usage.

In order to provide more flexibility within the limitations set by FORTRAN, many
library developers have chosen to provide a choice of calling sequences to the
same underlying algorithm. Commonly there is a "specialist" or "comprehensive"
routine with a long parameter-list, offering all the facilities that might be
needed by an expert user; and also one or more "easy-to-use" routines with much
shorter parameter-lists, sufficient for solving straightforward problems. Often
the easy-to-use routines call the specialist routine directly, with suitable
choices for the extra parameters of the specialist routines.

But this arrangement is still comparatively inflexible. Much greater flexibility
could be achieved if users could simply omit parameters which they did not need to
specify. Indeed it could become unnecessary to supply a separate "easy-to-use"
routine at all.

Note that even in FORTRAN 66 the Harwell Library has adopted an approach of putting
optional scalar "parameters" in COMMON blocks rather than in the parameter-list,
allowing users to access them or not as they wish, and supplying default values,
if required, in BLOCK DATA subprograms (Reid and Hopper (1980)). However portability problems involving BLOCK DATA are an obstacle to this approach. Also the
default values must be constants. ADA allows parameters of mode in - but no
others - to be omitted, and default values supplied, but again the default values
must be constants.

Facilities have been proposed for FORTRAN 8X, in particular by Reid (1980b) and
Herington (1981), which offer the range of flexibility that seems desirable. We
illustrate many of the possibilities by taking for an example the Mark 9 NAG
routine C05NCF for the solution of a system of non-linear equations $f(x) = 0$. The
routine is essentially equivalent to the MINPACK routine HYBRD (Moré, Garbow and
Hillstrom (1980)).

In both NAG and MINPACK there is an easy-to-use routine with the parameters:

    FCN,   subroutine defining f

    N,   the number of equations

    X, x at solution, dimension (N)

    FVEC, $f(x)$ at solution, dimension (N)

    XTOL,   convergence criterion

    W,   workspace array, dimension (LW)

LW,   dimension of W, $\geq$ N$*$(3$*$N+17)/2

IFAIL,   failure indicator

Apart from the workspace, these can be regarded as mandatory parameters.

The specialist routine has the following 14 additional parameters inserted after XTOL in the parameter-list:

MAXFEV, maximum no. of calls to FCN

ML,⎫
    ⎬   specifying band-structure of Jacobian
MU,⎭

EPSFCN, accuracy of FCN

DIAG,   scaling factors, dimension (N)

MODE,   specifies if DIAG is set by user

FACTOR, used to bound search region

NPRINT, frequency of monitoring

NFEV,   actual no. of calls to FCN

FJ,   Jacobian at solution, dimension (IFJ,N)

IFJ,   1st. dimension of FJ

R,   from QR-factorization of Jacobian, dimension (LR)

LR,   dimension of R, $\geq$N$*$(N+1)/2

QTF, $Q^T \underline{f}(\underline{x})$ at solution, dimension (N)

These can be regarded as optional parameters. Note that the parameter MODE is included solely to specify whether or not DIAG is preset by the user, and would become completely redundant if the user were simply able to omit DIAG.

The easy-to-use routine calls the specialist routine, supplying the additional parameters as follows. For scalar parameters, local variables are used and default values supplied for input parameters:

MAXFEV   200$*$(N+1)

ML   N-1

MU   N-1

EPSFCN   0.0

MODE   2

FACTOR   100.0

NPRINT   0

(Note that some of these are expressions involving other parameter values.) For array parameters, partitions of the workspace array are used. The array DIAG is an input parameter; the arrays FJ, R and QTF are output parameters. The elements of the array DIAG are all preset to 1.0.

The proposals for FORTRAN 8X would allow any of these optional parameters to be omitted from a call of the routine. If an array parameter was omitted the necessary array would be generated dynamically within the routine. The routine would be able to test whether a particular parameter was omitted and, if so, not merely supply a default value, but take default action (such as might be necessary to initialize an omitted array parameter).

Of course, if parameters are omitted, it is highly desirable, for clarity and reliability, to be able to specify them by keyword, the keywords being the same as the dummy names of the parameters. Then in our example, the "specialist" routine CO5NCF could be called:

a.)   as an "easy-to-use" routine with parameters:

      FCN=F,N=7,X=SOL,FVEC=RES,XTOL=1.E-5,IFAIL=IND

b.)   if, say subsequent "post-optimal" analysis is required, adding the parameters:

      FJ=JAC,R=COV,QTF=Y

c.)   if, say, the problem requires scaling and monitoring, adding the parameters:

      DIAG=SC,MAXFEV=1000,NPRINT=50,NFEV=NF

and there are many other possibilities, which would meet the needs of individual users far better than the present all-or-nothing choice between the easy-to-use and the specialist routine.

3. NUMERICAL PRECISION

Improvements in calling sequences could be of considerable benefit to non-numerical computation also. But peculiar and fundamental to numerical computation are matters concerning the precision of floating-point data. Current machines usually have hardware for two levels of precision ("single" and "double"); some have a third ("quadruple"). FORTRAN supports two levels of precision (with some limitations described below); Algol 60 supports only one level; Algol 68 potentially supports any number of levels, the actual number being dependent on the implementation.

Numerical computation requires access to more than one level of precision because:

a.)   Often it is desirable, and sometimes essential, to perform part of a computation at a higher precision than the rest.

b.)   A convenient, though crude, way to check the accuracy and stability of a computation is to re-execute it at a higher precision.

c.)   On some machines the lowest level of precision is simply inadequate for many computations.

For this last reason, NAG prepares two versions of the NAG FORTRAN Library: a single precision version and a double precision version. (A similar approach is adopted by most other library developers). On many machines the double precision version is the only, or at least the preferred, version available. Almost all the code in the library uses a uniform basic level of precision (single or double); the use of a higher level of precision (double or quadruple), for reason a.) above, is confined to two inner-product routines.

The double precision version is derived from the single precision version almost entirely by a global change from REAL to DOUBLE PRECISION (with corresponding modifications of constants, intrinsic functions etc.) as described by Du Croz, Hague and Siemieniuch (1977)). The transformations are straightforward in principle; but some difficulties arise in practice because not all single precision constructs have double precision analogues in standard FORTRAN 66, for example: F-format; the functions FLOAT and AINT; and COMPLEX data type and related constructs. In practice we do indeed encounter compilers in which the required double precision constructs are not available. FORTRAN 77 has brought an improvement: it has reduced the number of transformations required (by introducing

generic functions), and has supplied double-precision analogues of all single precision constructs, <u>except</u> those related to COMPLEX data type. Library developers
must therefore either avoid COMPLEX data altogether; or assume the existence of a
non-standard DOUBLE COMPLEX data type (as recommended by Reid (1980a)); or
simulate it.   The NAG Library contains 10 routines that have COMPLEX parameters.
For many implementations of the double precision library, DOUBLE COMPLEX
parameters are simulated by DOUBLE PRECISION arrays with an additional leading
dimension of 2. Note that a further 20 routines in the NAG Library solve problems
involving complex data, though without using COMPLEX data type, so the potential
benefits of clean facilities for COMPLEX data are considerable.

Difficulties also arise when the single precision source-text contains some
essential double precision computation; this needs to be transformed to quadruple
precision computation, which may not be available either in hardware or in extensions to the language. Hence any essential higher precision working must be carefully isolated and, if necessary, given special treatment (e.g. using machine-
language).

But most of these difficulties could be avoided; indeed the need to transform the
source-text at all could be avoided, since a global change of precision could in
principle be performed by the compiler during code-generation. The ICL F1 compiler
for the ICL 2980 has particularly good facilities: the source-text can be written
in terms of three distinct symbolic levels of precision: REAL*E (or, simply, REAL);
REAL*D (or DOUBLE PRECISION); and REAL*Q. A directive to the compiler allows each
of these symbolic levels of precision to be matched, at compile-time, to either
4-byte, 8-byte or 16-byte floating-point data. Surprisingly few FORTRAN compilers
offer anything like this remarkably useful facility. If universally available, it
would allow library developers to maintain only one version of their source-text,
instead of two, as at present.

Much more elaborate facilities for specifying the precision of floating-point data
have been included in ADA and proposed for FORTRAN 8X; these would make it generally possible to change the level of hardware precision without wholesale modi-
fication of the source-text.

One area of numerical software which is particularly sensitive to differences in
numerical precision is software for special functions. The methodology adopted in
the NAG Library has been described by Schonfelder (1976). It involves generating
distinct segments of code for each of a number of different levels of precision
(currently 6 different levels are catered for). The master version of the source-
text contains all the alternative segments of code embedded in comments. This must
be edited to "activate" the segments required for a particular machine. It would
be helpful if compiler directives were available to select the code to be compiled.
Thus the need for intermediate editing of the source-text could be avoided.

4. THE COMPUTATIONAL ENVIRONMENT

To write transportable numerical software we need to design adaptable algorithms,
that is, algorithms that adapt to the computational characteristics of the
environment in which they are being executed, in order to achieve the maximum
accuracy and efficiency that the environment allows. This principle requires that
certain "environmental parameters" be available to the algorithms.

The most important part of the computational environment for numerical software is
the floating-point hardware - the representation of floating-point numbers, and
the arithmetic operations upon them. There has been much discussion about the
choice and precise definition of these parameters, and also about how they should
be made available as a feature of programming languages, e.g. Ford (1978), Brown
and Feldman (1980). In the absence of built-in features, especially in FORTRAN,
library developers have had to make their own decisions and devise their own
mechanisms for making the parameters available, either by automatic insertion of

values in DATA statements as in the IMSL Library (Aird (1977)), or by providing library functions to return the required values as in the NAG Library, or the PORT Library (Fox, Hall and Schryer (1978)).

In fact this situation has had some advantages, in that it has allowed library developers to keep control of the choice and definition of parameters, and of the assignment of values to them. If environmental parameters are provided as built-in features of the language and if library developers are to make use of them, they need to be very sure that sufficient parameters are defined, that the definitions are rigorous, and above all, that correct values are assigned to them by compiler-writers. It is very important to have adequate means of verifying the values of the parameters, such as the test program developed by Schryer (1981) for the model of the environment proposed by Brown and Feldman (1980).

Certainly NAG has occasionally decided to make adjustments to the assigned values of the parameters (with implicit adjustments to their definition). For example on the Prime 400 it proved impossible to compute DSQRT(DOVFLO) without error, where DOVFLO is the overflow threshold for double precision numbers and its value had been determined with reference solely to the arithmetic hardware. In consequence, we reduced the value of DOVFLO (returned by NAG Library function X02ACF); the reduction - from 2^{32636} to 2^{32576} - was not likely to have undesirable side-effects. (We also complained to the manufacturer about the unexpected (and unnecessary) limitation on the arguments permitted by DSQRT). Similar problems arose when attempting to compute DLOG(DUNFLO). Thus we are faced with the choice between adding further conditions to the definitions of DOVFLO and DUNFLO, or requiring certain conditions to be satisfied by the implementation of DSQRT and DLOG. The latter seems preferable in the long term.

This example brings in another part of the computational environment - the library of mathematical functions provided with the compiler. Almost all writers of numerical software make use of these functions, yet they might be regarded, poten-tially at least, as the "Achilles' heel" of numerical software. They have suffered from: incomplete definition (compare the improvements in the definitions between FORTRAN 66 and FORTRAN 77); lack of agreed standards for implementation; inadequate implementation; and inadequate documentation (e.g. what accuracy is actually achieved, what are the permitted argument ranges).

In addition to the example just quoted, we mention the following details of NAG's experience:

a.) On the Prime 400, a routine to compute eigenvectors by inverse iteration failed, because the computed eigenvalues were insufficiently accurate; the inaccuracies in the eigenvalues were traced to inaccuracies in DSQRT. The problem was resolved by relaxing a tolerance factor, specifically for this implementation.

b.) On the Honeywell Level 66, a routine unexpectedly failed because it assumed that $|\cos(x)| \leq 1$ and this assumption was violated by DCOS. The library routine had to be modified so that it did not rely on that assumption.

c.) NAG has found it convenient to introduce two environmental parameters to specify the largest positive and negative permitted arguments for EXP, since it is not usually true that EXP can be evaluated for all arguments between ALOG(SUNFLO) and ALOG(SOVFLO).

d.) Similarly NAG has introduced an environmental parameter to specify the largest permitted arguments for SIN and COS. Both the argument ranges and the accuracy of SIN and COS crucially affect, for example, the argument ranges and accuracy of library routines for computing Bessel and Airy functions. On many machines, SIN and COS return results with some meaningful accuracy for arguments up to approximately the reciprocal of the machine precision; on some machines, however, the maximum argument is the square root of this. Moreover the actual limits are not always documented.

Therefore at present library developers must beware of making unjustified assumptions about the mathematical functions, and may choose to characterize some features of their actual behaviour by additional environmental parameters. It would be pleasant if progress could be made toward defining standard argument ranges and standard relations that must be satisfied by the values computed for the mathematical functions. But their most important characteristic is their accuracy, and this should be subject to thorough certification. We welcome, in this direction, the publication of test software by Cody and Waite (1980).

The weaknesses of the mathematical functions should not be exaggerated; in practice we hope for the best and only occasionally, after thorough testing, have been disappointed. But the situation is unsatisfactory. Moreover, while the mathematical functions are built-in features of FORTRAN, they are part of the standard prelude in Algol 68 and not part of the definition of the ADA language at all. There is thus a trend toward excluding the mathematical functions from the standardisation of programming languages. This trend is sensible: the relevant expertise is not normally possessed by those who design or implement programming languages. But we have been arguing that the standardization of routines for mathematical functions requires more attention, not less. It is therefore essential that other groups of experts undertake this responsibility.

A final lesson of this experience is, we believe, that great care should be taken before further, possibly more elaborate, computational features are in any sense built in to programming languages.

5. LIBRARY MECHANISMS

It is possible for a subroutine library to be made available in <u>source-text</u> form, both by developers to sites, and by sites to individual users. However there are widely recognised advantages if the library can be made available in <u>compiled</u> form: that is, if library routines can be compiled separately, and the compiled code stored in a library file for subsequent selective linking to a user's program: selection of the required routines is automated; repeated compilations of the same code are avoided; and the correctness of the compiled code, after it has once been rigorously tested, is not put at risk by minor modifications to the compiler. It has always been NAG's policy to make its libraries available in compiled form wherever possible.

It is sometimes argued that library mechanisms are not "part of the language". We take a broad practical view of what should be considered part of a programming language. As an analogy, the order of letters in, say, the Russian alphabet could be regarded as "not part of" the Russian language, in that it does not affect the grammar or meaning of the language. But, without knowing the order of the alphabet, it is almost impossible to use a dictionary to help translate a Russian text; and in practice the order of the alphabet is usually taught at an early stage.

We consider that the design of any programming language - certainly one intended for numerical computation - must take full account of the need to construct and use libraries of compiled routines in a portable manner.

On some machines, the software provided to create libraries of compiled routines or to extract routines from them may have undesirable restrictions, e.g. on the ordering of routines within the library, on the size of the library files or, more seriously, on the ability to satisfy references from a library routine to a routine loaded with the calling program (not possible on the Hewlett Packard HP3000). On some systems libraries can only be handled in source form, e.g. Burroughs B5700 or the WATFIV system on IBM machines. The important thing is that such restrictions should not be a consequence of features of the language.

There does not appear to be any direct reference to separate compilation of subprograms in either the FORTRAN 66 or FORTRAN 77 standard. However it was certainly contemplated by the original designers of FORTRAN: Backus (1978) reports
that a document, dated 1957, "describes how symbolic information is retained in
the relocatable binary form of a subroutine so that the 'binary symbolic subroutine loader' can implement references to separately compiled subroutines".

In practice the only feature of FORTRAN 66 which normally causes any difficultly
when creating and using library files of compiled routines, is the use of BLOCK
DATA subprograms. This may be implicitly acknowledged in the FORTRAN 77 standard
which requires that each BLOCK DATA subprogram must have a unique name; the Notes
(Appendix B) state that "the name of a block data subprogram ... is available ...
for possible use within a computer environment".

Nor is there any explicit mention of separate compilation in the original Algol 60
Report (1960) or its subsequent revisions (1963, 1977). But it was clearly
thought to be desirable by many of those involved in the design of Algol 60 - see
the reports by Perlis (1978) and Naur (1978) - and has been implemented in most
Algol 60 compilation systems. Note however that this has involved a slight extension to the language - the introduction of a special basic symbol (usually <u>algol</u>)
to denote the body of a separately compiled procedure.

Thus both FORTRAN and Algol 60 have in practice allowed independent compilation of
library routines.

However our experience with Algol 68 has been quite different. Since it has been
described in some detail by Hodgson (1980) and by Ford, Hodgson and Sayers (1979),
we shall only summarise the main points here. In order to exploit the full power
of the language in designing the interface to numerical algorithms, the NAG Algol
68 Library contains not merely procedures, but also operators, definitions of
data structures (modes) and global variables. The individual modules in the
library cannot be compiled independently of one another; the compiler may require
information about other modules, already compiled or yet to be compiled (i.e.
separate, but dependent, compilation). When NAG came to implement its Algol 68
Library on different machines - ICL 1900, CDC CYBER, ICL 2900 - we found that the
Library mechanisms, although adequate in themselves, were different from one
another. Moreover non-trivial modifications to the source-text were required to
adapt it from one mechanism to another. A standard library mechanism for Algol 68
has now been proposed by Lindsey and Boom (1978) - but is different again from the
three already implemented. Hodgson (1980) concludes:

> "while diverse library mechanisms remain, an algorithm has no unique 'best'
> source code - it must be modified for the different library mechanisms. This
> situation only arises where some of the more sophisticated features of the
> language have been used, but nevertheless where this does occur, the resulting
> codes are semantically very different. The existence of and adherence to a
> defined standard satisfying the above requirements for the library mechanism
> is just as important to library writers and users as is adherence to the
> definition of the language itself!"

It is encouraging that the designers of ADA have addressed this issue. However,
some of the difficulties encountered in Algol 68 may still lurk concealed by the
last sentence of the following quotation from the ADA Reference Manual (Section
10.5):

> "The elaboration of those [library] units is performed consistently with the
> partial ordering defined by the dependence relations imposed by with clauses
> .... The order of elaboration of library units that are package bodies must
> also be consistent with any dependence relations .... The program is illegal
> if no consistent order can be found .... If there are several possible orders,
> the program is erroneous if it relies on a specific order (among the possible
> orders)."

(Note that a program is called erroneous if it violates a rule of the language, but compilers are not required to check for violations of that rule. If an erroneous program is executed, its effect is unpredictable.)

6. CONCLUDING REMARKS

Sometimes the question is asked: why does NAG develop a subroutine library in more than one programming language? This is not the place to discuss the economic pros and cons, except to say that the task of translating, say, the NAG FORTRAN Library into a new programming language is enormous. It would indeed be a great advantage if a single language version of the library were sufficient to satisfy all users. But facilities for mixed language programming are far from adequate. It is certainly true that in certain specific environments many NAG FORTRAN library routines can be called from programs written in Algol 60, Algol 68, Simula or Pascal. It is also true that the Revised Algol 60 Report, the FORTRAN 77 standard and ADA Reference manual all explicitly mention the possibility of calling routines written in a different language. But, in practice, difficulties arise: most obviously, first, from 2-dimensional (or 3-dimensional) array parameters, which may be stored differently in different languages (this affects 48% of NAG FORTRAN routines); and second, from parameters which are themselves routines or functions (this affects 21% of NAG FORTRAN routines). Therefore, for most practical purposes, libraries and calling programs must be written in the same language.

In this paper we have discussed how the design and provision of numerical subroutine libraries may be affected, for better or for worse, by features of different programming languages. Yet the choice of a language for library development is likely to be governed more by the amount of use of a language than by the benefits it offers. We therefore hope that the design of programming languages will take account of the needs of library developers and library users.

ACKNOWLEDGEMENTS

Much of the material in this paper is based on work and discussion by many members of the NAG project, and the paper owes a large debt to them. However the opinions expressed are my own. I am grateful to B. Ford, S.J. Hague, G.S. Hodgson, W. Kahan and J.K. Reid for helpful comments on earlier drafts of the paper.

REFERENCES

Aird T.J. (1977), The IMSL FORTRAN Converter: An Approach to Solving Portability Problems, in: W. Cowell (ed.), Portability of Numerical Software, Lecture Notes in Computer Science 57.

Backus J.W. (1978), The History of FORTRAN I, II and III, SIGPLAN Notices 13, no. 8, 165-180.

Brown W.S. and Feldman S.I. (1980), Environment Parameters and Basic Functions for Floating-point Computation, ACM Trans. Math. Software 6, 510-523.

Cody W.J. and Waite W. (1980), Software Manual for the Elementary Functions (Prentice Hall).

Delves L.M. (1977), Algol 68 as a Language for Numerical Software, in: W. Cowell (ed.), Portability of Numerical Software, Lecture Notes in Computer Science 57.

Du Croz J.J., Hague S.J. and Siemieniuch J.L. (1977), Aids to Portability within the NAG Project, in: W. Cowell (ed.), Portability of Numerical Software, Lecture Notes in Computer Science 57.

Ford B. (1978), Parameterization of the Environment for Transportable Numerical Software, ACM Trans. Math. Software 4, 100-103.

Ford B., Hodgson G.S. and Sayers D.K. (1979), Evaluation of Numerical Software Intended for Many Machines - Is It Possible? in Fosdick L.D. (ed.), Performance Evaluation of Numerical Software (North Holland).

Fox P.A., Hall A.D. and Schryer N.L. (1978), The PORT Mathematical Subroutine Library, ACM Trans. Math. Software 4, 104-126.

Hague S.J. and Ford B. (1976), Portability - Prediction and Correction, Software - Practice and Experience, 6, 61-69.

Herington D.A. (1981), Proposals for Procedure Interface Extensions, X3J3 paper 77(8)DAH-1.

Hodgson G.S. (1980), Library Structures in Algol 68, in: Hennell M.A. and Delves L.M. (eds.), Production and Assessment of Numerical Software (Academic Press).

Lindsey C.H. and Boom H.J. (1978), A Modules and Separate Compilation Facility for Algol 68, Algol Bulletin 43, 19.

Moré J.J., Garbow B.S. and Hillstrom K.E. (1980), User Guide for MINPACK-1, Argonne National Laboratory report no. ANL-80-74.

Naur P. (1978), The European Side of the Last Phase of the Development of ALGOL 60, SIGPLAN Notices 13, no. 8, 15-44.

Perlis A. (1978), The American Side of the Development of ALGOL, SIGPLAN Notices 13, no. 8, 3-14.

Reid J.K. (1980a), Complex Double Precision in Association with FORTRAN 77, SIGNUM Newsletter 15, no. 3, 11-17.

Reid J.K. (1980b), Language Requirements for Numerical Software, paper presented to ISO/TC97/SC5 FORTRAN Experts Meeting, October 1980.

Reid J.K. and Hopper M.J. (1980), Production, Testing and Documentation of Mathematical Software, in: Hennell M.A. and Delves L.M. (eds.), Production and Assessment of Numerical Software (Academic Press).

Schonfelder J.L. (1976), The Production of Special Function Routines for a Multi-Machine Library, Software - Practice and Experience 6, 71-82.

Schryer N.L. (1981), A Test of a Computer's Floating-Point Arithmetic Unit, Bell Laboratories Computing Science Technical Report No. 89.

DISCUSSION

Kahan   I would like to amplify the discussion in this paper of the standardisation of elementary arithmetic functions. I believe we should establish a (presumably) small number of "Semantic Packages", each of which characterises exponent range, accuracy, exceptions, etc. of every arithmetic function for each floating-point format supported by a range or family of machines. This proposal is discussed in detail in my paper later in these proceedings. One instance of such a semantic package is available from IBM as its documentation for arithmetic operations and elementary functions; another instance is Brown's model for arithmetic operations when supplied with values for the pertinent parameters.

The reason behind my request, that the semantic package be not standardised as part of any programming language, is that the expertise required to specify arithmetic, elementary and matrix functions is not normally possessed by those who design and/or implement languages. Rather than dump our complaints upon their heads, IFIP Working Group 2.5 should help them out by pointing to good instances of semantic packages, and to good examples of elementary functions (e.g. IBM's 370 functions by Hirondo Kuki et al., or the book by Cody and Waite); we should try to specify how a semantic package can be framed so as to interact properly with appropriately framed syntax of any programming language. Most important, we should join with the language designers and implementors in persuading the providers and/or purveyors of computing systems that the ensemble of floating-point arithmetic operations, elementary functions, etc. should be jointly the responsibility of hardware implementors and experts in numerical software, both employed by that provider - and/or - purveyor. Otherwise the compiler writers and applications programmers will be obliged to patch the holes left by inexperienced hardware and software implementors, and some of those holes are much too big to patch.

Ford In the medium term Du Croz and I would agree with your proposal and where possible support such work. In the short term rather than attempt to rewrite inadequate elementary functions routines available to users on a number of different machine ranges we seek to persuade and cajole the manufacturer to provide adequate facilities.

Fox Has the NAG Library been implemented for the Cray-1?

Ford Yes, the Library is available for the Cray-1. But it is only the standard NAG Library for scalar-processing machines, compiled by the rather effective Cray FORTRAN compiler (which seeks to identify and handle any possible vector operations in FORTRAN source-text). Some work is in hand to prepare a version of the source-text that is partially optimised for the Cray-1, but still based on the same underlying algorithms. A separate source-text, from a common base library, will be required for the CYBER-205. In the medium term, in collaboration with other groups, we must invest many man-years in preparing new algorithms for vector-processing and parallel-processing machines.

Fox This leads to a more general question. Will FORTRAN now experience new pulls in opposing directions as users restructure programs to force vectorization or other adaptations to new hardware concepts? The emerging parallel, pipeline and vector computers clearly impose divergent language requirements.

Ford As indicated above optimal use of hardware will probably require different algorithms for the different systems. Some code optimisation can be achieved by automatic source text transformation by software tools such as the FORTRAN Intelligent Editor being designed by Hague.

Fox "The NAG Mark 8 FORTRAN Library has 466 (primary) routines and a total of 4380 parameters". If the error flag (e.g. "IFAIL") in the calling sequence were eliminated by using a central error-handling facility (as in the PORT Library), and the array parameters defining scratch space were cut out by using a dynamic stack allocation scheme, presumably 466 to say 1000 calling sequence parameters could be eliminated. Wouldn't the simplification be worth it?

Ford The problem with the proposed approach is that a programmer then requires to include the error-handling facility and dynamic stack allocation scheme in his program (for execution or for despatch to other users) even if he only wishes to use a single library routine. However the PORT approach does enjoy obvious merits.

Nelder Two restrictions on calling sequences are irritating to a statistician user working interactively.

(1) A statistical algorithm often produces, at least potentially, many output structures, but the user may frequently not wish to inspect more than a small fraction of them. It is highly desirable, therefore, that he not have to supply actual parameters for output structures not needed in a particular call.

(2) General-purpose languages, which are still essentially batch-oriented in spirit, fuse the allocation of actual parameters with the invocation of a procedure. For interactive use there is good reason to separate these two actions; more generally there is good reason to allow the user to change only the subsets of parameters between calls, instead of writing out the full list again.

We are unlikely to get these facilities in general-purpose languages, so that there will be a continuing need for problem-oriented (very-high-level) languages for solving specific problems. It should be a requirement in the specification of a general-purpose language that the language make it easy to write problem-oriented languages for specific areas.

Gentleman While I concur that calling sequences are a major problem, especially in languages that demand formal and actual conformance in number and type, there is another solution beyond the keywords you suggest.

This solution is to group parameters into one or more records, and pass these records instead. This has the same advantages as the keywords in that actual parameters are associated by symbolic name, and furthermore the same set of parameters can be used, as in common, in several procedure invocations. Variant records allow passing only the subset of parameters that are meaningful.

Unfortunately the machinery provided in many languages is inadequate for this. Some languages, e.g. FORTRAN, do not provide records. Some languages, e.g. Pascal, do not permit us to obtain references to arbitrary data entities, and consequently make it difficult or impossible to include output parameters in such passed records. Many languages, including both Pascal and Ada, make it impossible to pass procedures as elements of such records. And of course languages that do not support variant records still imply passing meaningless parameters, in the form of record fields that have no meaningful value in certain contexts.

Reid It is much easier to read a subroutine call with keywords, and ease of use is the fundamental user requirement.

Brown The comment in your paper about poor implementation of the double precision square root function on a well-known minicomputer raises the question of whether mathematical functions, packages and libraries should be included in language standards, and how and by whom they should be selected, defined and implemented. This question was suggested earlier in an exchange between Feldman, who asserted that libraries increase the power of a language, and Rice, who replied that such a conclusion is appropriate only to the extent that they are included in the language definition. Meissner, at the recent SIGNUM meeting (July 1981) expressed the concern that functions included in a language standard might be implemented poorly by many compiler writers, while functions omitted from the standard might be implemented poorly by many users.

Ford Kahan spoke earlier in this discussion session about elementary functions, arguing that the general availability, in the medium to long term, of "semantic packages", would obviate the need for the inclusion of such functions in language standards. In the long term I would support his approach - provided individuals (e.g. Cody, Kahan, ...) and groups (e.g. W.G. 2.5, IEEE, ...) could be found to complete the necessary "semantic packages". In the short term I would prefer to rely on the manufacturers (and hence upon a standard) rather than many individual users, many of whom might get the algorithms wrong and all of whom would be taking a methodological step backwards.

With regard to packages and libraries, and their inclusion in language standards we would certainly welcome the definition in all future standards of the programming environment (e.g. the APSE in ADA) of a language processor as well as the definition of the language itself. In particular this should include definition of library and package incorporation mechanisms. Hence the mechanism by which these broadening facilities could be included in the language/programming environment would be defined. This is vital. Whether specific libraries or packages should be included in a language specification is a moot point.

Meissner Schryer has specified a set of tests on environmental parameters. A similar test for the basic maths library has been suggested by Cody and White. A step towards standardization could consist of a testing procedure specified in an appendix to the standard.

George Paul I have two basic comments on the presentation. The talk focussed on the length of argument lists required by FORTRAN and ALGOL. I also believe you consider the linkage conventions required and their attendant overhead. With the addition of array processing features to FORTRAN 8X, the FORTRAN processors will also adopt dope vector descriptions for array-valued arguments passed to subprograms. While this will shorten the required argument lists to be comparable to ALGOL, it will add overhead for scalar implementations. For example vector-valued or array-valued arguments must be assumed to have an arbitrary stride in memory which will be provided via the dope vector at run time. As a result the indexing computation on a scalar machine will be forced to perform additional computation which can not be removed by optimization unless some additional mechanism is provided in the language which allows the user to specify that a particular argument will always be passed with a unit stride. Such a mechanism is under consideration for FORTRAN 8X.

The second remark has to do with maintaining precisely identical results for a given argument value between scalar and vector versions of the same intrinsic library function - say exponential. Assuming that one wishes to provide the highest possible performance in both scalars and vector modes it is likely that different approximations for the same intrinsic function will be optimal for the two modes. This is primarily the result of the overhead costs associated with the range reduction for the various elements of the array-valued argument in vector mode. The natural choice will be to select an approximation for vector mode which spans the broadest possible argument range to minimize the range reduction costs, the cost of selecting different coefficients for different approximation ranges and/or the costs of computing total different approximation formulas for the different components of the array-valued argument. Note also that if this approach is taken then the valid argument ranges for the scalar and vector versions of the library functions. I believe that both WG 2.5 and SIGNUM should raise this flag and make it clear that this is unacceptable and what the consequences to users might be.

THE RELATIONSHIP BETWEEN NUMERICAL COMPUTATION
AND PROGRAMMING LANGUAGES, J.K. Reid (editor)
North-Holland Publishing Company
© *IFIP, 1982*

EXPERIENCE WITH THE OLYMPUS SYSTEM

K.V. Roberts

Culham Laboratory
Abingdon,Oxford, OX14 3DB, UK
(EURATOM/UKAEA Fusion Association)

OLYMPUS is a programming system designed for use by computational
physicists using Fortran 66. The methods are however expected to
have wider application to other languages and application areas.
OLYMPUS programs are intended to have a quality and generality comparable to those of the published scientific literature, and software engineering techniques have been developed for achieving this
including word-processor facilities for the production of code and
documentation to uniform standards. The article discusses experience
in using the system at Culham and elsewhere over a 10-year period and
makes suggestions for future scientific programming languages.

1. INTRODUCTION

The OLYMPUS system [1] was started at the Culham Laboratory about 10 years ago as
a set of practical techniques for organizing the large simulation programs and
utility software used in computational physics and especially in fusion research
[2]. Its methodology is consistent with the structured programming concepts of
Dahl, Dijkstra and Hoare [3] but complementary, being concerned less with the
theoretical structure of languages or with programming style than with basic software engineering questions such as design, layout, construction, operation, documentation and portability as well as with the relation between different programs
belonging to the same project.

The central idea is to use our understanding of well-established disciplines such
as the physical and natural sciences, mathematics and engineering as a guide to
how the relatively new field of software engineering ought to develop, and several
analogies have been consciously exploited including those with the published
scientific literature, theoretical physics and mathematics, comparative anatomy,
architecture and engineering. Many of the OLYMPUS techniques can be used to advantage with any type of software, programming style or language and some were in
fact introduced first for Algol 60, but currently the published system [4-6] is
only available for Fortran 66 since that is the language commonly used by physicists at the present time.

2. THE STANDARD PROGRAM FILE

According to our first analogy [7] we think of the program source listing as a
document to be read and understood by many people over a period of years, often
passing through several versions. A great deal of expensive time and effort is
spent in trying to understand program listings and it is therefore worth making
them as intelligible as possible. But many of the necessary techniques were
already developed long ago, in connection with the publication of scientific and
mathematical books and journal articles, and these have therefore been carefully
examined, and exploited where appropriate for OLYMPUS program listings.

Books and articles generally have a standard structure which may include a title
page, preface, contents list and indexes; a division into chapters, sections and
subsections, each with a heading and often with a decimal numbering scheme; and a
clear layout which makes use of blank lines for separation,and indentation of
numbered equations from the explanatory text. Mathematical and scientific works
employ a standard notation and nomenclature with a concise symbolism which makes
good use of the available range of typefaces and character sets, including the
Greek alphabet and special signs. Explanatory tables and diagrams are used where
appropriate together with cross-references and footnotes. There is a well-developed
publication and distribution system for the scientific literature and a refereeing
and editing procedure to maintain quality and to minimize the possibility of
errors. Open publication also helps to remove errors as well as to encourage consistency. An efficient referencing scheme enables published works to be quickly
located. Finally, word-processing facilities are being increasingly used both by
authors and also by publishers to speed up the production process.

As a first step we adopt the same basic structure called the Standard Program File
(SPF) for every OLYMPUS program. The SPF listing is intended to have an organization and clarity similar to that of a scientific or mathematical textbook and consists of 4 main sections each headed by a control statement:

> 1.   C/ DOCUMENTATION
> 2.   C/ COMMON
> 3.   C/ FORTRAN
> 4.   C/ TEST DATA

These sections are further subdivided into modules headed by:

> C/ MODULE (name)

The SPF is system-independent and is converted by a utility program called the SPF
Processor into the particular type of input file required by a specific computer
system. There are no system-dependent JCL (Job Control Language) statements within
the SPF itself, the only control statements that are used being the OLYMPUS statements with C/ in columns 1 and 2, which are sufficient to inform the Processor of
the structure of the program, and which together with other features to be explained later considerably enhance its portability.

3.   DOCUMENTATION SECTION

The Documentation Section always contains the modules:

> TITLE   Title Page
> INTRO   Introduction
> INDSUB   Index of subprograms
> INDCOM   Index of COMMON blocks
> INDVAR   Alphabetic index of COMMON variables
> INDBLK   Index of COMMON variables by blocks
> MINDEX   Master index

with standard names and in standard order. The indexes are alphabetically ordered
with standard layout and enable the meaning and location of a subprogram, COMMON
block, variable or array to be looked up quickly. The layout had originally to be
set up by hand, but except for INTRO each of the first 6 modules is now constructed
and updated automatically by a word-processing utility called the GENSIS Generator
which uses the single free-format MINDEX master index as its input. GENSIS also
constructs the COMMON blocks themselves together with several of the standard subroutines. The programmer therefore has little difficulty in establishing and maintaining a clear standard structure for the SPF listing, and the user can readily
find his way about.

4. PROGRAM ARCHITECTURE

Many of the programs used in computational physics have similar tasks to perform and differ only in the details of the equations to be solved [2]. One important type of calculation is the initial-value problem, in which data are set at some initial instant, and then the evolution of a physical system is followed step by step as time proceeds. The OLYMPUS system was originally designed for programs of this type but was later found to have other applications. Both in theoretical physics and in mathematics it is usual to employ a common structure and notation wherever possible, an excellent example being classical dynamics in which the state of the system is defined by coordinates and momenta (q,p) and its evolution by the Hamiltonian function $H(q,p)$, and the OLYMPUS architecture has been largely based on this analogy.

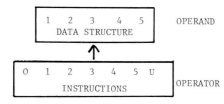

Fig. 1
Program Architecture

The architecture of an OLYMPUS program consists of two parts, the Data Structure and the Instructions, illustrated in Fig.1. The data structure is represented by Groups of decimally-labelled COMMON blocks and the instructions by Classes of decimally-labelled Fortran subprograms. Each group and each class has a similar function in all OLYMPUS programs:

Table 1.

Overall Program Structure

COMMON BLOCKS		SUBPROGRAMS	
Group	Purpose	Class	Purpose
		0.	Control
1.	General OLYMPUS data	1.	Prologue
2.	Physical problem	2.	Calculation
3.	Numerical scheme	3.	Output
4.	Housekeeping	4.	Epilogue
5.	I/O and diagnostics	5.	Diagnostics
		U.	Utilities

By analogy with Hamiltonian dynamics we think of the state of the calculation as defined by the current values of the COMMON variables and arrays and picture the Fortran subprograms as operators which act on this state to change it. This 'dynamical' architecture which is very familiar to physicists has proved highly successful in practice since it makes the working of the code transparent. Diagnostic facilities are available in OLYMPUS to display the values of individual COMMON variables and arrays or of complete blocks in a convenient annotated form so providing a 'snapshot' of the current state of the calculation. Only a limited use is made of local variables and subprogram arguments.

We have found in practice that programs written in an Algol block-structured language are harder to standardize since a wide variety of architectures is available. After some experimentation, we have found it appropriate in Algol 60 to use an outermost 'global' block to set array dimensions, and an inner 'common' block of data and procedures which is closely similar to the set of COMMON blocks and Fortran subprograms of Table 1, and to choose the same names as in OLYMPUS Fortran. This makes it easier to pass from one language to the other. Extensions of OLYMPUS to further languages will probably be handled in a similar way.

5. DATA STRUCTURE

A large part of OLYMPUS is concerned with the data structure, its organization, maintenance and initialization. A typical computational physics program may involve several hundred COMMON variable and array names together with a considerable number of internal names. To bring some order into this situation, initial letters are used not only to distinguish between real and integer types as is usual in Fortran, but also between COMMON and internal names, loop indices and formal parameters. Type LOGICAL uses 'L' as the second letter. These conventions greatly reduce the chance of name clashes and enable the reader to understand the listing.

In general an OLYMPUS program contains only one copy of each labelled COMMON block. This is constructed automatically by GENSIS and forms part of the COMMON section of the SPF. It is inserted into a subprogram by the Processor control statement.

C/ INSERT <name>

Implicit equivalencing of COMMON storage areas is therefore not normally used, although it can be employed in special cases with adequate explanation. Similarly, a COMMON name normally has a unique meaning, (explained in the indexes), and duplicates are not allowed. COMMON array dimensions are defined symbolically in MINDEX and the numerical values which are inserted by the Processor may quickly be changed.

The COMMON names are organized into Groups as indicated in Table 1, and then further into blocks, each block having a decimal number as well as a module name which is chosen to be the same as its Fortran name, and a purpose which is defined by an entry in the INDCOM index, e.g.

C   COMESH   Mesh and auxiliary variables   C3.2

The general OLYMPUS data blocks of Group 1 are intended to be independent not only of the computer system but also of the particular program. At present they include I/O channel numbers, timestep control, date and time information, labels, and diagnostic control parameters. There are many possibilities for extending this group, and at Culham there are blocks containing fundamental physical constants, physical coefficients, standard printing and plotting variables and so on, although these have not yet been officially included in the first group.

The remaining groups are organized by the programmer, with the general aim of making the data structure as rational and intelligible as possible. During the development of a program it is often necessary to reorganize COMMON, but this is readily done by editing the MINDEX master index online, and by adding or removing C/ INSERT statements as appropriate.

6. SUBPROGRAM STRUCTURE

All OLYMPUS programs share the same overall subprogram structure. This is achieved quite simply by running them under the control of a Class 0 library subprogram <0.3> COTROL which references a particular set of subprogram names, and which therefore requires that subprograms with these names should be provided for the loader. Some of these 'principal' subprograms are constructed automatically by GENSIS and others are written by the programmer. These can in turn call any other user-written subprograms, but the framework enforced by COTROL is enough to make the overall structure of the program clear.

There is a useful analogy here with comparative anatomy. In biology there are millions of species which are organized according to the Linnaean system into phyla, classes, orders, families and genera. It is possible to study these as a whole because of the evolutionary relationship between them, which means that different organisms share common features which can be given the same names. Clearly a similar degree of organization would be useful in computer science, but

because of the lack of a common origin, programs which perform the same task often
have arbitrarily different structures. OLYMPUS attempts to avoid this as far as
possible, so reducing programming effort and making it easier for a user to move
from one program to another.

Most of the common framework is concerned with initialization of the data structure, which is carried out by the Class 1 subprograms of the Prologue to be discussed in the next section. Classes 2 and 3 are represented by their leading
members which are intended to control these parts of the calculation:

<2.1>	STEPON	Step on the calculation
<3.1>	OUTPUT	Control the output

As written by the programmer, subroutine STEPON should consist of an annotated
list of calls to subprograms <2.2>,<2.3>,... which perform the various stages of
the timestep cycle in turn, while subroutine OUTPUT calls subprograms that deal
specifically with various types of output. Since they always appear at the same
places in the decimally-ordered SPF listing they provide 'contents lists' for these
important sections of the program. Classes 4 and 5 do not contain control features
as such, but their principal subprograms ensure that run termination and diagnostics are organized in a standard way.

The main program is always a dummy, its purpose being to define COMMON storage
areas and to call the standard control routines. Each module has a decimal name,
like a chapter number, e.g., <2.1> STEPON is C2S1.

7.   INITIALISATION OF THE DATA STRUCTURE

Although Class 2 contains most of the physics, much of the complexity of a program
can appear in Class 1, and since similar tasks have to be carried out for any
calculation it is worth organizing these once-for-all. At the beginning of a new
run subroutine COTROL calls the following principal subprograms:

Table 2.

Principal Class 1 Subprograms

<1.1>*	LABRUN	Label the run
<1.2>*	CLEAR	Clear variables and arrays
<1.3>	PRESET	Set default values
<1.4>*	DATA	Define data specific to run
<1.5>	AUXVAL	Set auxiliary values
<1.6>	INITAL	Define physical initial conditions
<1.8>	START	Start to restart the run

Those marked* are automatically provided by GENSIS. A somewhat different pattern
is used for a restart, involving <1.7>* RESUME, which reads the previous state of
the COMMON storage from disc.

Subroutine LABRUN reads data cards and labels the run in a standard way, while
CLEAR sets COMMON variables and arrays to zero or to logical .FALSE. as appropriate.
It is constructed by GENSIS and is updated automatically when new variables or
arrays are added to MINDEX or the dimensions changed.

A very straightforward pattern is used for setting data values and control parameters in OLYMPUS programs. This is done in three stages. First, all those
values that can be independently specified are given default values in <1.3>
PRESET. This subroutine is divided into decimally numbered sections corresponding
to the groups of Table 1, and then into decimally-numbered sub-sections corresponding to the blocks within each group, each section or sub-section having a
heading which is the same as the description in the index INDCOM. If required
this framework can be constructed automatically by GENSIS. Within each block the
variables and arrays are assigned values in alphabetical order. It is often convenient to choose default values that define a typical run or an explanatory test
case.

Next, any values that the user wishes to change are overwritten by NAMELIST data
read in (1.4) DATA. The procedure is so standard that only a single input statement is needed:

$$READ(NREAD,NEWRUN)$$

where NREAD is the input channel and NEWRUN is the NAMELIST name corresponding to
the data for a new run, (a similar procedure being used for a restart with namelist RESET). The programmer indicates in MINDEX names that are to appear in these
namelists, and then the namelist declarations are constructed automatically by
GENSIS. Although the NAMELIST facility is not part of ANSI Fortran 66 or 77 it has
many advantages; only those few values that are to be changed need be specified
by the user, they are defined mnemonically in the form (name)=(value), the order
and format are arbitrary, and the single input statement is simple both to implement and to understand. As a result NAMELIST is increasingly used by computational
physicists, and I would certainly recommend that it should be incorporated not only
into a new version of Fortran but also into other languages.

The final stage is carried out by (1.5) AUXVAL, which again has the block by block
framework of PRESET. The purpose of this subroutine is to set auxiliary values
that are used throughout the program but depend on the input data: for example if
NX is the number of mesh points, we might set NXP1=NX+1.

Because this standard 3-stage pattern is used by all OLYMPUS programs it is largely
unnecessary to explain the data setup for a particular case, although an individual
program may have specific features, such as the input of a large block of physical
coefficients from a disc file or the definition of an analytic formula by a function subprogram, which are not covered by the NAMELIST method.

The purpose of INITAL and START is to separate the set-up of the physical initial
conditions from any housekeeping that is needed before beginning the first timestep.

8.   STRUCTURE AND LAYOUT OF A SUBPROGRAM

We think of a subprogram as analogous to a chapter in a textbook. It has a decimal
number within the program as a whole, and a purpose which is defined in the index
INDSUB and on a heading card at the beginning of the subprogram itself, but the
hierarchical structure is carried further down, and each of the major subprograms
is organized into decimally-numbered sections and sub-sections with definite tasks
to perform, and these are also explained by headings. Finally, individual points
of detail are explained by ordinary comments.

Experience has shown this hierarchical form of structure and documentation to be
very necessary for large programs, and that the mere addition of a large number of
comments is not enough to make the working clear. This is especially true for
programs written in assembly language which often have a comment on every line and
yet remain difficult to follow.

To make the subprogram listing as readable as possible, definite columns have been
chosen for the decimal numbering, headings, subheadings, comments, executable statements, and declarations. Blank lines are used for spacing, and ruled lines ------
are drawn across the page at the beginning of each section. The Fortran statement
numbers are in ascending order and are correlated with the decimal numbering of the
sections and sub-sections.

Although it is found as expected that this arrangement greatly improves the readability of the code, and does indeed save time for experienced OLYMPUS programmers,
it takes time to explain the precise conventions to newcomers to the system, and it
also more difficult to select the correct columns on-line than on a coding sheet
although the tab facility of an editor can often be used for this purpose. To
remove these problems, and to provide other advantages,a word-processor called the

COMPOS compositor has been written. This accepts a free-format on-line input in which the sections and sub-sections are simply indicated by # or ## respectively and arbitrary statement numbers can be used. It then automatically organizes the numbering and generates the desired OLYMPUS layout. The output can in turn be 'proof-corrected' with additional free-format corrections and then reprocessed by COMPOS, when the numbering will if necessary be updated.

So far COMPOS is only in use at Culham, but it has been written in OLYMPUS Fortran 66 and will be published in the Computer Physics Communications Program Library [8] in due course. It can to a large extent be used for the conversion of existing programs to OLYMPUS form and this facility will be extended. The construction of well-formated code is a natural application for word-processors and it seems certain that their use will become widespread.

References are included in OLYMPUS subprograms to the equation numbers of an associated paper or report, which itself refers to the decimal numbering of the subprograms. This cross-referencing is of considerable assistance in understanding a large program. Another way of improving the readability of the listing is to put the comments in lower case in order to distinguish them from the Fortran statements and this is now done in OLYMPUS programs.

As already mentioned, distinct initial letters are used for internal real (Z) and integer (I) variables and arrays as well as for loop indices (J) and formal parameters (P,K) and this also helps to improve readability.

Symbolic names rather than numerical values are used for I/O channels, array sizes DO-loop ranges, character codes, physical and mathematical constants and so on. This has two advantages: the meaning of the constant is clearer to the reader, and the value need only be assigned in one place (e.g. in PRESET) where he will know where to find it, so that it can be changed consistently throughout the program by altering this one statement.

A general rule is to think of each subprogram as a document to be read by other people and to design it for maximum readability, just as one would write a chapter in a mathematical or scientific book.

9.   PORTABILITY

Another rule is that the program should be as portable as possible, that is, it should run on different types of computer system with as few changes as possible to the source code. Ideally no changes whatever should be needed to the SPF. So far this ideal has not usually been completely attained, but typically it may only be necessary to change two or three statements in a program of several thousand lines.

Apart from minor exceptions an OLYMPUS program is written in standard ANSI Fortran 66 [9,10]. The main exception is NAMELIST which is available in most implementations but can if necessary be simulated, e.g. by replacing the NAMELIST data by Fortran statements in a modified version of subroutine DATA which is recompiled for each run (the 'Fortran Data' technique). Facilities which are not available in Standard Fortran are usually provided by library subprograms, an equivalent version written in non-standard Fortran or in assembly language being provided for each type of system on which OLYMPUS programs are run.

Many of these library subprograms are collected together in the OLYMPUS Control and Utility Package. This contains a dummy main program, the control subroutine COTROL, dummy versions of all the principal subprograms, and a package of utility subprograms called CYCLOPS. Versions have been published for 3 types of computer system [4-6] but the package has been implemented for a wide range of systems in several countries. The package contains a working program called CRONUS which may be used for test and demonstration purposes, and also provides a template on which a real program is gradually built up by progressive replacement of the dummies.

The programmer thus has a working program from the start which is often an advantage during the development stage.

An OLYMPUS SPF is not a complete program in itself, but runs within a standard environment which is provided by the OLYMPUS package. This defines parameters such as channel numbers that are specific to the type of computer or to the local installation, and also 'translates' system routines such as those dealing with the time and date into a standard form. The standard parameters are at present contained in block [C1.1] COMBAS (basic system parameters) and could with advantage be considerably extended to include several other features of interest at run time e.g., type of computer, store size, word length, accuracy and so on.

The main areas of difficulty in making OLYMPUS programs fully portable have been input/output and character manipulation. This is not surprising since random access storage devices were not available when ANSI Fortran 66 was defined, and since Fortran was not originally intended as a language for text manipulation although it turned out to be quite good for this purpose and is widely used. These problems should largely be removed in Fortran 77. However, one interesting way of avoiding the problems of dialect Fortran is to provide alternative versions of statements within the SPF which are marked by suitable codes, the appropriate version then being selected by the SPF Processor. Such a facility has not yet been implemented for OLYMPUS, but it has been used by Crees [11], while the LRLTRAN language available at the National Magnetic Fusion Energy Computer Center (NMFECC) at Livermore [12] has a sophisticated macro-preprocessor that can be used in this way.

It was found necessary to make special provision in the SPF Processor for automatic conversion to double-length real arithmetic. Provided that care is taken in the writing of the code, single-precision Fortran can be converted to double precision by introducing the declaration

IMPLICIT DOUBLE PRECISION (A-H,O-Z)

together with a set of statement functions that convert the single-length Fortran functions to their double-length equivalents. This can be achieved by a pair of C/ INSERT statements which the Processor then expands.

One application for this facility is to enable test results to be as nearly as possible identical on the different computer systems so that conversion from one type of system to another can be checked. A 32-bit word length is usually insufficient to give high accuracy, so that the results change in going to 60-bit or 64-bit words. On the other hand if the code is written in double precision for a 32-bit machine space will be wasted if it is run without alteration on machines with the longer word lengths. Another reason is that conservation identities are often not satisfied with sufficient accuracy when 32-bit words are used, so that they cannot be used as a check on the validity of the code. The conversion facility of the SPF Processor enables the code to be written in single length and used in this form for production runs, but switched to double length during testing or for the production of Test Run output. Of course this facility should really be an integral part of the language and the compiler.

Used with care, these facilities make it possible to write OLYMPUS programs in SPF form that will run on any type of computer system with adequate capability. One advantage of this is that computational physicists frequently exchange programs or transfer them from one type of machine to another. Having to make unnecessary changes to the code wastes time and effort and causes delays in implementation, especially since such programs are often under continual development and pass through many versions. Rather than convert several versions one after another, or to stay with an old version, it is much better to be able to run a new version as soon as it is received.

Two major difficulties remain. One is the introduction of array and vector processing machines that require a generalized version of Fortran to achieve their full performance. The solution here should be to standardize this generalized version and to implement it also for the conventional type of machine, otherwise these will not be able to run many of the programs that will be written in the future even at reduced efficiency. The other difficulty is the increasing use of subprogram packages and libraries. Clearly it is efficient for a program to be built up so far as possible from prefabricated subprograms since this reduces programming effort and delays as well as the likelihood of errors. Such subprograms may in practice be regarded as an extension of the Fortran language, so that unless they are generally available the effect is to make programs non-portable between installations even if they have machines of the same kind. Here the solution must be to encourage the development and dissemination of standard subprogram packages to cover as wide a field of application as possible. These should be regarded as part of the regular scientific literature like books and journals, the more important ones being available at all major installations and the others obtainable by a distribution scheme equivalent to a National Lending Library, perhaps via a national or international computer network provided that a suitable charging scheme can be worked out.

10. THE PUBLICATION OF COMPUTER PROGRAMS

One of the problems in the development of good software standards is that computer programs are not 'visible' like other products of art, science and engineering. In most fields of endeavour, examples of excellent past work are available for all to see, or at least, all students who are working in the field. This tends to set both standards and conventions. In computing on the other hand there are very few examples of large programs whose source code is published in visible form, and publication is mostly confined to short algorithms. As a result, most people who begin work on large programs have not seen such programs before and techniques progress only slowly and remain fragmented. Illustrative material is not available for advanced programming courses.

In aiming to bring the readability of OLYMPUS listings up to the standard of scientific and mathematical textbooks and papers, one of our intentions has been to publish them in a similar way. There is already a vehicle for the publication and distribution of computer programs in digital form in the Computer Physics Communications (CPC) International Physics Program Library [8], and several OLYMPUS programs [13-17] as well as the OLYMPUS Control and Utility Package [4-6] are available from this library. The CPC Library is associated with the Computer Physics Communications journal which publishes the program writeups including selected output from one or more Test Runs. Both the program and writeup are edited and refereed, the procedure being similar to that for a regular scientific paper. The purpose of the printed test run output is to enable a subsequent user, who may need to carry out some conversion to run the program on a different type of machine, to check that the numerical results remain unchanged. Of course this cannot provide a complete check but it may help to signal major errors. With OLYMPUS programs very few such changes if any should be required.

The CPC scheme enables physics programs to be published in digital form and thus made generally available to any would-be user, whether or not he subscribes to the journal, since there is a subscription scheme for individual programs [8]. However it does not yet solve the problem of the lack of visibility of computer programs since the source listings themsleves are not normally printed in the journal unless to illustrate some programming method or documentation technique. Indeed a typical large Fortran program is usually much too long and diffuse to be suitable for publication in a journal: for example the OLYMPUS code ATHENE 1 [16] contains 6698 lines which would correspond to about 120 A4-size pages, and many programs are considerably longer than this.

One solution is to publish the program listing together with a copy of the writeup as a Laboratory Report which would supplement the publication in CPC, and this will be done at Culham for some OLYMPUS examples. Provided that the report is referenced in the journal article it will then be available to all interested readers. It has been found that the listings can be reduced in size by a linear factor 2 and still remain sufficiently readable, so reducing the number of pages by a factor of 4.

In the past many published listings have been photographic reproductions of ordinary line printer output and therefore of rather poor quality. However now that the text of scientific reports and papers is increasingly being produced on word-processors which use high-quality printers or computer type-setting, there is no reason why the source-listing should not be printed in the same way. It is particularly recommended that Fortran comments should make full use of lower case since this allows them to be much more readable.

Nevertheless ANSI Fortran 66 or 77 has many disadvantages as a medium for human communication and it is useful to consider some of these and to ask how they might be removed. First, there is the obvious point that much of the area of a printed listing is wasted: it simply remains blank. Algol 60 and related languages allow several statements and comments on one line, as well as multiple assignment statements. Even assembly languages allow a comment on the same line as an instruction. Standard Fortran does not allow this flexibility so that the listing is considerably more bulky than it need be, although the restrictions are removed in some dialect versions of Fortran [12,18]. However, it is a simple matter to incorporate these generalizations into a preprocessor and this may be done in a future version of the OLYMPUS system.

A more serious problem is the diffuseness of the Fortran language itself compared to the symbolism of analytic mathematics. In a study [19] of a 3-dimensional magnetohydrodynamics code it was found that the difference formulas expressed in Fortran required 100 times as many characters as did the original differential equations when written in the usual vector notation. Such a diffuse language is not a good medium for the communication of algorithms between people, and indeed it is well known that computer programs are much harder to understand than mathematics.

One might try to improve the intelligibility of programs in two different ways. Either the programming language itself might be made more powerful and symbolic and therefore more concise, as has been attempted with Algol 68, or one might develop an algebraic preprocessor that constructs and codes the difference formulas automatically, so that it is not necessary to publish the program but only the input to the preprocessor, which although much more concise nevertheless contains all the essentials of the algorithm. Both approaches have been used at Culham in a symbolic system based on Algol 60 [20], although this has not been incorporated into OLYMPUS since Algol 60 is no longer widely available in physics laboratories.

In planning at the present time any new language for scientific use, or any extension of an existing language such as Fortran, there is one important point on which attention should be focussed. Word processors are now being increasingly used for the preparation and printing of scientific documents that require a full range of mathematical notation, including Greek letters, subscripts, superscripts, italics, under- and over-lining, special mathematical signs and so on. It takes some years to design and implement a new programming language, and one must expect that well before this can be done, most laboratories will have been equipped with terminals and printers that can handle this extended character set. It would therefore be a mistake to restrict a new language or language dialect to the 49 characters used in Fortran 77 (upper-case letters, digits, space, and 13 mathematical signs), or even to the 95 ASCII printing characters. Rather, one should now consider whether any significant advances could be made in programming methodology and in the readability and conciseness of program listings if the full range of mathematical symbolism were employed.

There is also no reason why a program listing should not include comments in the form of explanatory diagrams, since a printer such as the Versatec is capable of combining text and graphics, and a preprocessor can readily strip off the graphics before sending the source file to the compiler.

According to the OLYMPUS methodology there are three aspects to a source file as indicated in Fig. 2. It must be created and updated by a programmer, possibly with the aid of a code-generation facility (A); it is a document to be read both by the original programmer and also by later users (B); and it is a set of instructions for the compiler (C).

In the early Fortran systems, based on cards, the formats of all three aspects were the same. That is, the symbols that were punched on the cards were printed on the listing and were also sent to the compiler, although the compiler could ignore the comments and details of layout that were intended to make the listing more intelligible to a human reader. When OLYMPUS began to be used online, it was however realised that what a programmer finds it most convenient to type at the

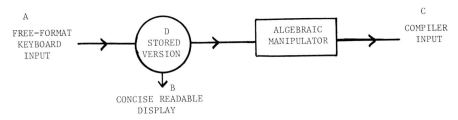

Fig.2 Generation and display of scientific program

keyboard (A) and what he likes to read (B) are not necessarily the same. This led to the COMPOS facility, by which the input (A) is made as simple as possible, and automatically converted into a readable form (B) which obeys the OLYMPUS conventions. However it is still true that B and C are the same, except for the minor changes carried out by the SPF Processor.

Now that general mathematical formulas can be displayed on screen terminals and printed on hardcopy, one must ask whether B and C should indeed remain the same in future. Experience seems to suggest that 2-dimensional formulas which make a liberal use of subscripts and superscripts, sometimes at more than one level, are easier for people to understand than the 1-dimensional linear formulas so far used by programming languages. On the other hand it will probably be necessary to retain a 1-dimensional format for the compiler input since this is likely still to be a string of bytes on a particular channel. Nevertheless it would be a mistake to think that it is therefore restricted to 128 or even 256 distinct symbols since quite arbitrary information can be represented in this form.

To sum up, therefore, we expect that the published versions (B) of future algorithmic and programming languages will make full use of 2-dimensional mathematical symbolism and graphics; that they will be constructed and proof-corrected from the keyboard (A) using both word-processing facilities such as COMPOS and program generation systems such as GENSIS, but they will continue to be transmitted to the compiler as a string of bytes (C). The stored version of the program will also be a string of bytes (D), not necessarily the same as C. In many cases the published version B may be in a highly condensed symbolic form which is converted to version

C by an algebraic manipulation system each time it is to be compiled, and that the programmer and other users would not generally see the expanded version that is sent to the compiler.

11. THE EXPERT FACILITY

In software engineering a computer program is thought of as an information proces-sing machine, and standard engineering techniques in areas such as design, optimi-zation of efficiency, construction, testing, tuning, control, documentation and maintenance are so far as possible applied. One important technique in hardware design is to build in a set of diagnostic outlets from which information can be transmitted to the test engineer without disturbing the working of the machine itself. In many large devices such as the JET fusion apparatus now under construc-tion at Culham, such information is continuously monitored by the control computers during the operation stage of the machine.

A similar scheme is provided in OLYMPUS programs by the EXPERT facility. At criti-cal points within the program the designer includes a statement

CALL EXPERT(ICLASS,ISUB,n)

where (ICLASS,ISUB) is the decimal number of the subprogram and n is the serial number of the diagnostic point within the subprogram. Using COMPOS he can do this by typing CALL EXPERT in column 1, when the statement will be aligned on column 40 for better visibility and the argument list automatically constructed.

This call transmits the exact position from which it is made to the EXPERT subrou-tine, which then codes the position into a jump statement to a point within EXPERT at which the programmer can include appropriate diagnostic reports. A call from point 5 of subprogram (2.13) is for example coded into a jump to statement label 21305 of EXPERT, and by arranging these labels in numerically ascending order the programmer can maintain a convenient list of the diagnostic reports that he is currently using.

An important advantage of this scheme is that for many purposes it is not necessary to alter or to recompile the program itself, but simply to update and recompile EXPERT. Most of the diagnostic reports can be obtained by examining the values either of individual COMMON variables or arrays or of complete COMMON blocks, and CYCLOPS subroutines are provided for doing this in neat form. These are also use-ful during the development phase since they make it possible to test the working of the program before the final output format has been implemented.

Of course many computer systems now provide quite sophisticated trace and debug facilities, and where these are available they can be used to supplement or replace the simpler diagnostic facilities of the OLYMPUS system.

Another important use for the EXPERT facility is to modify the working of the pro-gram for specific runs. In computational physics it is often necessary to inves-tigate the consequences of different physical assumptions or numerical methods, or to add extra forms of output, so that the program is in a continual state of modi-fication. If this is done by changing the main source code itself the situation can quickly get out of hand, with many slightly different versions of the complete program or of individual subprograms, whose validity is in doubt since they have probably been altered without repeating the full series of initial tests. EXPERT allows sections of code to be inserted at pre-defined points without recompilation of the main source code, so that only a single version of this is required, and all the changes needed for a specific set of runs can be concentrated in the EXPERT listing which, in effect, serves as a generalized form of input data file. This method of working has been found at Culham to be very effective.

12. CONCLUDING REMARKS

The philosophy underlying the OLYMPUS system is that the methods available to computational physicists should be comparable to those used in theoretical and experimental physics. The programs should be as universal and comprehensible as mathematics, and should at the same time be as well-designed and engineered as scientific apparatus. A number of techniques have been worked out and put into practice for achieving this, initially based on Fortran 66 because that is the language that physicists mostly employ at present. Although originally designed for initial-value problems, they have proved equally effective for the text manipulation programs COMPOS, GENSIS and the SPF Processor and appear to be of rather general application.

OLYMPUS has been extensively used by the Computational Physics Group at Culham on many different computers and has proved very successful. Recently it has been implemented on the Culham PRIME computer system whose Command and Abbreviation facilities [21] enable a very neat OLYMPUS Command Language to be introduced. Several OLYMPUS programs [13-17] together with the Control and Utility Package [4-6] have been published in CPC and used in various countries on a wide range of machines. The main difficulty until now has been defining and explaining the detailed conventions that are needed to achieve a consistent layout, and that initially had to be followed at the keyboard. This has limited the number of people using the system and has made it hard to maintain the conventions with complete rigour as one would do in a book. With the introduction of COMPOS and GENSIS these difficulties should disappear since the conventions are satisfied automatically without the programmer having to remember them.

The quality and readability of OLYMPUS listings is high by current Fortran programming standards and the next step will be to publish several of these as Laboratory Reports for training purposes. However it is also intended to experiment with reducing the bulk of the source code by improved layout and symbolism as suggested earlier.

13. ACKNOWLEDGEMENTS

Many people have contributed to the development of the OLYMPUS system in addition to the author including J.P.Christiansen, R.L.Dewar, J.W.Eastwood, M.H.Hughes, G.C.Lister, A.P.V.Roberts, P.D.Roberts and M.L.Watkins. Dr.M.H.Hughes has especially been responsible for implementing the COMPOS,GENSIS and SPF Processor facilities, and Dr.J.P.Christiansen for the OLYMPUS Control and Utility Package.

14. REFERENCES

[1] Roberts,K.V., An Introduction to the OLYMPUS System, Computer Phys.Commun. 7 (1974) 237-243.
[2] Hockney,R.W. and Eastwood,J.W, Computer Simulation Using Particles, Chapter 3: The Simulation Program, McGraw-Hill, New York 1981.
[3] Dahl,O.J., Dijkstra,E.W. and Hoare,C.A.R., Structured Programming (Academic Press,New Yrok,1972).
[4] Roberts,K.V. and Christiansen,J.P., OLYMPUS: A Standard Control Package for Initial-Value Fortran Programs, Computer Phys. Commun. 7 (1974) 245-270.
[5] Hughes,M.H.,Roberts,K.V. and Roberts,P.D., OLYMPUS and Preprocessor Package for an IBM 370/165, Computer Phys. Commun. 9 (1975) 51-58.
[6] Hughes,M.H.,Roberts,K.V. and Lister,G.C., OLYMPUS Control and Utility Package for the CDC 6500, Computer Phys. Commun. 10 (1975) 167-181.
[7] Roberts,K.V., The Publication of Scientific Fortran Programs, Computer Phys. Commun. 1(1969) 1-9.
[8] Jackson,C., The CPC Program Library, Computer Phys. Commun. 19 (1980) 11-15.
[9] USA Standard Fortran, USA X3.10-1966, (USA Standards Institute, New York, March 19-6).
[10] Standard Fortran Programming Manual,(National Computer Centre,Manchester,1970).

[11] Crees,M.A., ASYPCK, A Program for Calculating Asymptotic Solutions of the
      Coupled Equations of Electron Collision Theory, §10.5 Activation and De-
      Activation of Machine-Dependent Statements, Computer Phys. Commun. 19
      (1980) 103-137.
[12] The LRLTRAN Language for the CHATR Compiler, (National Magnetic Fusion Energy
      Computer Center, Livermore, California, current version online on NMFECC Net-
      work).
[13] Christiansen,J.P.Ashby,D.E.T.F., and Roberts,K.V., MEDUSA, A One-Dimensional
      Laser Fusion Code, Computer Phys.Commun. 7(1974) 271-287.
[14] Appert,K.,Berger,D.,Gruber,R.,Troyon,F., and Roberts,K.V., THALIA - a One-
      Dimensional Magnetohydrodynamic Stability Program using the Method of Finite
      Elements, Computer Phys. Commun. 10(1975) 11-29.
[15] Dudder,H.D., and Henderson,D.B., RAMSES, a Two-Dimensional, PIC-Type, Laser
      Pulse Propagation Code, Computer Phys. Commun. 10 (1975) 155-166.
[16] Christiansen,J.P., Roberts,K.V., and Long,J.W., ATHENE 1: A One-Dimensional
      Equilibrium-Diffusion Code, Computer Phys. Commun. 14 (1978) 423-445.
[17] Christiansen, J.P. and Winsor,N.K., CASTER 2: A Two-Dimensional Laser Target
      Code, Computer Phys. Comm. 17 (1979) 397-412.
[18] Fortran IV Reference Manual: Mark III Foreground (Honeywell), Order No.3102.
      01E, March 1974).
[19] Roberts,K.V. and Boris,J.P., The Solution of Partial Differential Equations
      in a Symbolic Style of Algol, J. Comp. Phys. 8 (1971) 83-105.
[20] Petravic,M., Kuo-Petravic,G. and Roberts,K.V., Automatic Optimization of
      Symbolic Algol Programs. I. General Principles, J. Comp. Phys. 10 (1972)
      503-533.
[21] PRIMOS COMMANDS Reference Guide (PRIME Computer, FDR3108-101B, 1980).

DISCUSSION

Hull   Would you say more about how you deal with the subroutines in a package
that come from a library, such as the NAG Library, but that are apparently not
part of the Olympus program for distribution purposes?

Roberts The problems of routines that are part of a proprietory library package,
or of a local installation library, is one that affects a considerable number of
the programs published in the Computer Physics Communications Program Library,
and not only OLYMPUS programs.   Our present intention is that the subscriber
should be able to run the program even if these particular routines are not
available to him.   Usually we require that the action of the routines and their
arguments should be very carefully defined, for example on comment cards, so that
the subscriber can if necessary replace them by versions available at his own
installation. Graph-plotting is normally omitted unless it employs only routines
at a very low level.

Cowell   There would be no proprietory question if the EISPACK routines were used,
since they are available in the public domain.

Roberts   Even routines that are in the public domain are not yet generally
available to all our users, many of whom are in the developing countries.
However, I agree that packages such as EISPACK and NAG should be as readily
available to scientists as are the leading scientific books and journals, or
tables of mathematical functions. Installations should buy these packages. They
are not expensive compared to the cost of the computer itself.

Feldman In what form do you publish programs?

Roberts CPC programs are made available as source decks on magnetic tape in two
ways; to subscribers who take the complete library, and to people who wish to
purchase one or more individual programs.   The complete library is available on-
line on research networks in several countries. Program writeups are published
in the CPC journal.

Feldman How many people need the source text of the programs?

Roberts Many physics programs are very complex and it is necessary to 'read' the listing in order to understand the working and to verify that it is correct, just like a mathematical or theoretical paper. Also, these programs are often modified and extended by users. I am not yet satisfied with the documentation standards and we could make greater use of word-processing techniques for improving program readability.

Meissner I agree with the idea that use should be made of word-processing for "publicising" computer programs. A general utility should be made available which does at least two things
(1) put paragraph-sized blocks of comments into well-formatted, readable text
(2) align the characters of source-code blocks.
This is well within the current state-of-the-art for word-processing.

Roberts Yes. But even comments on every line are not enough. Many of the comments in OLYMPUS programs are decimalized section and subsection headings, and we find indexes very useful too.

Feldman Word-processing is now very easy, and there are many software tools available in this and related areas. I question your continuing use of the analogy for publishing programs with the publishing of papers. You don't publish an engineering drawing unless its content falls down!

Roberts I think that one should distinguish between library subroutines, which are normally accepted by the user as correct, and large complex physics programs which are essentially research tools. One has no guarantee that these large programs will operate correctly under all circumstances, and the more readable they are, the more likely it is that errors will be weeded out.

Rice What is the procedure that the journal Computational Physics Communications uses in refereeing programs? Specifically:
(1) What is the acceptance vote?
(2) Are programs run by the referees and, if so, what kind of performance evaluation is expected?
(3) How are the OLYMPUS conventions and standards enforced?

Roberts The author submits a CPC manuscript to a Specialist Editor in the appropriate physics field. This manuscript consists of a Program Summary and Long Writeup, together with the compile-load-go job output for one or more test runs. The editor or referee does not run the code himself, but checks the writeup and listing like an ordinary scientific paper for requirements such as accuracy, originality, relevance, efficiency, completeness, readability and so on.Short extracts from the test run are reproduced in the journal as part of the writeup to allow a subsequent user to verify that he obtains the same results as the author. The final acceptance rate is currently fairly high, but programs are often returned to the authors for improvements first.
     We have found that the most effective way of enforcing the more mechanical parts of the OLYMPUS conventions and standards is to have a word-processor that constructs and maintains them automatically. Others are best learnt by example, in the same way that one learns to use a standard set of conventions in mathematics and physics by reading books and papers. This means that good-quality listings should be distributed fairly widely, which has been difficult in the past with standard line printers but should become easier now that word-processor output devices are being used for ordinary scientific papers.

Nelder Which machines have OLYMPUS available?

Roberts   Versions of the basic OLYMPUS package have been published in CPC for 3
different machines, but probably more than 20 versions have been implemented by
various user groups in a number of countries.   The word-processing facilities
have been implemented for several systems but not published as yet, and it is
hoped to do this soon.

Brown   I applaud your analogy between publishing programs and   publishing
scientific papers.   However most scientific papers contain gaps or errors of a
kind that would be intolerable in programs.   Further very few programs are read
by more than a handful of human readers.   I was also surprised by your assertion
that implicit typing rules in OLYMPUS are similar to notational conventions in
mathematical writing.   Many mathematical papers define all symbols in prose.   I
suggest that the declarations in a computer program are similar in function to
such prose.

Roberts   I agree that programs must be more meticulously correct than scientific
papers, but this surely means that the techniques that have been found useful for
organising the scientific literature will be even more essential for programs. I
disagree with the suggestion that programs are not read.   My experience is that
source listings are constantly read and that some of the large scientific
programs continue in use for many years and pass through many hands and many
editions.   This is surely also true for system software.   Often it is expensive
and even dangerous if their working cannot be understood.
   There is quite a good analogy between OLYMPUS programs and mathematical
writing, in fact, because the physical meaning of the symbols is defined in
'prose' in the alphabetic indexes, while the type is apparent to the reader from
the kind of symbol that is used.   Declarations define the type but not the
meaning, and it may be inconvenient to search through lists of declarations each
time to determine the status of each identifier.   Some of these large programs
have more than a thousand different symbols in COMMON alone.

Ford   How do you handle errors (and their correction) in the OLYMPUS programs?

Roberts We aim to get the programs widely distributed and read.   The CPC library
has a scheme for publishing erratum notices and correction decks, and for
informing users of changes.

Ford   So you adopt the computer manufacturer's implicit policy of squeezing out
errors by wide use.

Roberts   Yes, but this is also the way in which errors are eliminated from the
scientific and mathematical literature.   The initial refereeing is supplemented
by continual reading and re-reading.   If a published result is wrong, sooner or
later some teacher, student or research colleague will notice.   Errors are
inevitable with software too but it should be possible to remove them in the same
way.

Snoek   Rice emphasised consideration in language definition of how easy it is to
write a program.   Roberts emphasised how easy it is to read a program.
Inevitably there is a conflict between the two points.   I would add a third
consideration.   Language designers should consider how easy it is to learn the
language. FORTRAN gives an acceptable compromise between these three aims.

Rice   There is not necessarily a conflict between the three aims.

Snoek   APL is bad in all three ways.

Rice   But there are good things in APL.

**SESSION 2
ARITHMETIC**

Chair
H.J. Stetter

Discussant
R.P. Brent

*THE RELATIONSHIP BETWEEN NUMERICAL COMPUTATION
AND PROGRAMMING LANGUAGES, J.K. Reid (editor)*
North-Holland Publishing Company
© *IFIP, 1982*

FLOATING-POINT PARAMETERS, MODELS AND STANDARDS

W. J. Cody[1]

Applied Mathematics Division
Argonne National Laboratory
Argonne, Illinois 60439

This paper is a critical survey of recent proposals for environmental parameterization for floating-point arithmetic. The proposals discussed are from IFIP WG 2.5, Brown and Feldman, the ANSI X3J3 committee, Ada and the draft IEEE Standard for Floating-Point Arithmetic.

1. Introduction

There are several reasons for parameterizing the details of the computational environment. Obviously, parameterization helps us to write programs that can be moved between disparate computing systems with little loss of performance. When properly supported by axioms and models of arithmetic, parameters are also useful in error analysis and in formal proofs related to program performance. Finally, the existence of parameters encourages programmers to think about the numerical environment, thus subtly influencing the way computation is done.

There have been several major attempts at environmental parameterization for these purposes. These range from simple lists of static parameters to more complicated proposals that include detailed idealized models of floating-point arithmetic (including systems of axioms), programs for manipulating components of floating-point numbers and formal extensions to language standards. IFIP WG 2.5 has proposed both a set of static parameters [Ford, 1978] and a set of intrinsic functions for manipulating floating-point data [Ford, Reid and Smith, 1977]. Similar proposals by Brown and Feldman [1980] are based on a detailed model of floating-point arithmetic [Brown, 1981]. The model is also formally embedded in Ada[2] [Ada, 1980]. Still another proposal is being considered by the ANSI X3J3 Fortran Standards Committee for inclusion in the next Fortran standard [X3J3, 1981]. Even the draft IEEE Floating-Point Standard [IEEE P754, 1981] contains suggestions for environmental facilities.

This paper is a critical survey of all the above efforts. The next section is a brief historical summary of work in this area. Sections 3 through 5 are individual discussions of the IFIP WG 2.5 parameterization, Brown and Feldman's suggestions, and the Ada and X3J3 proposals. We describe each effort in enough detail that the advantages and disadvantages can be understood, and discuss the fundamental issues involved. In Section 6 we discuss intrinsic functions supporting parameterization, including those for manipulating components of floating-point data. The final section is a short, critical summary.

2. History

Peter Naur launched our interest in environmental parameters with a visionary paper [Naur, 1967] proposing that common programming languages be extended to include "... Environment Enquiries ... designed to place a carefully chosen set of information about the available equipment at the disposal of the programmer ...." Recognizing that "... an Environmental Enquiry can only be meaningful on the background of an explicitly described machine structure.", he prefaced his discussion with a simple model for the representation of integer and real data.   He specifically suggested we need language mechanisms, such as intrinsic subprograms, for determining the representation of reals, integers, characters, bit patterns and strings.

Responses to his requests, especially recent responses, have often taken the forms he foresaw. All are based explicitly or implicitly on an underlying model of the arithmetic, most suggest language mechanisms for accessing the parameters of the model, and many seek to formally embed these mechanisms in language standards. In later sections we will compare proposals against this 'blueprint'.

Some of Naur's suggestions became part of Algol 68 (see, e.g., [Delves, 1977]), but his eloquent pleas were ignored for other languages until Redish and Ward [1971] made a few specific suggestions for the definition and implementation of environmental facilities suitable for Fortran.   Their main contribution, however, was to suggest subprograms for manipulating the components of a floating-point number, such as determining or specifying the unbiased exponent. Similar subprograms were later discussed by Sterbenz [1974], and have now become a part of most proposals for environmental facilities.

Most efforts of the early 1970's, however, were simply compilations of parameter values for selected machines to satisfy specific needs.   Three such compilations in use for several years were summarized at the *Portability of Numerical Software Conference* at Oak Brook in 1976.   Aird [1977] discussed parameter names and values used with the IMSL Fortran Converter, a system designed to extract machine-specific versions of a Fortran program from a master version.   Fox [1977] and Du Croz, Hague and Siemieniuch [1977] discussed similar compilations accessible through subroutines in the PORT and NAG libraries, respectively.   The PORT programs were eventually published [Fox, Hall and Schryer, 1978] and are now widely used.

The Oak Brook conference and the IFIP WG 2.5 meeting that immediately followed were major milestones for parameterization. In addition to the papers just mentioned, Ford [1977] presented WG 2.5's preliminary list of parameter names and definitions, Brown gave the first first public discussion of his model for floating-point arithmetic [Brown and Hall, 1977], and Cody [1977] discussed the philosophy behind parameterization and refocussed attention on functions for manipulating components of floating-point numbers.   Each of these presentations bore fruit. A final report on the WG 2.5 proposals appeared about two years later [Ford, 1978], serving primarily as a catalyst for other efforts, while a polished version of Brown's model [Brown, 1981] became the foundation for a major new environmental proposal [Brown and Feldman, 1980].   Finally, sparked by the Oak Brook papers, the WG 2.5 meeting drafted a proposal for intrinsic Fortran functions to manipulate floating-point data [Ford, Reid and Smith, 1977].   This was forwarded with other proposals to the ANSI X3J3 committee then putting finishing touches on what became the Fortran 77 Standard [ANSI, 1978].   The suggestions arrived too late to be included in the new standard, but motivated X3J3 to recruit a member of WG 2.5 to help draft the next standard. We discuss these matters in more detail later.

In most of this work, requisite parameter values are determined ahead of time and made available in some static way, but several attempts have also been made to determine parameter values dynamically. Malcolm [1972] pioneered with Fortran programs that determine three important floating-point parameters on many machines. Unfortunately, they seriously malfunction on others. Gentleman and Marovich [1974] explained these malfunctions and modified the programs to extend their coverage. George [1975] provided additional programs that determine the representation of characters, integers and floating-point numbers by cycling data between the CPU and an external I/O unit. Cody and Waite [1980] extended Malcolm's program to determine thirteen floating-point parameters, but the new program malfunctions on the same machines as Malcolm's original. Finally, Schryer [1981] produced a small suite of portable Fortran programs that test support for a particular version of Brown's model. The programs can also be used to determine arithmetic parameters within certain limitations.

3. WG 2.5 Conventions

It took three years for WG 2.5 to reach consensus on their proposed conventions for names and definitions of parameters useful in preparing transportable numerical software. When they finally appeared, the editors of *Transactions on Mathematical Software* immediately adopted them as standards for algorithms published in the journal [Editor's note prefacing Ford, 1978]. Despite this endorsement, to date they have not been used in published algorithms. They are important primarily because they were first, and because of their catalytic influence on other efforts.

In the WG 2.5 view the environment to be parameterized includes not only the hardware but also the supportive compilers and operating systems. This suggests that a complete set of parameters for numerical programs should include specification of the I/O facilities and certain compiler and system characteristics in addition to the usual arithmetic parameters. The group's goal was to produce a language independent list of useful parameters together with definitions and suggested *characteristic* names for algorithm writers and one or more *explicit* names for inclusion in source code. Neither the parameter values nor the mechanism for making them available was to be specified. Thus the group had modest but useful goals.

Unfortunately, the language in the final report is imprecise and not appropriate for a 'standard'. While the intuitive purpose for each parameter is clear, the exact definition is sometimes vague. Feldman's [1979] critique points out some of the problems. We do not detail the objections here, but instead direct the interested reader to Feldman's paper.

Table I lists the characteristic names for all parameters included in the final report. We consider only the six parameters related to the floating-point system, however. As Smith [1978] has pointed out, most of the arithmetic parameters are defined for the static representation of numbers in the host system; only the definitions of relative precision and symmetric range refer to results of arithmetic operations. The underlying model for the representation is

Table I

WG 2.5 Parameter List

Arithmetic Set	I/O Set	Miscellaneous Set
radix	standard input unit	characters per word
mantissa length	standard output unit	memory page size
relative precision	standard error	number of decimal
overflow threshold	message unit	digits (converted
underflow threshold	characters per input	by compiler)
symmetric range	record	
	characters per output	
	record	

never explicitly given, violating the first requirement of Naur's blueprint. This weakens the report, leaving the reader to interpret the parameter definitions according to his own vague and informal model.

The parameters are defined for the unrestricted set of representable floating-point numbers and not for some artificially constrained subset. Such unrestricted parameters permit computations to probe the limits of the hosting system when it is safe to do so, but they do not necessarily provide enough warning of the limits when it is not safe to approach them. For example, the underflow threshold is informally defined as the smallest positive floating-point number x such that both x and $-x$ are representable as elements of the system. What is not said is that computations involving x may not always yield the expected results. There are small, representable and *computable* floating-point numbers on some popular computers such that

$$x \neq 0,$$

but either

$$1.0 * x = 0$$

or

$$x + x = 0.$$

Here the underflow threshold reflects reality (x is representable and even computable) but not practicality (x is not too useful). Parameters derived from the unrestricted floating-point system are treasured by skilled numerical analysts seeking to squeeze the ultimate in performance from their software, but they may be too dangerous for less skilled colleagues seeking only to write safe, transportable programs.

In retrospect, the main value in the WG 2.5 proposals has been their catalytic effect on other efforts, although that was probably not what WG 2.5 originally intended. The final report may have helped Brown identify difficulties in parameterization, and thus influenced his work. Certainly vestiges of the IFIP proposals remain in the X3J3 work.

4. Brown and Feldman's Proposals

In contrast with the primarily static approach of WG 2.5, Brown and Feldman [1980] discuss arithmetic parameterization relative to the dynamic behavior of an arithmetic system. Following Naur's blueprint, they base their work on a specific detailed model of floating-point arithmetic [Brown, 1981]. The model combines a representation scheme similar to that previously used for error analysis by Wilkinson [1963] and Forsythe and Moler [1967] with a workable axiomatization of computer arithmetic (see [Dekker, 1981] for a comparison of this axiomatic system with others). It thus meets all the goals of parameterization by providing a framework for writing transportable software and for proving certain mathematical properties of the software.

Brown and Feldman's complete environmental proposal includes both parameters (see Table II) and manipulative functions (see Table V). We discuss only the parameterization here, delaying the function discussion until later. A system of model numbers is defined by four basic parameters, all integers, and three derived parameters, all floating-point numbers. A non-zero model number is represented as

$$x = b^e f$$

where the significand f is a normalized, signed fraction, $1/b \leq |f| < 1$, containing p base-b digits, $e_{min} \leq e \leq e_{max}$, and e is an integer. Under ideal circumstances, model parameters would be identical to those describing the representation of floating-point numbers in the host system. Ordinarily, however, model parameters are adjusted to guarantee that four axioms governing operations on model numbers are satisfied, and the model numbers are therefore a proper subset of the machine numbers. The axioms impose an interval arithmetic by introducing model intervals, i.e., small intervals of machine numbers whose endpoints are model numbers.

Exact statements of the axioms are too complicated to give here. For our purposes, the essence of the axioms is that arithmetic operations on model numbers are at least as accurate as chopped arithmetic, that errors in operations

Table II

Environmental Parameters Proposed by Brown and Feldman [1980]

Name	Definition
Basic Model Parameters	
base	b
precision	p
minimum exponent	e_{min}
maximum exponent	e_{max}
Derived Model Parameters	
maximum relative spacing	$\varepsilon = b^{1-p}$
smallest positive number	$\sigma = b^{e_{min}-1}$
largest number	$\lambda = b^{e_{max}}(1-b^{-p})$

involving non-model numbers are consistent with errors in operations involving the model intervals containing them, that comparisons between model numbers are exact, and that comparisons involving non-model numbers are consistent with comparisons involving the model intervals containing them. The model parameters differ from those for the host system to the extent that operations in the host system fail to satisfy the axioms. For example, the model parameter e_{min} is penalized on machines implementing comparisons with floating-point subtraction. On such machines, e_{min} must be large enough that the difference between two model numbers does not underflow, hence must differ from the e_{min} for machine numbers by at least the model precision p. That, in turn, may be less than the machine precision on machines lacking guard digits, having a peculiar rounding algorithm, or on which multiplication is not commutative.

Thus, the four basic parameters define a 'well-behaved' arithmetic system embedded in the host arithmetic system. The axioms permit mathematical proofs of certain properties, including error bounds, for programs whose computations stay within the model. Because the proofs depend only on the model, they hold abstractly, i.e., they hold for any valid model arithmetic regardless of the hosting system.

Brown's model is an excellent abstract model of what computer arithmetic does. Given a particular machine, it is always possible to choose the parameters to obtain an optimal model of the arithmetic system, i.e., one in which the parameters are as close to the machine parameters as possible under the constraints imposed by the axioms. Then all computational results are explainable by the axioms. Unfortunately, this does not relieve the practical programmer of the need to thoroughly understand the arithmetic foibles of his system. When some machine numbers are not model numbers, the axioms must be correspondingly 'fuzzy' in places. For example, Axiom 4 allows that, when doing logical comparisons of arithmetic quantities, under certain conditions the computer may report $x<y$, $x=y$, or $x>y$, regardless of the facts. The model may explain this puzzling behavior, but the practical programmer must still be aware of the idiosyncrasies of his computer if he is to write a flawless program. He must still know under what conditions the computer is likely to misrepresent the facts, and why.

In this important sense the model does not make the preparation of transportable software that much easier. Previously existing difficulties assume new forms. It is now necessary to determine the model and to design programs that stay within model limits when it is important to do so. Thus the user must now understand both the foibles of his machine and the model. He must be aware, for example, that the system does not automatically signal underflow when he transgresses the lower boundaries of the model, and he must know when this matters. This is not a criticism of the model, because the model is a good one; it is a criticism of *blind reliance* on the model for programming purposes.

Fortunately there is some help in determining the model for a particular host environment. Schryer [1981] has produced a portable suite of Fortran programs for checking that a particular system supports a particular model. The programs test the dynamic behavior of an arithmetic system by comparing the results of computations using carefully selected operands with exact mathematical results. Within limits, the programs can also be used to discover the 'best' model parameters form a cold start. This is an incredibly difficult task because it is necessary to probe the 'dirty corners' of an unknown arithmetic system without stumbling over an abortive error such as overflow.

In summary, Brown and Feldman's proposals follow Naur's blueprint closely by including a detailed model, and by suggesting language mechanisms for accessing the parameters of the model. They do not formally propose embedding these facilities in a standard language, but things are phrased so that this would not be difficult to do. Indeed, Ada already includes large portions of the proposal. The success of this approach is demonstrated by the portability of the PORT library [Fox, 1981].

5. Ada and Fortran

The efforts just described have influenced Ada [Ada, 1980] and recent work on the next ANSI standard for Fortran [X3J3, 1981]. There are some striking similarities in this work considering the differences in the origins of the proposals, and also some significant differences. Both efforts follow Naur's blueprint closely.

Ada is a powerful language supporting many different data types. For floating-point computations, the type FLOAT is predefined with precision determined by the host hardware [Wichmann, 1981]. In addition to other pre-defined types corresponding to different hardware precisions, the language permits user-defined types and subtypes. The attributes of each defined type or subtype are made available at run time through an intricate system of environ-mental parameters. The following discussion only hints at the richness of the environmental facilities in Ada. Again we restrict our attention to floating-point facilities.

Ada has explicitly embraced a restricted version of Brown's model for user-defined floating-point types. The user must specify a minimum decimal accuracy in declaring a real data type, although language implementations are free to provide greater accuracy. Unlike Brown's flexible model, the type declaration automatically determines all the associated environmental parameters according to rigid rules involving only the declared integer number of decimal digits of significance in stored data. A similar model applies to predefined types with their inherent precisions. Ada assumes that the machine is binary and that nonzero floating-point data are representable as

$$\pm 2^e * f,$$

where e is an exponent and f is a normalized binary (fraction) significand.

Model parameters for a particular floating-point data type F are given in Table III. Similar parameters exist for each defined or predefined floating-point data type and subtype. They are accessed by replacing the 'F' in the above parameter names with the appropriate type name.

For each defined floating-point type there are additional parameters that report machine-dependent attributes. For example, F'MACHINE_ROUNDS reports whether the machine operations supporting type F data round or not, and F'MACHINE_ RADIX reports the radix for the machine representation of type F data. Thus Ada includes both the artificial parameterization of Brown's model for use when portability is paramount in importance, and the machine parameterization sought by WG 2.5 for use when full exploitation of specific hardware is more important than portability. The most serious objection to this scheme is the rigidity of

Table III

Ada Intrinsic Parameters for Floating—Point Data Type F

Name	Definition
F'DIGITS	D, the integral number of decimal digits specified in the type declaration
F'MANTISSA	binary length of the model significand, defined as the next integer greater than $D*\log(10)/\log(2)$
F'EMAX	maximum value of the model exponent, defined as $4*F'MANTISSA$
F'SMALL	smallest positive model number, defined as $2^{-F'EMAX-1}$
F'LARGE	largest model number, defined as $2^{F'EMAX}*[1.0-2.0^{-F'MANTISSA}]$
F'EPSILON	the distance between 1.0 and the next greater model number, defined as $2^{1-F'MANTISSA}$

the model. In particular, the restrictive ratio of the maximum exponent to the significand length has implications for the types of numerical problems that can be treated portably, and for the complexity of the programs that treat them.

Any discussion of proposed environmental facilities in Fortran is necessarily tentative because the X3J3 committee is still debating the issues. The approach being considered has evolved more from WG 2.5 ideas than from Brown's model, although parameterization is based on a model. Floating-point numbers are considered to be represented as Brown has described, with the same four basic and three derived parameters characterizing the representation. Again the basic parameters are integers and the derived parameters are reals. Instead of selecting parameter values to force the model to satisfy axioms, however, the values are chosen so that the model 'best fits' the host machine, where the concept of 'best fits' is left undefined. The sense in which the model fits the host is therefore implementation-dependent, and presumably a given host could support more than one model at different times. Although the vagueness of the model as described in X3J3 documents will disturb some, it does not appear to be an important limitation on what is proposed. It is intended that specific implementations include complete semantic details of the particular model used.

The basic approach to environmental facilities is through generic intrinsic functions (as suggested by Brown and Feldman [1980]), that is, functions in which, unless otherwise specified, the type of the result is inherent in the argument. The proposal includes both functions for ascertaining parameter values (see Table IV), discussed here, and functions for manipulating components of data (see Table V), discussed later. In each case, an enquiry function reports the parameter value appropriate for the data type of the argument X, and reports the same values for all arguments of a given data type. The proposal is general in including a model of integer arithmetic and valid returns for certain functions

Table IV

**Proposed Fortran Intrinsic Functions
for Environmental Enquiry**

Name	Definition
RADIX(X)	b
PRECISION(X)	p
MINEX(X)	e_{min}
MAXEXP(X)	e_{max}
HUGE(X)	$b^{e_{max}}(1 - b^{-p})$
TINY(X)	$b^{e_{min}-1}$
EPSILON(X)	b^{1-p}

for integer data types as well.   Thus the precision reported by PRECISION(X) depends only on whether X is a real or an integer data type, and on the number of RADIX(X) digits in the stored significand for that type data.   Because PRECISION(X) is a basic parameter, the result is always an integer.   On the other hand, the value of EPSILON(X) is a real of the same type as X.

The intent here is to provide useful facilities without imposing burdensome details.   The simplicity and flexibility of this approach contrast sharply with the complexity and rigidity of the Ada approach.   Because model arithmetic is not constrained to satisfy axioms, the model cannot be used to formally prove properties of programs, nor to do formal error analysis.   On the other hand, we ought to be able to write reasonably transportable programs with intelligent use of these parameters.   After all, there now exist large collections of programs that can be transported between machines by specifying a handful of machine-dependent constants that depend simply on the proposed parameters.   At the least, these programs can be made completely portable when the parameters are supplied automatically instead of being explicitly specified in source code.

6. Manipulative Functions

We now turn our attention to facilities for the manipulation of floating-point data, including the decomposition of data into components and the reconstruction of data from components.   At first glance these capabilities do not appear to be directly related to parameterization, but we will see shortly that many parameter values can be recovered through simple application of manipulative functions.   Table V summarizes the manipulative functions found in Brown and Feldman [1980], the X3J3 proposals [X3J3, 1981], and discussions of the draft IEEE Floating-Point Standard [IEEE P754, 1981; Kahan, 1981].   No such functions are included in Ada.   Modifications of the three functions in the WG 2.5 proposal that originally focussed attention on manipulative functions [Ford, Reid and Smith, 1977] are included in the X3J3 list.

Table V
Summary of Manipulative Functions

	Name		Purpose
Brown/Feldman	**X3J3**	**IEEE**	
$exponent(x)$	EXPONENT(X)	$logb(x)$	extracts integer exponent e from x
$fraction(x)$	FRACTION(X)		extracts fraction f from x
$synthesize(x,e)$	SETEXPONENT(X,E)		returns b^e fraction (x)
$scale(x,e)$	SCALE(X,E)	$scalb(x,e)$	returns xb^e
$\alpha(x)$	ABSSPACE(X)		returns absolute spacing, b^{e-p} for $\|x\| > \sigma/\epsilon$, and σ for $\|x\| < \sigma/\epsilon$
$\beta(x)$	RECSPACE(X)		returns reciprocal relative spacing, $\|f\|b^p$
	NEAREST(X,Y)	$nextafter(x,y)$	returns nearest machine neighbor to x in direction determined by y

Each function is intended to be a generic intrinsic function suitable for inclusion in language standards. This means that, except for exponent extraction, the function results are always floating-point data with the same attributes as the floating-point argument. (The exponent e is always an unbiased, signed integer.) Our discussion of these functions is necessarily cavalier. Obviously, the definitions must accommodate situations where requested manipulations would give out-of-bounds results. The cited sources have been careful to specify what happens under such conditions, but we ignore such things here. There are also some subtle differences between the functions that are not apparent from the informal definitions in the table. We ignore most of those in this discussion, as well.

The definitions in the table use the terminology of Brown's model, but the parameter values and functional properties must be interpreted in the context of each particular proposal. The Brown/Feldman functions are based on Brown's model arithmetic, the X3J3 functions are based on the X3J3 model in which parameters are 'machine' parameters, and the IEEE functions are based on the draft IEEE binary floating point standard. In this latter scheme, the significand of a stored, nonzero normalized number x consists of a fraction f and an implicit bit to the left of the binary point. That is,

$$x = \pm 2^e(1+f).$$

A normalized significand therefore satisfies the condition $1 \leq 1+f < 2$, which is different from the conditions imposed on significands in the other schemes. The IEEE scheme also includes special representations for $\pm\infty$, ± 0, non-numbers and numbers resulting from 'gradual underflow' (see [IEEE P754, 1981] for details). Thus the parameter b in the table refers to the radix for model numbers when discussing $scale(x,e)$, probably refers to the machine radix when discussing SCALE(X,E), and is 2 when discussing $scalb(x,e)$.

The most striking feature of this table is the repetition of functions. As with the environmental parameters, the functions proposed by X3J3 include the functions proposed by Brown and Feldman, attesting to the importance and utility of this particular selection. The IEEE set is a minimal spanning set, however. Most of the other functions and many of the parameters can be obtained in a simple way from $logb$ and $scalb$, modulo model differences, or their counterparts in the other systems. (For convenience, Cody and Waite [1980] augment these with a synthesize-like function, but that is not necessary.) For example, in the Brown/Feldman set

$$synthesize(x,e) = scale(x,e - exponent(x)),$$
$$fraction(x) = scale(x, -exponent(x)),$$

and

$$b = scale(1.0,1).$$

The IEEE function $nextafter$ is necessary to recover the remaining functions and most of the remaining parameters. Thus,

$$\sigma = nextafter(0.0,1.0),$$

and

$$e_{min} = exponent(\sigma).$$

Unfortunately, there is no way to recover e_{max} or λ from these functions without special representations of ∞ such as those in the proposed IEEE arithmetic. When such representations are available,

$$\lambda = nextafter(\infty,1.0).$$

and

$$e_{max} = logb(\lambda).$$

To show the problems that arise in defining these functions, we compare $nextafter(x,y)$, loosely defined as the nearest neighbor to x in the direction of y, with NEAREST(X,Y), loosely defined as the nearest neighbor to X in the direction of *the infinity with the same sign* as Y. Thus

$$nextafter(2.0,1.0) \neq \text{NEAREST}(2.0,1.0).$$

This disconcerting situation arises because the model behind the X3J3 proposal lacks an ∞. Regardless of the size of x, we can use $nextafter(x,\infty)$ in the IEEE system to find the next larger number, modulo exceptions. But when x is large, say close to the overflow threshold, and the system lacks the special element ∞, then it is cumbersome to determine an appropriate y between x and the missing ∞ for use with this function. The X3J3 proposal avoids this difficulty by letting the sign of the second argument, rather than its size, define the direction. Of course, Y=0 now becomes exceptional, because 0 is unsigned; that was not a problem under the IEEE scheme. Thus the X3J3 proposal has exchanged one troublesome problem for a more tractable one.

The Brown/Feldman functions also have some surprising problems. Each is defined for all nonzero real x, model number or not, and can be evaluated whenever the result is representable in the machine. Results are required to be exact when arguments are model numbers, but may be in error when arguments are non-model numbers. Thus, the sequence

$$y := scale(x,1)$$
$$z := scale(y,-1)$$

may return a z different from the original x when the model radix differs from the machine radix. Similarly, the results of $scale(x,e)$ and $synthesize(x,e)$ are unpredictable when they are outside the limits imposed by the model, but not necessarily outside the limits imposed by the host system. Specification of these functions has been left purposely vague in this situation to permit unencumbered inline implementations.

7. Summary

We consider the WG 2.5 work to be primarily catalytic. Its main contribution was to focus attention on parameterization and to 'blaze trail' for the proposals that followed. Our understanding of the difficulties involved and of possible strategies has improved with each new attempt at parameterization. There remain today three viable proposals for parameterization and three for intrinsic manipulative functions. Facilities provided in the Brown/Feldman and X3J3 proposals are almost identical. Each suggests a parameterization supported by intrinsic functions. The main difference between the two proposals is that the former is based on a conservative model for floating-point arithmetic while the latter is based on a model that 'best fits' a host system.

Use of the conservative model permits the writing of transportable programs that perform properly as long as computations do not transgress model boundaries. Brown [1981] claims that "Once the model parameters are available, programmers can forget about floating-point anomalies and think about the model machine instead of the actual machines on which their programs will run." Unfortunately, that is not completely true, because software environments do not automatically warn of model boundary transgressions. Such transgressions, often subtle, may have startling consequences, as when comparisons report falsely partially because the quantities being compared have more precision than the model acknowledges. To be sure, the situation is no worse with the model than it would be without the model. The point is that the model lends a false sense of security unless it is thoroughly understood, and that requires more of the user. The model will not do as Brown claims until all hardware anomalies disappear. At that time the model will no longer be an abstract model; it will be a parameterization, without penalties, of the computer arithmetic itself.

The model behind the X3J3 proposals is intended to be just that: a convenient parameterization, without penalties, of the host system with no guarantees attached. The user must still protect himself against system vagaries. All the model and its supportive intrinsic functions do is provide a means for specifying and treating machine dependencies.

Ada is a language rich in parameterization of data types. If anything, parameterization is overdone. In contrast with the Brown/Feldman and X3J3 models, the Ada model for user-defined data types is too rigid and unyielding, being completely specified by only one parameter. Of necessity, such a model is extremely conservative and overly restrictive for certain important numerical purposes. We cannot advocate it in its present form.

Of all the proposals discussed, the 'cleanest' is the IEEE proposal supported by the draft IEEE binary arithmetic standard. The three IEEE intrinsic functions duplicate or surpass the capabilities of the intrinsic functions in other proposals. For example, they can be used to recover any of the proposed environmental parameters, modulo model differences. The careful design of IEEE arithmetic permits complete specification of the functions. Beyond the response to exceptions, nothing is implementation dependent or ambiguous.

We believe the IEEE scheme to be the best proposal of the lot. Unfortunately, it is useful only on 'IEEE systems' and is not the universal answer to environmental parameterization. That answer lies somewhere between the mathematical preciseness of Brown's model and the utility of the X3J3 proposals.

8. Acknowledgements

Our interest in this subject has been nourished by interactions with many colleagues. The members of IFIP WG 2.5, the drafters of the proposed IEEE Floating-Point Standard, participants in various meetings and symposia, and our colleagues at Argonne National Laboratory have all contributed ideas to this work. We accept full responsibility for contorting their wisdom into the form presented here. We especially want to thank J. Reid and B. Smith for their constructive criticism of an early version of this manuscript.

9. References

Ada [1980]. *Ada Reference Manual*, Honeywell, Minneapolis; reprinted in H. Ledgard, [1981]. *ADA, An Introduction, Ada Reference Manual*, (Springer-Verlag, New York, July 1980).

Aird, T. J. [1977]. The IMSL Fortran converter: an approach to solving portability problems, in: W. Cowell (ed.), *Portability of Numerical Software, Lecture Notes in Computer Science, Vol. 57*, (Springer-Verlag, New York: 368-388).

ANSI [1978]. *American National Standard Programming Language FORTRAN, ANSI X3.9-1978*, American National Standards Institute, New York.

Brown, W. S. [1981]. *A Simple but Realistic Model of Floating-Point Computation*, Computing Science Technical Report No. 83, Bell Laboratories, Murray Hill, N.J.

Brown, W. S. and Feldman, S. I. [1980]. Environment parameters and basic functions for floating-point computation, *TOMS 6*: 510-523.

Brown, W. S. and Hall, A. D. [1977]. FORTRAN portability via models and tools, in: W. Cowell (ed.), *Portability of Numerical Software, Lecture Notes in Computer Science, Vol. 57*, (Springer-Verlag, New York: 158-164).

Cody, W. J. [1977]. Machine parameters for numerical analysis, in: W. Cowell (ed.), *Portability of Mathematical Software, Lecture Notes in Computer Science, Vol. 57*, (Springer-Verlag, New York: 49-67).

Cody, W. J. and Waite, W. [1980]. *Software Manual for the Elementary Functions*, (Prentice-Hall, Englewood Cliffs, N.J.).

Dekker, T. J. [1981]. *Program Correctness and Machine Arithmetic*, Report 81-07, Mathematisch Institut, Amsterdam.

Delves, L. M. [1977].   Algol 68 as a language for numerical software, in: W. Cowell, (ed.), *Portability of Numerical Software, Lecture Notes in Computer Science, Vol. 57* (Springer-Verlag, New York: 95-126).

Du Croz, J. J., Hague, S. J. and Siemieniuch J. L. [1977]. Aids to portability within the NAG project, in: W. Cowell (ed.), *Portability of Numerical Software, Lecture Notes in Computer Science, Vol. 57* (Springer-Verlag, New York: 389-404).

Feldman, S. I. [1979].   Comments on the IFIP Working Group 2.5 proposal on 'Parameterization of the environment for transportable numerical software', *SIGNUM Newsletter 14(2)*: 18-20.

Ford, B. [1977].   Preparing conventions for parameters for transportable software, in: W. Cowell (ed.), *Portability of Numerical Software, Lecture Notes in Computer Science, Vol. 57*, (Springer-Verlag, New York: 68-91).

Ford, B. [1978]. Parameterization of the environment for transportable numerical software , *TOMS 4*: 100-103.

Ford, B., Reid, J. K. and Smith, B. T. [1977].   Three proposed amendments to the draft proposed ANS Fortran standard, *SIGNUM Newsletter 12(2)*: 18-20.

Forsythe, G. E. and Moler, C. B. [1967].   *Computer Solution of Linear Algebraic Systems* (Prentice-Hall, Englewood Cliffs, N.J.).

Fox, P. A. [1977].   PORT - a portable mathematical subroutine library, in W. Cowell (ed.), *Portability of Numerical Software, Lecture Notes in Computer Science, Vol. 57* (Springer-Verlag, New York: 165-177).

Fox, P. A. [1981].   The PORT mathematical subroutine library, to appear in: W. Cowell (ed.), *Sources and Development of Mathematical Software* (Prentice-Hall, Englewood Cliffs, N.J.).

Fox, P. A., Hall, A. D. and Schryer, N. L. [1978]. Algorithm 528, framework for a portable library, *TOMS 4*: 177-188.   (Algorithm headings only.   See *Collected Algor. ACM* for complete programs.)

Gentleman, W. M. and Marovich, S. B. [1974].   More on algorithms that reveal properties of floating-point arithmetic units, *CACM 17*: 276-277.

George, J. E. [1975]. Algorithms to reveal the representation of characters, integers and floating-point numbers, *TOMS 1*: 210-216.

IEEE P754 [1981]. A proposed standard for binary floating-point arithmetic, Draft 8.0 of IEEE Task P754," *Computer 14(3)*: 51-62.

Kahan, W. [1981]. Why do we need a floating-point standard?, to appear.

Malcolm, M. A. [1972]. Algorithms to reveal properties of floating-point arithmetic, *CACM 15*: 949-951.

Naur, P. [1967]. Machine dependent programming in common languages, *BIT 7*: 123-131. (See also Proposals for a new language," *Algol Bulletin 18*, October 1964: 26-31.)

Redish, K. A. and Ward, W. [1971]. Environment enquiries for numerical analysis," *SIGNUM Newsletter 6(1)*: 10-15.

Schryer, N. L. [1981]. *A Test of a Computer's Floating-Point Arithmetic Unit* Computing Science Technical Report No. 89, Bell Laboratories, Murray Hill, N.J.

Smith, B. T. [1978]. A comparison of two recent approaches to machine parameterization for mathematical software, *Proceedings of a Conference on the Programming Environment for Development of Numerical Software*, JPL Publication 78-92, Jet Propulsion Laboratory, Pasadena CA: 15-17.

Sterbenz, P. H. [1974]. *Floating-Point Computation*, Prentice-Hall, Englewood Cliffs, N.J.

Wichmann, B. A. [1981]. *Tutorial Material on the Real Data-Types in Ada*, Final Technical Report, U.S. Army European Research Office, London.

Wilkinson, J. H. [1963]. *Rounding Errors in Algebraic Processes* (Prentice-Hall, Englewood Cliffs, N.J.)

X3J3 [1981]. Floating Point Intrinsics, included in *X3J3 S6/78.1*.

---

[1]This work was supported by the Applied Mathematical Sciences Research Program (KC-04-02) of the Office of Energy Research of the U.S. Department of Energy under Contract W-31-109-Eng-38.

[2]Ada is a trademark of DoD.

DISCUSSION

Boyle   What do you mean by "tie to programming languages"?

Cody   The use of intrinsic functions and keywords.

Brown   Wichmann intends that "packages" to manipulate floating-point numbers will be added to Ada.

Moler   A key point of your talk is that we need the clean arithmetic provided by the IEEE standard.   The clowns are at the manufacturers, not here.   By using the Brown-Feldman model we are condoning shoddy arithmetic.   Ideally the restriction to "model numbers" should be unnecessary.   Are we influencing the hardware designers?

Cody   Some manufacturers are committed to ignoring the IEEE standard.   Others are following it.   The IEEE is working on a radix-independent standard.   If it is as successful as the radix-2 standard then progress will have been made.

Moler   NaNs, infinities etc. are nice but not necessary for parameterization of the machine precision, radix, overflow threshold, etc.

Cody   With existing machines there are problems in the middle of the range too, e.g., $1.0 * X \neq X$ for some real X.

Reinsch   Your claim that Schryer's program can determine the parameters of Brown's model is not completely correct, because the model for a particular machine is not unique.   For example, on IBM machines the base could be 2 or 16.   Thus, one of the most important machine parameters, the radix, cannot be derived from the model.

Cody   This is true: see the written version of my paper.

Brown   I agree with Cody's comments on the limitations of my model, but not with Moler's remark that it condones sloppy arithmetic.   In fact, I believe it clarifies the high price that must be paid for sloppy arithmetic. Incidentally, I have stopped working on the model, and a paper on it is to appear in the December 1981 issue of TOMS [ACM Transactions on Mathematical Software].

Smith   One of your criticisms of the X3J3 proposal was that the characterization of "best fit" was not specified by the X3J3 proposals for an environmental enquiry feature.   This was done on purpose so that the semantics of the enquiry feature could be taylored to fit the many disparate machine architectures. As machine architectures become more regular (as exemplified by the IEEE binary floating point standard) the semantic packages which describe how the model fits the machine will converge to an acceptable one, which hopefully will be as simple as the IEEE description of its environmental enquiry facility.   My intention, though, as a member of X3J3, is to propose that X3J3 provide an Appendix to the Standard to describe the ways that the X3J3 model can "best fit" a machine;   these ways could include Brown's axioms;   or a requirement that the model radix match the machine radix;   and so on.

Cody   Well said.

Kahan   Brown and I disagree much less than might appear.   Brown is a manager; he must and does see the world as it is.   I am an academic and see the world as it might be.

One of the motivations for the short list of environmental enquiry operations offered in the proposed IEEE standard was intellectual parsimony: a minimal set of notions is introduced, complete in the sense that all others can be obtained from them, and convenient in the sense that we obtain as directly as possible the functionality we need. So we do not introduce directly any words to describe the radix or word size; these two attributes are fundamentals in the model of arithmetic but are rarely used except inside the ln and exp subroutines. More often we want

Overflowthreshold := Nextafter (+∞,1)

Tiniest positive number := Nextafter (0.0,1)

Relative uncertainty due to round off

$$:= \begin{cases} \text{Nextafter } (1.0,2) - 1.0 \text{ if arithmetic is chopped,} \\ (\text{Nextafter } (1.0,2) - 1.0)/2 \text{ if arithmetic is} \\ \text{rounded} \end{cases}$$

Smallest positive number unaffected by underflow

$$:= \frac{\text{(Tiniest positive number)}}{\text{Relative uncertainty due to roundoff}}$$

Number of ulps (units in the last place) from x to y

$$:= \frac{y-x}{\text{Nextafter } (x,y) - x}$$

On computers which lack the symbols ±∞, we might let "overflowthreshold" be a reserved word with the meaning

largest positive magnitude representable without anomaly.

On computers with arithmetic no better than W.S. Brown's model requires, we might distinguish between

Nextafter (x,y) := representable number after x towards y

and

Modelnextafter (x,y) := model number after x towards y.

Provided compilers can evaluate constant expressions of modest complexity at compile time, the functions provided in the IEEE proposal, possibly augmented as indicated above, seem to be more nearly intellectually parsimonious than some other proposals, with no loss in functionality. The functions logb and scalb use the radix implicitly without burdening anyone with the words "radix" (or "base"), "maxexponent", "minexponent", or their equivalents. Similarly, we avoid words like "eps" whose meanings have already been debased through widespread use for disparate purposes.

Brown Feldman and I agree completely about the importance to users of the overflow and underflow thresholds - that is the reason for derived parameters.

Meissner Will the X3J3 set of parameters do the right things if IEEE arithmetic is available?

Cody Yes, because then Brown's model numbers will be the representable (normalised) numbers.

Lake   I would like to speak as one of the "clowns" who have been shot at
earlier.   The first point is that if you ladies and gentleman care to put up
the money and let a development contract you will get the arithmetic that you
want.   I am perhaps the only person here to have coded these operations at
the bit level for practical use - I have coded multiply and divide for the
ICL Distributed Array Processor. We were careful in this work, though not
conforming to the Brown-Feldman model with full mantissa length.   The multiply
is symmetric, multiplies by 1 correctly, is correctly ordered, and comparisons
are correct and never overflow. We have stressed a feature not in the Brown-
Feldman model - the average bias which is of importance in physical modelling
and can lead to drift of total energy etc. Has anyone looked at this question
of average bias?

Brent   Yes, Von Neumann and many others: see my paper "On the Precision
Attainable with Various Floating-point Number Systems", IEEE Trans. Comp.
C-22 (1973), 601-607.

Ford   NAG has a practical interest in "environmental parameters". I stress
"environmental parameters". We find that the required parameters for
transportable software are a growing set.   To accommodate to the many (often
dirty) arithmetics that we have to work with, there is a need for the
parameters to be restricted (for the sake of safety) as compared with the numbers
suggested by manufacturers in their hardware reference manuals.   How is this
slack handled with the Brown model?

     Recently three people (some 5000 miles apart) evaluated the "Bell"
parameters for the Cray. Each got different numbers. Was this evidence of the
slack?

Cody   Each of the three might be correct. Brown's model penalises the
parameters to account for floppy hardware, so there are infinitely many correct
models.

Ford   I believe that the various models that you discussed are seeking to
handle the problem of this slack, each in its particular way.   This is a
fundamental problem in developing a suitable set of environmental parameters.
I note that the non-uniqueness of parameter values from the Brown model appears
to be an acceptable fact.   Surely we must seek a reliable (presumably unique)
set - hence we need to know where the slackness (due to recognizing the problems
from many machines, dirty arithmetic, etc.) is taken up in the model.

     It is rumoured that Schryer's package will determine a set of parameters
for Brown's model. Is this package available; if so, how do we get it?

Schryer   My test program is complete and has been run successfully in about a
dozen different computing environments. Also, I have written a paper that
describes the program and presents most of the test results.

Brown   Schryer's paper is available from Bell Labs, and the program is
currently being considered for licensing.   There is a possibility that it will
be protected by copyright rather than by trade secrecy, so that the algorithm
it uses can be subjected to full public scrutiny.

Blum   From the programming and numerical analysis standpoint, I suggest that
the parameter problem be viewed as a problem in specifying precisely a family
of floating-point data types that will suffice for numerical software and a
family of mappings between the types to define conversions for transportability
purposes.   Now, parameters in themselves do not constitute a definition of a
family of floating-point types.   They merely assist us in defining the carriers -
or underlying domains - of the various types in the family.   We must also specify

precisely the semantics of the operations (+, *, round, etc.) and relations (=, <, >, etc.) in terms of the parameters. This can be done in several ways, for example, (i) axiomatically, using formal or informal axioms consisting of equations and logical assertions; and (ii) operationally, by choosing some canonical idealized machine representation of numbers and expressing the parameterized types in terms of the canonical one. Brown's axioms are an attempt at method (i) and the IEEE standard appears to be an attempt at method (ii).

The choice of which parameters to use is primarily a question of whether they provide sufficient structure to define all reasonable floating-point types that arise in numerical computation and whether the mappings between the types, both operands and operators, are readily specifiable in terms of the parameters. Without such a complete specification of the semantics of types and conversions between them, transportability will continue to be the vague pragmatic matter it has always been.

Questions: does transportability imply close agreement of arithmetic results or simply precise specification of the disagreement of results? What is the role of exceptions in transportability? Is some degree of non-determinism in arithmetic results tolerable in relations as well as operations? If we do not allow such nondeterminism, do we impose undue constraints on the implementations?

Cody Transportability does not imply close agreement of arithmetic results, nor does it imply precise specification of the disagreement of results. It merely implies computational results consistent with the host environment with minimal change in source code, e.g., proper assignment of parameter values.

Exceptions are frequently host-dependent. Robust software attemps to avoid exceptions but it is difficult to write "failsafe" programs.

Realistically, we must live with a certain amount of nondeterminism in relations that involve arithmetic operations. Ideally, we would like relational results to be as crisp as arithmetic results. This crispness is one of the great advantages of the proposed IEEE arithmetic.

Blum What are we gaining? When is something transportable and when isn't it?

Cody To start, we need to know the underflow and overflow thresholds.

Kahan Would you regard a program as transportable if the specifications were satisfied on both machines even if the results were very different? We all believe that we could write programs to tell what machine (Cray, IBM,...) they are running on. Except for accounting applications it doesn't matter if the results are not exactly the same. We can't expect identical results for trigonometric functions, matrix inversion, etc.

Ford We talk of the transportability [Hague S.J. and Ford B., Software-Practice and Experience 6, 61-69 (1976)] of the source program; generally not of the computed results. A program may be converted between precision types in Fortran (single and double) in a transportable manner. However the results in such a case will be computed to different precisions.

*THE RELATIONSHIP BETWEEN NUMERICAL COMPUTATION
AND PROGRAMMING LANGUAGES, J.K. Reid (editor)*
North-Holland Publishing Company
© *IFIP, 1982*

THE USE OF CONTROLLED PRECISION

T.E. Hull

Department of Computer Science
University of Toronto
Toronto, Ontario, M5S 1A7
Canada

More appropriate programming language facilities
are needed for the description of numerical processes. A proposal for new facilities involving
clean arithmetic, complete precision control, and
convenient exception handling capabilities is
outlined. A number of examples are given to
illustrate the use of the new facilities, particularly of the precision control. Experience with
two preprocessors, the design of a complete language, and the construction of an arithmetic unit
are referred to briefly. It is claimed that good
language facilities of the kind proposed in this
paper can be provided in a reasonably convenient
and economical way.

1. INTRODUCTION

Existing programming languages tend to restrict our ability to describe the numerical processes we would like to implement. We believe
that the situation would improve very significantly if three particular capabilities could be provided in the language in a straightforward and convenient way.

The capabilities we have in mind are associated with precision control and exception handling, and will be described in more detail in
section 2.

Then, in section 3, language facilities which provide these capabilities are outlined. But implementation questions are not considered
at this stage.

In section 4, a number of examples are presented. It is intended
that these examples show how convenient the proposed new facilities
are for describing a very wide variety of numerical processes in a
convenient and flexible way.

A few further points about the language facilities are discussed in
section 5, and then, in section 6, experience with some attempts to
implement various aspects of the new facilities is described briefly.
Finally, the main ideas are summarized in section 7.

2. NEEDED CAPABILITIES

We identify three floating-point capabilities which we believe are
needed if we are to describe adequately the variety of numerical pro-

cesses we would like to have.

   (1)   The first capability is to carry out <u>different parts of a</u>
<u>calculation in different precisions</u>. Typically this arises when one
part of a calculation needs to be carried out in higher precision.
An obvious example is the accumulation of a dot product. Another is
computing the "residual" vector in solving a system of linear equations.

We may also need a wider exponent range as well as a greater number
of significant digits. One simple example is in the calculation of
a Euclidean norm.

   (2)   The second capability is to carry out <u>a portion of a calcu-</u>
<u>lation in more than one precision</u>. Determining the effect of round-
off error on a calculation is just one situation in which this need
arises. A more general situation is one in which part of a calculation is to be repeated in higher and higher precision until an
estimate of the error is small enough. (If the error estimate is to
be a guaranteed bound on the error, it turns out to be convenient to
use interval arithmetic for its computation.)

   (3)   The third capability is to <u>have control over what is done</u>
<u>in case of exceptions such as overflow or underflow</u>. If underflow
occurs we may wish to set the result to zero and carry on with the
calculation. On the other hand, we may instead wish to repeat the
calculation with a wider exponent range.

3.   LANGUAGE FACILITIES

The purpose of this section is to outline very briefly some language
facilities which provide these capabilities. The work is based on
earlier ideas described in [4-8]. The discussion is intended to be
entirely at the language level, but does include both the syntax and
the semantics. Implementation questions are not considered in this
section.

We assume that the floating-point numbers have a normalized, sign-and-
magnitude, decimal representation, and that all arithmetic operations
on them are properly rounded. Precision is to be under the control
of the programmer and will be denoted by the number of decimal
digits in the significand. (The exponent range will be under control
of the programmer as well. For simplicity here, we assume that the
exponent range is tied to the precision by requiring the exponent
range for precision p to be $[-10p, 10p]$.)

We also assume that real variables can have two "values" in addition
to the floating-point numbers, namely "not-yet-assigned" and
"indeterminate". And we assume that wrap-around results are left
after overflow and underflow, except that "indeterminate" is left
after zero-divide.

We assume that complex, interval, and interval complex arithmetic are
all available. As mentioned earlier, interval arithmetic is convenient in the calculation of upper bounds.

An essential feature of the language facilities we propose is that
the programmer specify the precision of the variables <u>separately</u>
from the precision of the operations performed on them.

(1) <u>The precision of the variables</u> is declared as illustrated in the following:

```
real(10) x, y, z      /* x, y, z have 10-digit significands */
real(18) u, v         /* u, v have 18-digit significands   */
real(p) r(100)        /* r is an array                     */
real(p+4) s           /* s has a (p+4)-digit significand   */
```

(2) <u>The precision of the operations</u> is specified by the programmer as in:

```
begin precision(16)        or        begin precision(p)
   real(16) temp                        real(p) temp
   _____                                _____
   z = x + sin(y) + v
   call solve(z,...)                  end
   _____
end
```

All operations in the "precision" block are done in the "current precision" specified for the block, including not only the arithmetic operations, but also the elementary functions, and the procedures called from within the block (unless the procedure itself contains a precision block with a different precision, which is then nested within the original block and has precedence in determining the precision to be used in the inner block).

Thus, if x and y have precision 10, they will, in effect, have to be padded out with zeros before they can enter into the precision 16 calculation specified above. The sine, and the addition in x + sin(y), will both be done in precision 16. If v has precision 18, it must be rounded down to 16 before entering into the calculation. Finally, the 16-digit value of the resulting expression must be rounded to 10 digits before being assigned to z, assuming z has been declared to have precision 10.

(3) <u>Exception handling</u> is specified by the user as illustrated in the following:

```
begin
   on(overflow)        /* the block is the scope of the handler */
   _____
   (underflow)
      result = 0       /* "result" is a reserved word           */
   end on              /* other possible exceptions             */
   _____               /* include "inexact" and "error"         */
end
```

In this example, underflow would lead to a value of 0 being assigned to "result" (in place of the wrap-around value provided by the arithmetic unit before trapping), and then the calculation resumes from the point at which the interruption took place.

Special functions to determine the precision of a specific variable, or of the environment, will be needed. We will use "precisionof(w)" to provide the precision of the variable w, and the parameterless function "currentprecision" to provide the precision of the environment.

To avoid having to include dimension information in calling sequences
we also need special functions to determine subscript limits for
arrays. We use, e.g., lowvalue(1,v) to provide the lowest value of
the first subscript of v, and highvalue(2,A) to provide the highest
value of the second subscript of A.

4.   EXAMPLES

We now consider a few examples to illustrate how the language facilities of the preceding section can be used.

The example in Figure 1 shows how a Euclidean norm can be obtained.
Here the precision of the environment is used to determine an appropriate precision for the intermediate calculations (double the precision of the environment in this case). Besides accumulating a
more accurate sum, unnecessary problems with overflow and underflow
are avoided because the higher precision implies a wider exponent
range as well. After the square root is taken (and the magnitude of
the exponent is correspondingly restricted) the value is assigned to
"norm", whose precision is expected to be the same as the precision
of the environment.

```
procedure Euclidnorm(v,norm)
   real(*) v(*), norm
   integer n, p
   n = highvalue(1,v)
   p = precisionof(v)
   begin precision(2*p)
      real(2*p) temp
      integer i
      temp = 0
      for i = 1 to n
         temp = temp + v(i)*v(i)
      end for
      norm = sqrt(temp)
   end begin
end procedure
```

Figure 1. A procedure for determining the Euclidean norm
of a vector v. Higher precision is used here to provide
a wider exponent range, as well as a larger number of
significant digits during the accumulation of the sum.

The example in Figure 2 is for finding the least squares solution
to an overdetermined system of linear equations Ax = b. Forming the
normal equations and solving them in double precision, which is
quite straightforward with the precision control facilities being
discussed, may also be more efficient than an alternative method
such as one based on the modified Gram-Schmidt technique.

The two examples considered so far show how the first capability described in section 2 can be exploited. In both cases a part of a
calculation is being carried out in higher precision than the precision of the environment. For the Euclidean norm, the higher preci-
avoids unnecessary problems with intermediate overflow or underflow,
as well as providing more accuracy. For the least squares problem,

```
procedure leastsquares(A,x,b,ind)
    real(*) A(*,*), b(*), x(*)
    integer ind, p, n
    p = currentprecision
    n = highvalue(2,A)
    begin precision(2*p)
        real(2*p) C(n,n), d(n)
        call procedure to set C = A^T A
        call procedure to set d = A^T b
        call procedure to factor C, or return an
            error indication through ind
    ...exit begin block if ind ≠ 0
        call procedure to solve Cx = d for x
    end begin
end procedure
```

Figure 2.   A procedure for finding the least squares
solution of Ax = b by forming and solving the normal
equations in double precision.

the higher precision avoids unnecessary difficulties that might arise
because forming the normal equations might seriously worsen the condition of the problem.

Other examples of where the first capability might be exploited
include the following:

(1)   Calculating residuals.

(2)   Summing alternating series, such as the series for $\exp(-x)$
when x has a moderate size. Of course, there is a trick to avoid the
difficulty caused by the cancellation of terms in this case, which is
to sum the series for $\exp(x)$ and take the reciprocal. But, such
tricks are usually not available. For example, consider the series
for the Bessel function $J_0(x)$.

(3)   Solving stiff systems of ordinary differential equations
where the right sides typically suffer from cancellation between
relatively large terms, as in chemical kinetics. Here the right
sides may need to be evaluated in higher precision.

(4)   Random number generators. It is a trivial matter to
generate random numbers "in line" if an intermediate calculation can
be done in higher precision. For example, if x has an initial value
in (0,1) and has precision 10, while m = 3**17 and also has precision
10, then repetition of the following will produce a sequence of
pseudo-random numbers:

```
begin precision(20)
    x = m*x - floor(m*x)
end
```

Another feature of the examples in Figures 1 and 2 is that they show
how natural it is to write procedures that "inherit" the precision
of the environment from which they are called. There is no need for
a separate version for each precision. This idea can be extended

T.E. Hull

in a natural way to any kind of arithmetic, complex, interval, etc.,
and is not restricted to real arithmetic.

A final example to illustrate the use of the first capability described in section 2, is given by the procedure for solving quadratic
equations in Figure 3. Once again, unnecessary trouble with intermediate overflows or underflows (in evaluating the discriminant is
avoided.

```
procedure quadratic(a, b, c, r1, r2, ind)
    real(*) a, b, c, r1, r2
    integer ind, p
    p = precisionof(a)
    begin precision(2*p)
        real(2*p) disc
        if a = 0
            if b = 0
                ind = 3
            else
                ind = 2
                r1 = -c/b
            end if
        else
            disc = b*b - 4*a*c
            if disc < 0
                ind = 1
                r1 = -b/a
                r2 = (sqrt(-disc))/(2*a)
            else
                ind = 0
                if b > 0
                    r1 = (-b-sqrt(disc))/(2*a)
                else
                    r1 = (-b+sqrt(disc))/(2*a)
                end if
                if r1 = 0
                    r2 = 0
                else
                    r2 = c/(r1*a)
                end if
            end if
        end if
    end
end procedure
```

Figure 3. A procedure for solving quadratic equations.
If ind = 3, there are no roots (or, if c = 0, an infinite number). If ind = 2, there is one real root
r1. If ind = 1, there are complex conjugate roots
with real and imaginary parts r1 and r2 respectively.
If ind = 0, there are two real roots r1 and r2.

```
initialize precision p
set acceptable = false
while(not acceptable)
    begin precision(p)
        attempt solution
        determine error bound or error estimate
        if error is small enough
            set acceptable = true
        end if
    end
    increase p
end while
```

Figure 4. Paradigm for repeatedly solving a problem in higher and higher precision until some error criterion is satisfied.

We turn now to examples in which the second capability described in section 2 is exploited. This is the ability to repeat a portion of a calculation in higher and higher precision.

The paradigm shown in Figure 4 illustrates the use of this capability. Here a problem is solved repeatedly, in higher and higher precision, until some error criterion is satisfied. More detail will of course be needed in a practical application. For example, there will usually have to be some other way of terminating the loop, in case the error criterion can never be satisfied.

One can imagine such a scheme being used in finding solutions to polynomial equations, in which case the higher precision may be needed primarily to cope with multiple, or near-multiple roots. Alternatively, a quadrature program could be designed so that it is able to resort to higher precision in any circumstances that seem to warrant its use.

A specific example of this general idea is shown in Figure 5, where linear equations are solved in higher and higher precision (increasing by 6 each time) until an error bound is no larger than a user-supplied tolerance.

Figure 5 also illustrates the use of exception handling facilities in a rather straightforward way. The handler simply causes the calculation to proceed immediately to the next higher precision. Of course overflow or underflow may still occur at maxp, in which case there is a return to the calling program with ind = -1.

It has already been mentioned that the determination of a guaranteed error bound can involve the use of interval arithmetic. An example that illustrates this is given in Figure 6, where a bound on the max norm of the residual b-Ax is calculated, such as would be needed in the linear equation solver of Figure 5.

A number of other examples are given in [6].

```
procedure linearequations(A,b,tol,maxp,x,ind)
    real(*) A(*,*), b(*), tol, x(*), error
    integer maxp, ind, p, n
    p = currentprecision
    n = highvalue(1,A)
    ind = -1
    while(ind ≠ 0 & p ≤ maxp)
        begin precision(p)
            real(p) C(n,n), d(n), y(n)
            on(overflow or underflow)
        ......exit begin block
            end on
            assign arrays C = A, d = b
            call procedure to factor C
            call procedure to solve Cy = d for y
            x = y
            determine error in x
            if error ≤ tol
                ind = 0
            end if
        end begin
        if p = maxp, increase p, else set p = min(p+6,maxp)
    end while
end procedure
```

Figure 5. A procedure for solving the linear equations Ax=b.
A trial solution, along with a measure of the error, is computed in higher and higher precision until either error ≤ tol
or the precision exceeds maxp; the value of ind is set according to whether the accuracy requirement is achieved or not.

```
function resbound(b,A,x)
    real(*) b(*), A(*,*), x(*), resbound
    interval(*) r
    integer p, n, i, j
    p = currentprecision
    n = highvalue(1,b)
    resbound = 0
    for i = 1 to n
        begin interval precision(2*p)
            interval(2*p) temp
            temp = b(i)
            for j = 1 to n
                temp = temp - A(i,j)*x(j)
            end for
            r = temp
        end
        resbound = max(resbound,right(abs(r)))
    end for
end function
```

Figure 6. Each component of the residual vector is computed
in double precision interval arithmetic, and "rounded out"
before being assigned to the single precision interval r.
Right(abs(r)) bounds the absolute values of all the values
in r, so resbound is a true upper bound for the max norm
of the residual vector b-Ax.

5. FURTHER POINTS ABOUT THE LANGUAGE FACILITIES

A few points that did not show up in the examples will be mentioned briefly in this section.

(1) It should be pointed out that the facilities proposed in this paper will allow the user to design procedures intended to return values to approximately the accuracy of the current precision. Thus, for example, the linear equation solver in Figure 6 need not be supplied with a tolerance. The procedure could have been designed to take tol to be a small multiple of $10**(-p+1)$.

(2) In most cases, the precision could have "at least" rather than "exact". However, there are some situations in which "at least" is not good enough, such as when the user wants to measure the effects of roundoff, or when such a measure is to be used to estimate what value of p should be used next (instead of simply increasing p by an arbitrary amount and trying again). In any event, it is better to know exactly what your program is supposed to do, and to know only that its precision is "at least" so many decimal digits is still "implementation dependent", and hence not portable.

(3) There may be occasions in which a user would like to use the result of one iteration to be the starting point of another iteration in higher precision. This would obviously lead to the need for something like the "own" variables in Algol, along with the resulting implementation problems.

(4) What has been discussed so far has been to point out the advantages of using higher precision under certain circumstances. But sometimes there can also be an advantage in using lower precision. For example, this would occur in storing large amounts of data that needs to be known only to 2 or 3 decimal digits, as is the case in some graphics applications.

(5) Adoption of language facilities such as are proposed here would eliminate most of the problems associated with transportability of programs, for example in connection with environmental enquiry functions, and in coping with the different precisions on different machines.

6. EXPERIENCE SO FAR

All of the examples given in this paper have been written in a pseudo-code for which no processor exists as yet. However, some experience has been gained in three different areas that are closely related to what such a processor would provide, and it is our belief that the results are promising enough to justify working towards the development of such a processor.

(1) The first area in which some experience has been gained is in the use of two preprocessors which implement the essential features of the precision control facilities used in the examples of this paper.

One preprocessor [7,8] is for a language now called Algol-H, which is an extension of Algol-W that provides precision control in terms of word lengths, rather than decimal digits, but which makes no provision for exception handling. The programs produced by this system

are very slow because each arithmetic operator is translated into a
procedure call, even for single and double precision. Nevertheless,
a large number of examples have been run successfully, including
some that require higher and higher precision until a prescribed
error requirement is met. A typical result was one in which triple
precision was needed to obtain the zeros of the famous polynomial
$(x-1)(x-2)...(x-20)$ to within a few decimal places of accuracy.

The other preprocessor has been completed only recently, although it
was first proposed a number of years ago in [8]. It is for a language called Fortran-X. It extends Fortran in two ways, one in the
direction of reasonable structuring facilities and the other in order
to provide precision control. As with Algol-H, the precision is
specified in terms of word lengths, but the single and double precision arithmetic is implemented at normal machine speeds, subroutine
calls being used only if the precision is three or higher. Results
obtained with this preprocessor have been very similar to those
obtained with Algol-H.

      (2) Experience with the design of a complete language has been
obtained recently by Curley [2]. The language is called PNCL (for
Prototype Numerical Computation Language), and is based on Pascal.
It provides almost all of the features proposed in this paper. The
main exception is that it uses binary arithmetic, and precision must
be expressed in multiples of 8 bits. The compiler has not yet been
completed, but the language has been tested on a number of examples,
and this testing has already led to some ideas for further refinement of the language.

      (3) Some further experience has also been obtained with the
design and construction of an arithmetic unit called CADAC (for
Clean Arithmetic with Decimal base And Controlled precision). This
unit has been designed to support all of the proposed language
facilities, with reasonable efficiency. (There is a minor exception
in that the number of decimal digits of precision must be even, but,
otherwise, the unit will support arbitrarily high precision decimal
arithmetic, with provision for easy implementation of interval
arithmetic, and complete exception handling capabilities. The precision of the operations is specified separately from the precision
of the operands.)

The design of CADAC was completed in December, 1980, and the unit is
now being constructed. Running at 10 mHz, it will perform an 8-digit
multiply in 4 μs. More details are given in [1].

The primary purpose of this unit is to demonstrate the feasibility
of supporting the language facilities proposed in this paper in a
reasonably convenient and economical way.

7.   SUMMARY AND CONCLUSIONS

We began by identifying three capabilities which are not currently
available in most programming languages but which should be provided
for if we are not to be restricted in the numerical processes we are
able to implement. One need is to carry out part of a calculation
in higher precision. Another is to be able to repeat a part of a
calculation in higher and higher precision. And the third is to let
the user have control over what happens when an exception occurs.

Language facilities to meet these needs were proposed. A key feature of these facilities is the separation of the precision of the operands from the precision of the operations performed on them. Otherwise the proposals depend on the availability of clean, decimal, floating-point arithmetic, along with the ability to specify precisely what to do in case an exception arises.

Most of the paper has been devoted to examples which illustrate the use of these facilities. It was intended that these examples demonstrate the ease with which a very wide variety of numerical processes can be implemented with the aid of the proposed facilities.

Experience with preprocessors, the design of a complete language, and the design and construction of a special hardware unit, have provided support for the claim that the proposed language facilities can be provided in a reasonably convenient and economical way.

Most programming languages could be extended relatively easily in the direction of the facilities proposed here. Reasonable upward compatability with precision declarations and precision blocks is quite straightforward, as long as "at least" is required instead of "exact". (The default in case no environment precision is specified could obviously be whatever is now required in the language being extended.)

In the meantime, it also appears that more appropriate hardware units could be developed to support these facilities in a reasonably economical way.

REFERENCES

[1]   Cohen, M., Hamacher, V.C. and Hull, T.E., CADAC: An Arithmetic Unit for Clean Decimal Arithmetic and Controlled Precision, Proceedings 5th Symposium on Computer Arithmetic (IEEE Computer Society, Ann Arbor, Michigan, 1981), 106-112.

[2]   Curley, A., PNCL: A Prototype Numerical Computation Language, M.Sc. thesis, Department of Computer Science, University of Toronto (1981).

[3]   Dongarra, J.J., Bunch, J.R., Moler, C.B. and Stewart, G.W., LINPACK Users' Guide (SIAM, Philadelphia, 1979).

[4]   Hull, T.E., Semantics of Floating-Point Arithmetic and Elementary Functions, Portability of Numerical Software, edited by Wayne Cowell (Springer-Verlag, 1977), 37-48.

[5]   Hull, T.E., Desirable Floating-Point and Elementary Functions for Numerical Computation, Proceedings Conference on the Programming Environment for Development of Numerical Software, SIGNUM Newsletter 14 (1979), 96-99, and a similar paper in Proceedings 4th Symposium on Computer Arithmetic (IEEE Computer Society, Santa Monica, California, 1978), 63-69.

[6]   Hull, T.E., Precision Control, Exception Handling and a Choice of Numerical Algorithms, Proceedings Dundee Conference on Numerical Analysis (1981), to appear.

[7] Hull, T.E. and Hofbauer, J.J., Language Facilities for Multiple
     Precision Floating-Point Computation, with Examples, and the
     Description of a Preprocessor, Technical Report 63 (Department
     of Computer Science, University of Toronto, 1974).

[8] Hull, T.E. and Hofbauer, J.J., Language Facilities for Numerical
     Computation, Proceedings ACM-SIAM Conference on Mathematical
     Software II (Purdue University, 1974), 1-18.

DISCUSSION

<u>Roberts</u> What is the timescale?   Users would like to use your
ideas, but they may take several years to standardise and several
more years to implement.   Wouldn't using a preprocessor or an
Algol 68 implementation be faster?

<u>Hull</u> As mentioned in my talk, we have one preprocessor that extends
Algol-W, but it is very slow. Another which has just been completed
extends Fortran and is much faster.   There is also a preprocessor
using R. Brent's package. Perhaps some members of the present
audience will start using them.

<u>Roberts</u>   Perhaps we should use preprocessors, but they need to be
universally available.

<u>Brent</u>   The preprocessor for my MP package is described in [TOMS
6 (1980), 146-149].   However, preprocessors can be awkward to use.
I don't believe that variable-precision arithmetic will catch on
(except for very special applications) until there is an implementation
which is as elegant as Hull's proposal and also reasonably efficient.

<u>Tienari</u>   How do you represent program constants like π, e etc.
in the varying precision blocks?   I think you would need a pre-
determined upper limit to the computing precision.

<u>Hull</u> I should have mentioned Tienari's early work on the use of
variable precision in numerical computation at Stanford in the '60s.
We allow "constants" in precision blocks (but, strictly speaking,
they are no longer constant).   I have not contemplated use of the
symbol "π", but we need to provide the value π somehow
(dynamically).

<u>Kahan</u> You can use $\pi = 4 \arctan(1)$, and $e = \exp(1)$; it's someone
else's responsibility to provide arctan and exp.

<u>Brent</u>   Using my MP package with the Augment precompiler, you can
simply write CTM('PI') for π, and statements like

$$X = \text{'PI'} * X \quad \text{(or similar expressions)}$$

where X is declared as MULTIPLE.   The precompiler generates a
subroutine call to compute π to the required accuracy. If the value
of π were saved, the computation would only need to be repeated if
the precision were increased, but this idea has not been implemented
because the computation of π by Machin's identity is very fast
compared to other multiple-precision computations (e.g., the
computation of trigonometric functions).

<u>Moler</u> Your example of solving linear equations is bad because the
errors in the solution might be due to errors in the matrix elements,
which would need to be recomputed to higher precision.

Hull   That could be. I assumed that the matrix was known exactly
(e.g., a scaled Hilbert matrix).   My point is that varying precision
enables you to get a more accurate answer if you want to, and if
your data justifies it.   You don't have to do so if you don't want
to.

Moler   I was quibbling over your example, not over the use of
varying precision.

Dekker   A distinction should be made between static variation of
precision allowing variations only at compile time, and dynamic
variation at run time as well.   In my opinion, static variation of
precision suffices to achieve several (though not all) goals mentioned
in Hull's paper.   It admits, however, of a much more efficient
implementation;   the compiler can for each block select the
operations and (elementary) function routines working in the
appropriate static precision and determine appropriately rounded
constants occurring in the block.

Boyle   How widely would dynamic precision be used?   Is it something
to be used in all programs all the time, or is it just highly useful
sometimes?   For example, would it be used in all NAG routines?

Hull   It would certainly be useful to writers of subroutine
libraries.   For example, it would be useful in a number of NAG
routines (but I don't know what percentage).

Meissner   Assuming you had a sufficiently precise arithmetic unit
and enough sufficiently precise registers,

> 1) could you evaluate the effects of varying
> precision by introducing "noise" at the appropriate
> level?
>
> 2) is it sufficient to reduce precision only at
> "assignment"?

Hull   Neither suggestions appears to me to be particularly
appealing.

Gentleman   1) Noise is not reproducible and the errors don't
            correlate as they should;   this tends to give unduly
            pessimistic estimates.

            2) Rounding on stores is fine except that you can't
            tell when the high-level language implementor will
            store temporaries.

Kahan   Dekker's question has not really been dealt with.   Fast
arithmetic, even if sloppy, drives out slow arithmetic.   Static
precision is likely to be faster than dynamic.   Thus NAG routines
could not use dynamic precision without the risk of being too slow.
Most users will run a "fast" program first, then use Brent's MP or
Macsyma's BIGFLOAT if necessary.

Brent   Two general comments:

1)   I have used dynamically variable precision extensively inside
the MP package and have found it very convenient.   Typically one

increases the precision slightly inside a special function routine and
then rounds the result at the end, so the result is correct to at
most a few ulps (and usually it is the correctly rounded result).
The relative cost is low because multiple-precision operations are
expensive in any case (but the expense is only a low-degree polynomial
in the precision).

2)   Usually "at least" precision p rather than "exact" precision
p is adequate, provided one knows what the precision actually is.
"At least" is easier to implement because the base does not have to
be ten.   "Exact" precision could be useful for pedagogical or testing
purposes.

*THE RELATIONSHIP BETWEEN NUMERICAL COMPUTATION
AND PROGRAMMING LANGUAGES, J.K. Reid (editor)
North-Holland Publishing Company*
© *IFIP, 1982*

A SYNOPSIS OF INTERVAL ARITHMETIC

FOR THE DESIGNER OF PROGRAMMING LANGUAGES

Christian Reinsch

Institut für Mathematik der
Technischen Universität München
D-8000 München 2, FRG

The straightforward technique of basic interval arithmetic is
recalled and two generalizations of Kahan [9] and Kulisch
[12],[13] are outlined. A discussion of the syntax extensions is given which are necessary for the implementation
of interval arithmetic within a numerically oriented high
level programming language.

1. Basic Interval Arithmetic

A powerful digital computer accomplishes per second more than a million rounding
errors. In most cases their effect will be negligible, but rarely can a programmer
make rigorous claims to the accuracy of the computed results. Also, most practical
computations have input data of limited accuracy and again, it might be difficult
to judge their influence on the solution. Where careful analysis of both effects
is impractical, some experimenting might help - like switching from single to
double precision in order to see the influence of rounding errors, and a systematic parameter variation to get a feeling for the "condition" of the problem.

Another possibility is to introduce a strict lower and upper bound for each real
number occurring in the course of computation. Obviously, the four arithmetic
operations are monotone so that bounds for the result follow from bounds for the
two operands. The finite precision of the computer can be taken into account by
directed rounding: towards $-\infty$ for the lower bound and towards $+\infty$ for the upper
bound. The same can be done for the customary unary operations like sign inversion,
squaring, and taking the square root. In short, a digital computation can be done
with *inclusion intervals* as input data, operands, and final results instead of the
ordinary real number approximations. The interval bounds are traced through the
entire calculation resulting in final intervals that contain the wanted answers.
At least, this should be possible as long as there are no arithmetic relations
involved as with maximum, minimum, or program branching.

We underline symbols to denote intervals and use the suffixes - and + for the corresponding endpoints; thus $\underline{x} = [x_-, x_+]$ with $-\infty < x_- \leq x_+ < +\infty$ is a closed and
bounded (compact) interval on the real line $\mathbb{R}$. It is important that in the computer
intervals should be stored and represented by the pair of their endpoints and *not*
by midpoint $(x_- + x_+)/2$ and half-length $(x_+ - x_-)/2$. Mathematically, this would
be equivalent, but the finite precision of machine-representable numbers renders
the second method useless whenever $|x_-| \ll |x_+|$ or vice versa so that the same
machine value is stored for midpoint and half-length and the interval effectively
extends to zero. The same deficiency applies to circular arithmetic in the complex
plane [7] where the inclusion sets are disks specified by midpoint and radius.

According to the above, Minkowsky's principle (each-value-rule) is used to extend to intervals the four arithmetic operations and certain unary operations or elementary functions:

$$\underline{x} * \underline{y} = \{x * y : x \in \underline{x}, y \in \underline{y}\} \quad \text{for } * \in \{+,-,\times,/\},$$

$$(0 \notin \underline{y} \text{ in case of division}),$$

$$\underline{f(x)} = \{f(x) : x \in \underline{x}\} \quad \text{for } f : \mathbb{R} \to \mathbb{R}, \text{ continuous}.$$

There results always an inclusion interval; its endpoints can be computed from the endpoints of the operands for the four arithmetic operations and for continuous, monotone functions f. Moore [14] calls $\underline{f}$ the *united extension* of f (extension by union). Note that $\underline{f(x)}$, the united extension of an arbitrary arithmetic expression, must be carefully distinguished from $f(\underline{x})$, the evaluation of this expression in interval arithmetic; e.g., in general, $\underline{x}^2 \neq \underline{x} \cdot \underline{x}$ and $\underline{x}/(\exp(\underline{x}) - 1) \neq \underline{x}/(\exp(\underline{x}) - 1)$. Implementations of the above definitions must take into account the computer's finite precision by rounding the lower endpoint towards $-\infty$ and the upper endpoint towards $+\infty$; for a more general definition see Section 4. Also required are the intersection $\cap$ and the convex hull (union with fill-in) $\overline{\cup}$ of two intervals:

$$\underline{x} \cap \underline{y} = [\max(x_-, y_-), \min(x_+, y_+)] \quad \text{(if not empty)},$$

$$\underline{x} \overline{\cup} \underline{y} = [\min(x_-, y_-), \max(x_+, y_+)].$$

There is no useful linear ordering for intervals and partial orderings (without the trichotomy of <, =, >) are not sufficiently close to the real model. For example, one can use Minkowsky's principle (the each-value-rule) also for the definition of $<, \leq, =, \neq, \geq, >$:

$$\underline{x} < \underline{y} \quad \text{if and only if} \quad x < y \quad \forall x \in \underline{x}, y \in \underline{y} \quad \text{i.e.} \quad x_+ < y_-$$

etc. Then, $\underline{x} = \underline{y}$ can be true only for degenerate intervals ($x_- = x_+ = y_- = y_+$) and $\underline{x} \neq \underline{y}$ means that the intervals are disjoint; these relations differ from the familiar set theoretic equality and its logical complement. Note also, that $\underline{x} \leq \underline{y}$ is not equivalent to $(\underline{x} < \underline{y})$ or $(\underline{x} = \underline{y})$.

2. Applicability

Interval arithmetic is always a case study and can never provide *general* insight. Its statements concern the accuracy of the computed result in the underlying case, never the numerical stability of the employed algorithm. Thus, it is not a substitute for a true rounding error analysis as presented in the famous book of Wilkinson [16]. However, for an application only the accuracy of the result matters and the general behavior of an algorithm is unimportant. For the user any method of validation is good enough and interval arithmetic has probably the greatest applicability.

Basically, it is possible to execute any numerical algorithm in interval arithmetic instead of ordinary floating-point arithmetic, although some care might be necessary with the arithmetic comparisons as outlined above. Also, exceptions like the division by an interval containing zero might occur more often. With this proceeding one gets rigorous error bounds for each computed result including round-off errors as well as uncertainties in the input data. One is not restricted to the use of numerically stable algorithms but can try also the more risky ones. If the inclusion intervals for the final answer are acceptably small then interval arithmetic was a success and the user has the desired validation.

Frequently, however, the intervals so obtained are too wide. In particular, the well-established algorithms from linear algebra, when transliterated naïvely into interval arithmetic, often give only trivial intervals judged by practical standards. This is especially unsatisfactory since from the *a priori* error analysis we conclude that these algorithms are numerically stable and know the worst case error bounds.

The reason for this discrepancy lies in the correlations of the errors: certainly, intervals are the ideal means to represent the uncertainty of a given value. However, they do not provide the possibility to take into account the existing correlations among any pair of variables and their errors. Most of our established algorithms rely just on such correlations. They do not attempt to keep all of the errors small, rather they accept arbitrarily large errors in intermediate results as long as they are correlated such that the final results will be unaffected. A typical example is Givens' or Householder's reduction of a symmetric matrix to tri-diagonal form, T, for the computation of the eigenvalues. The computed $\tilde{T}$ can have many totally wrong elements and yet have eigenvalues correct to working accuracy.

Hence it is not true that the good algorithms of numerical analysis, naïvely done in interval arithmetic, would provide useful error bounds. In order to be valuable, interval arithmetic needs its own algorithms where the error correlations must not play a vital role for the numerical stability. Thus the problem of numerical stability in conventional arithmetic has its counterpart in interval arithmetic: to arrange for the final intervals to be as narrow as possible, while still providing rigorous bounds.

Error correlations can also prevent acceptable results in other applications of interval arithmetic. A case in point are parameter studies where one wants to determine the influence of one or several input values on the final result. In other words, which interval contains the solution if the set of feasible arguments is a given multi-dimensional interval? In the limit of differentially narrow intervals, this is the question about the *condition* of the posed problem. Again, the answer provided by naïve interval arithmetic may be much to pessimistic.

The interesting applications of interval arithmetic thus require more sophisticated methods than a simple transliteration of conventional algorithms. For example, many iteration methods performed in normal floating-point arithmetic can be supplemented with a last step where interval arithmetic gives a bound for the error of the correction. A survey of such methods can be found in [1], [14]. The following example is from [9], Kahan's method and proof have been rediscovered several times.

Assume that Newton's iteration has been used to produce the approximation $\tilde{x}$ to the zero ξ of a function $f(x)$; x and f can be either scalars or vectors. It is assumed that f and its derivative, f', can be evaluated in interval arithmetic. An interval is to be found containing ξ. To this end, the following three intervals are computed:

$$\underline{x} \;=\; [\tilde{x}-\varepsilon,\, \tilde{x}+\varepsilon] \qquad \text{where } \varepsilon = 1 \text{ or } 2 \text{ units in the last place of } \tilde{x},$$

$$\underline{y} \;=\; -\,G\,f(\underline{x}) \qquad \text{where } G \doteq (f'(\tilde{x}))^{-1} \text{ in floating-point arithmetic,}$$

$$\underline{z} \;=\; -\,G\,f([\tilde{x},\tilde{x}]) + \underline{S}\,\underline{y} \quad \text{where } \underline{S} = 1 - G\,f'(\tilde{x}+\underline{y}) \text{ in interval arithmetic.}$$

Two assumptions can be made: (1) $0 \in f(\underline{x})$, otherwise more Newton iterations should be done, (2) $\|\underline{S}\| < 1$, otherwise the problem is too ill-conditioned to be treated in this way. Both assumptions can be readily checked. If $\underline{z}$ is contained in $\underline{y}$ then $\tilde{x}+\underline{z}$ is the inclusion interval, otherwise $\underline{y}$ can be replaced by $\underline{y}\,\overline{\cup}\,\underline{z}$ and the last step repeated. To prove this, Kahan considers the function $h(t) \equiv t - G\,f(\tilde{x}+t)$ $\equiv -\,G\,f(\tilde{x}) + t - G\,(f(\tilde{x}+t) - f(\tilde{x}))$ and shows with the help of the mean value theorem that h maps the compact set $\underline{y}$ into $\underline{z}$. Thus, if $\underline{z}$ is contained in $\underline{y}$, it follows from Brouwer's fixed point theorem that h has a fixed point in $\underline{z}$, or equivalently that f has a zero in $\tilde{x}+\underline{z}$ since G is non-singular if $\|\underline{S}\| < 1$.

As mentioned, x and f can be n-dimensional vectors. Two special cases, which have also been discussed several times in the literature, are linear systems where $f(x) \equiv b - Ax$, $G = -A^{-1}$, and the eigenvalue problem where the first n components of $f(x,\lambda)$ are $\lambda x - Ax$ and the last component is $c^T x - 1$ with c^T a fixed row vector used for normalization.

3. Kahan's "More Complete" Interval Arithmetic

Interval arithmetic as discussed above does not admit the division by an interval containing zero. This operation occurs rather often, especially with zero as an endpoint, so that in many cases the computation comes to a premature end. To counter this effect, Kahan [9] introduced also *unbounded* and *outer* intervals:

(i) $\underline{x} = [x_-, x_+] = \{x : x_- \leq x \leq x_+\}$ if $x_- \leq x_+$,

(ii) $\underline{x} = [x_-, \infty] = \{x : x_- \leq x\}$ any x_-,

(iii) $\underline{x} = [-\infty, x_+] = \{x : x \leq x_+\}$ any x_+,

(iv) $\underline{x} = [x_-, x_+] = [x_-, \infty] \cup [-\infty, x_+]$ if $x_- > x_+$,

In order to represent machine intervals of type (ii) and (iii) by pairs of floating-numbers, it is necessary that the computer has a special bit pattern reserved to encode ∞. Intervals of type (iv) are identifiable by the reverse arithmetic order of their endpoints. Kahan recommends visualizing this enlarged set of intervals by the projection of the real axis onto the unit circle and thereby identifying $-\infty$ with $+\infty$:

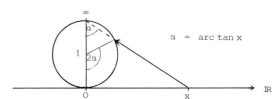

$\alpha = \arctan x$

Figure 1
Projection of the real axis onto
the unit circle (diameter = 1)·

In this system, the basic formula for the operations on intervals (each-value-rule),

$$\underline{x} * \underline{y} = \{x * y : x \in \underline{x}, y \in \underline{y}\} = \bigcup_{\substack{x \in \underline{x} \\ y \in \underline{y}}} x * y \quad \text{with} \quad * \in \{+,-,\times,/\}$$

has to be extended so that *all* operations become feasible and free of exceptions. To this end, the reals $\mathbb{R}$ are extended by ∞ to $\mathbb{R}^* = \mathbb{R} \cup \{\infty\}$ with the following arithmetic tables declared on $\mathbb{R}^*$:

ADD SUB	y	∞
x	$x \pm y$	∞
∞	∞	$\mathbb{R}^*$

real x,y

MUL	0	y	∞
0	0	0	$\mathbb{R}^*$
x	0	$x \cdot y$	∞
∞	$\mathbb{R}^*$	∞	∞

real $x \neq 0$, $y \neq 0$

DIV	0	y	∞
0	$\mathbb{R}^*$	0	0
x	∞	x / y	0
∞	∞	∞	$\mathbb{R}^*$

real $x \neq 0$, $y \neq 0$

The Arithmetic Tables on $\mathbb{R}^*$

Naturally, $\underline{x} * \underline{y} = \mathbb{R}^*$ if according to the tables $x * y = \mathbb{R}^*$ for at least one pair $x \in \underline{x}$, $y \in \underline{y}$. For example, from row 1,2 and column 2,3 of the multiplication table it follows that $[0,1] \times [1, \infty] = [-\infty, \infty] = \mathbb{R}^*$ while a simple consideration of limits would have given $[0, \infty]$ for the product because obviously $[\sigma,1] \times [1,\lambda] = [\sigma,\lambda]$. Kahan goes further and admits also open intervals, at least for 0 and ∞ as endpoints where this feature has its greatest practical value; examples:

$$(0,1] \times [1, \infty] = (0, \infty] \qquad \text{(from row 2, column 2 \& 3),}$$
$$[0,1] \times [1, \infty) = [0, \infty) \qquad \text{(from row 1 \& 2, column 2),}$$
$$(0,1] \times [1, \infty) = (0, \infty) \qquad \text{(from row 2, column 2),}$$

$$[0, \infty] + [0, \infty] = [-\infty, \infty] \qquad \text{(from row 1 \& 2, column 1 \& 2),}$$
$$[0, \infty) + [0, \infty] = [0, \infty] \qquad \text{(from row 1, column 1 \& 2).}$$

Thus, open intervals behave as the corresponding limiting cases. They are either input or generated by over/underflow of exponents.

In the same way the arithmetic relations, if based on the each-value-rule, can be extended to include the intervals of type (ii), (iii), and (iv). This requires the definition of the six relations $<$, $\leq$, $=$, $\neq$, $\geq$, and $>$ on the extended reals $\mathbb{R}^*$ and could be done with the help of six tables. A summary is that $x \neq \infty$, $\infty \neq y$, $\infty = \infty$, $\infty \leq \infty$, $\infty \geq \infty$ are true while every other relation involving ∞ is false. The open intervals do not involve ∞.

The distinction between closed and open intervals is easy to code for the special endpoints 0 and ∞ provided there are *two* internal representations for these quantities available, e.g., $+0$ and -0, $+\infty$ and $-\infty$. Some special combinations of these symbols can then be used to add a few more intervals to the system, e.g., Φ, $\mathbb{R}$, $\mathbb{R}^*$, $\mathbb{R}^* \setminus \{0\}$. In this system, the intersection of an inner with an outer interval or of two outer intervals need not be an interval but no other exception is possible.

Occasionally the midpoint and the width of an interval enter the computation. With outer intervals, the linear definition of Section 1 might not be very useful. Kahan suggests measuring along the circle in Figure 1, i.e., using the expression $\tan((\arctan x_- + \arctan x_+)/2)$ as midpoint and $\arctan x_+ - \arctan x_-$ for the width. The first formula can be simplified although a careful treatment of rounding errors and over/underflow of exponents is necessary.

The accurate and efficient implementation of this extended scheme is a formidable task and requires unusually good hardware [15]. The proposed IEEE floating-point standard [4] contains all provisions to make this possible [6], [10], among other features (i).   Encodings for $+0$, -0, $+\infty$, $-\infty$,

(ii)   Hardware support for the arithmetic and relational tables on $\mathbb{R}^*$,

(iii)   Directed rounding without over/underflow of exponents,

(iv)   Denormalized numbers and the normalizing mode for their processing.

4. Interval Arithmetic in Product Spaces and Ordered Sets

We have already used n-tuples of intervals, defined as the direct product of their components:

$$(\underline{x}_1, \ldots, \underline{x}_n) \quad = \quad \{(x_1, \ldots, x_n) : \ x_i \in \underline{x}_i, \ i = 1, \ldots, n\}.$$

Strictly speaking one should not call them "interval vectors" (or "interval matrices" etc.) because they do not form a linear space in the algebraic terminology; more appropriate might be *structured* intervals (vector intervals, matrix intervals etc.). Let $\text{II}\,\mathbb{R}^n$ denote the set of all n-tuples of real intervals and $\text{II}\,\mathbb{F}^n$ its subset where the endpoints are floating-point numbers of the computer. For any algebraic structure in $\mathbb{R}^n$ Kulisch [12], [13] defines a corresponding interval structure in $\text{II}\,\mathbb{R}^n$ resp. $\text{II}\,\mathbb{F}^n$ in two different ways.

The first possibility is to replace each real operation in the definition of the structured operation by the corresponding interval operation as defined previously by Minkowsky's principle. The following examples are complex interval arithmetic in $\text{II}\,\mathbb{R}^2$ and matrix operations in $\text{II}\,\mathbb{R}^{n \times n}$:

$$(\underline{a} + i\underline{b}) \cdot (\underline{x} + i\underline{y}) \quad = \quad (\underline{a} \cdot \underline{x} - \underline{b} \cdot \underline{y}) \ + \ i\,(\underline{b} \cdot \underline{x} + \underline{a} \cdot \underline{y}),$$

$$(\underline{a} + i\underline{b}) / (\underline{x} + i\underline{y}) \quad = \quad (\underline{a} \cdot \underline{x} + \underline{b} \cdot \underline{y}) / (\underline{x}^2 + \underline{y}^2) \ + \ i\,(\underline{b} \cdot \underline{x} - \underline{a} \cdot \underline{y}) / (\underline{x}^2 + \underline{y}^2),$$

$$(\underline{A} \cdot \underline{B})_{ik} \quad = \quad \sum_j \underline{a}_{ij} \cdot \underline{b}_{jk}$$

The other possibility applies Minkowsky's principle immediately to the operation so that only *one* rounding occurs in the compound operation. Thus, if $x * y$ is any operation declared for the n-tuples of reals, x and y, then the corresponding operation on a pair $\underline{x}$ and $\underline{y}$ of n-tuple intervals is defined as the smallest interval containing all $x * y$ with $x \in \underline{x}$, $y \in \underline{y}$:

$$\underline{x} * \underline{y} \quad = \quad \rho(\{x * y : \ x \in \underline{x}, \ y \in \underline{y}\}) \quad \text{where} \quad \rho(Z) \quad = \quad \bigcap_{Z \subset \underline{z} \in \text{II}\,\mathbb{F}^n} \underline{z} \,.$$

The function ρ is the outward rounding of an arbitrary subset of $\mathbb{R}^n$ to the next larger feasible interval illustrated in Figure 2:

Figure 2

Outward rounding in the power set of $\mathbb{R}^2$

As always with interval arithmetic, rounding increases the inherent inclusion set. Contrary to the basic interval arithmetic, however, the enlargement is now primarily caused by the correlation effect and not so much by the finite machine precision. For example, multiplication of the complex interval $[x - \varepsilon, x + \varepsilon] + i[y - \varepsilon, y + \varepsilon]$ by the singleton exp(iα) increases its width by the factor $|\cos\alpha| + |\sin\alpha|$ whose average value might be near 1.2. Clearly, several complex multiplications could blow things up. This problem is still present, but less severe, when complex intervals (rectangles) are replaced by disks in the complex plane, see [7].

One wonders therefore whether it is worthwhile to invoke multiprecision algorithms to keep the rounding error of all basic operations combined below half a unit in the last place. Nevertheless, the arithmetic with complex, vector, and matrix intervals *has* been implemented on a microprocessor system (Zilog Z80) at the University of Karlsruhe; the required algorithms are discussed in [13]. An enhanced Pascal compiler is completed and documented in [11], and work on a corresponding compiler for FORTRAN 66 is in progress, see [2]. It seems that the extended precision of a few guard digits in the arithmetic registers as suggested in the proposed IEEE floating-point standard is enough to keep the influence of finite precision negligible beside the correlation effect.

It should be mentioned that Kulisch has generalized his approach. Let $\{M, \leq\}$ be a partially ordered set and S an arbitrary subset of M. Rounding is defined as any mapping $\rho : M \to S$ with the properties

(1) $x \in S \;\;==>\;\; \rho(x) = x$ (Projection),

(2) $x \leq y \;\;==>\;\; \rho(x) \leq \rho(y)$ (Monotonicity).

A third rule renders this mapping unique, e.g.,

(3) $x \leq \rho(x)$ (Directed rounding).

The subset S is called an *upper semiscreen* if such a mapping ρ exists. This is the case if M and S are complete lattices and if $\inf A \in S$ for all $A \subseteq S$, i.e., if S is a *complete inf-subnet*. Now, if some operation * is defined on M, then a corresponding operation $\hat{*}$ on S is defined by $x \hat{*} y = \rho(x * y)$ for all $x, y \in S$. Interval arithmetic as discussed above is here included as the special case where M is the power set of $\mathbb{R}^n$ (the set of all subsets of $\mathbb{R}^n$) and $S = \Pi \mathbb{F}^n$ and inclusion as the order relation $\leq$. The operation * thus combines arbitrary subsets of $\mathbb{R}^n$ and might be either directly defined like the union $x \cup y$ or by the Minkowsky rule $x * y = \{\xi * \eta : \xi \in x, \eta \in y\}$. Kulisch calls this correspondence between operations * on the ordered set M and operations $\hat{*}$ on its upper semiscreen a *semimorphism*, because only part of the algebraic structure remains valid, essentially

$$x \in S, \;\; y \in S, \;\; x * y \in S \;\;==>\;\; x * y = x \hat{*} y,$$

$$a \in S, \;\; b \in S, \;\; x \in S, \;\; y \in S \;\; \text{and} \;\; a * b \leq x * y \;\;==>\;\; a \hat{*} b \leq x \hat{*} y,$$

$$x \in S, \;\; y \in S \;\;==>\;\; x * y \leq x \hat{*} y.$$

Vice versa, if these properties hold true, then $x \hat{*} y = \rho(x * y)$ for some mapping $\rho : M \to S$ with the above properties (1), (2), and (3).

5. Interval Arithmetic and High Level Programming Languages

Interval arithmetic is currently used at very few installations. This might be due to intrinsic problems of that method which too often produces results with little practical significance. On the other hand ardent supporters frequently blame the missing support from hardware and high level programming languages. Hardware support will be almost perfect with new arithmetic units like the INTEL 8087 chip which fully conform to the proposed IEEE standard for floating-point arithmetic. It might thus be worthwhile to improve the access to interval arithmetic in programming languages which, like FORTRAN, are primarily used for numerical computations.

If available at all to the user, interval arithmetic is mostly offered through calls upon subroutines, one for each basic operation, where all operands, intermediate results etc. must be explicitly dimensioned by the user. This reduces programming to the level of writing machine instructions. Much better is the situation at places where Crary's AUGMENT precompiler [5] and Yohe's adapted subroutine package for interval arithmetic [17] are available. With that support little is left wanting for the user as long as no run time error occurs; in such a case, however, the user has the burden of studying the output of the precompiler which is tedious even with a table of corresponding line numbers [18].

FORTRAN is the most appropriate host for interval arithmetic since

(i)   most numerical software is written in FORTRAN,

(ii)   from the language point of view, interval variables behave almost identically to the complex and double precision variables of FORTRAN,

(iii)   the evolution of FORTRAN is open for discussion.

In the following, the terminology is therefore slightly biased towards FORTRAN. Note, however, that a PASCAL compiler is available for the Zilog Z80 microprocessor with a full implementation of real and complex interval arithmetic with vectors and matrices as discussed above. For ALGOL 68 see [8].

6. Data Types

The numeric data types are defined by three independent attributes:

(i)   single/double precision,

(ii)   singleton/interval,

(iii)   real/complex.

The scheme can easily be extended to more than two levels of precision. All combinations of attributes should be allowed and the given ordering has only formal significance, e.g., the data type identifiers could be derived by prefixing the word symbol FLOAT first with D, then with I, and finally with C so that REAL, DOUBLE PRECISION, and COMPLEX are synonyms for FLOAT, DFLOAT, and CFLOAT resp. FLOAT needs one memory unit and each prefix doubles the number of units. If the storage sequence is to be specified then the most significant part comes before the least significant part, the left endpoint before the right endpoint, and the real part before the imaginary part in this ranking. The eight numerical data types might be visualized as the vertices of a cube with transitions along the edges. The corresponding coercion (conversion) rules are listed below with descending priority:

```
single pr.   →  double pr.:   expand  (no rounding is assumed),
singleton    →  interval:     create a degenerate interval,
real         →  complex:      set imaginary part to zero,
complex      →  real:         discard the imaginary part,
interval     →  singleton:    take midpoint,
double pr.   →  single pr.:   proper rounding for singletons and
                              outward rounding for intervals.
```

For example, the transition from a single precision interval to a double precision singleton first converts both endpoints to double precision and then computes the midpoint. As mentioned above the midpoint must be carefully defined and implemented; in particular, the computed midpoint of a degenerate interval must be its endpoints.

The integers do not fit nicely into this scheme. It is customary to do the conversion to and from integers via the single precision reals. Then the ranking of the numerical data type leads to the following lattice of Figure 3:

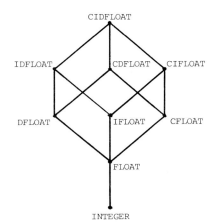

Figure 3

The order diagram (lattice)
of the nine numerical datatypes

For the reasons explained above the complex intervals do not have the importance of the single and double precision real intervals.

7. Constants and Input/Output Conventions

The language requires a representation of interval constants. The use of square brackets has been suggested for this purpose. Thus, a real interval is represented as [left endpoint, right endpoint] where the endpoints are single or double precision constants (same data type) and the decimal to binary conversion must be done with directed rounding. Complex intervals are pairs of real intervals, separated by a comma and enclosed in parentheses as usual. A few minor points should be mentioned.

1) Representation of infinity: If Kahan's more complete system of interval arithmetic is to be implemented, then infinity must be admitted for either endpoint. Due to the type conversions as discussed above and some of the intrinsic functions, infinity can then also be assigned to singletons and should therefore be a feasible value for these data types. This requires that infinity is a legal unsigned constant and that a special symbol must be reserved for it. Instead of ++, or **, or the overworked Dollar sign (which have been suggested), it might be preferable to use the word symbol .INF. much like .TRUE. and .FALSE.

2) Open intervals, at least at 0 and ∞, are also envisaged in Kahan's approach and need a proper syntax specification. Using the reverse brackets is a proposal close to the mathematical convention; abusing the unary minus sign is another proposal, probably inspired by the most likely internal encoding. The latter method is bound to cause confusion since in the underlying mathematical system, $+\infty$ and $-\infty$ mean the same as $+0$ and -0.

3) Special intervals like Φ (the empty interval), $\mathbb{R}$ (all reals), $\mathbb{R}^*$ (all reals and ∞), and $\mathbb{R}^* \setminus \{0\}$ ($\mathbb{R}^*$ except 0) have been mentioned and should be available as constants and for input/output. Again, the source language representation need

and probably should not follow the internal representation   (perhaps some otherwise unused combination of ±0 and ±∞).   More suggestive would be [],]*[, [*],]0[.

4) Some proposals for language extensions include the suffixes Δ, ∇ with decimal constants in order to denote decimal to binary conversion with directed rounding. It seems that this is unnecessary because the rounding direction should always be clear from the context.

5) The data type of a constant is traditionally determined from its syntax, e.g., INTEGER if no decimal point or scale factor with E or D is present. Constants in arithmetic expressions and data statements may then need a *type* conversion after the radix conversion as discussed in the next section. Thus, if X is an interval variable, 0.1 * X and [0.1, 0.1] * X must be distinguished. On the other hand, it would be possible to define the syntax so that a constant gets the data type required by context and that no second conversion need be invoked.

8. Arithmetic and Relational Expressions

The directed rounding which is needed for interval arithmetic can be made available to the programmer by allowing the operations +∇ and +Δ besides +, etc. In all cases, a +∇ b and a +Δ b must agree with the endpoints of [a,a] + [b,b] etc. This can be regarded as the precise definition of ∇ and Δ, applying also to complex variables. Notice that the proposed IEEE standard for floating-point arithmetic encourages the use of extended precision within the arithmetic unit so that the assignment operation can also require rounding. Thus, =∇ and =Δ should be included.

Intervals as additional data types do not change very much the syntax of arithmetic expressions. However, there are two additional set theoretic operations, inter-section and convex hull, for which symbols and priority rules must be chosen. Sym-bols already in use are .IS. and .CH. respectively and it is suggested that in-tersection ranks behind addition/subtraction and before convex hull, the smallest interval containing the union. Note that intersection is an invalid operation if it is empty (*in the basic interval system of Section 1*) or if it consists of two intervals (*in Kahan's more complete system of Section 3*). Thus, there should be some means of testing for these conditions.

Operands of different data type should be allowed in the arithmetic operations ("mixed modes"). In this case the operands should be automatically converted to the next higher common data type as indicated in the order diagram for the lattice of numeric data types (see Figure 3). Note that in contrast to ordinary FORTRAN this might require the conversion of *both* operands because the ordering is not total. Intersection and convex hull should be undefined if both operands are singletons.

The semantics of compound operations can be reduced to that of the elementary ones only if the sequence of the operations is uniquely defined. For example, if a and b are single precision variables and c double precision, then (a + b) + c and a + (b + c) do not mean the same. FORTRAN 77 has the from-left-to-right rule so that unfortunately a + b + c ≠ c + b + a *by definition* and not only through rounding errors. One should feel that this is an unnecessary and deplorable deviation from the al-gebraic rules. It is preferable to require for compound operations that all *factors* in a *term* be converted to the lowest common data type as defined by lattice theory, similarly all *terms* in an *arithmetic expression*, etc. In fact, the simplest solu-tion would be the *dominance rule* where all operands in an expression are coerced to the same data type, the lattice theoretic supremum. Here function arguments should participate if and only if the function is generic like ABS() and SQRT(). This should rarely cause a loss of efficiency in practical computations.

The power function deserves a comment. Contrary to the above rule for mixed data types, an integer exponent should not be converted to the type of the base. For real intervals its value is defined as the smallest machine interval containing the united extension:

$$\underline{x}^y \;\; = \;\; \rho(\{\, x^y \,:\, x \in \underline{x},\, y \in \underline{y}\,\}) \qquad\quad \text{(for } \rho : \mathbb{R} \to \mathbb{F} \text{ see Figure 2).}$$

It requires exceptionally good hardware to achieve this goal. For complex intervals it is not practical to require so demanding a definition. Except for squares one should allow several roundings and define $\underline{x}^y$ as $\exp(\underline{y}\cdot\log\underline{x})$ where each operation is done in interval arithmetic. An example shows the difference:

$$\underline{x} \;\; = \;\; [1.950,\; 2.050] + i\,[2.000,\; 2.000]\,,$$

$$\underline{x}^3 \;\; = \;\; [-16.00,\; -15.98] + i\,[14.81,\; 17.22]\,,$$

$$\exp(3\cdot\log\underline{x}) \;\; = \;\; [-17.29,\; -14.81] + i\,[14.79,\; 17.26]\,.$$

Note that the widening of the interval is not an effect of the machine precision (four decimals here). Needless to say that contrary to current practice, a complete algorithmic definition must be provided as part of the language specification.

Yohe [17] uses the symbols .VLT. etc. for the arithmetic relations < etc. based on the each-value-rule and .EQ. , .NE. for the set theoretic equality/inequality as discussed in Section 1. For complex intervals, <, ≤, ≥, > do not apply. Important for all data types are the relations of inclusion, $x \in \underline{y}$ and $\underline{x} \subset \underline{y}$, for which the same symbol, say .IN., could be used, see [2]. The operand on the right-hand side of .IN. should always be an interval. It should be noticed that in Kahan's more complete interval arithmetic no finite real compares with ∞ (since this is an identification of −∞ and +∞) so that evaluation of X .LT. Y etc. for real variables can cause an arithmetic exception. Also $[0,1] < [2,\infty]$ is false while $[0,1] < [2,\infty)$ is true.

9. Intrinsic Functions

Only a few points are mentioned because no details are required in the design of a programming language. Some of the generic functions like REAL, IMAG, CMPLX, CONJ, ABS, SQRT, and LOG should be available for intervals as the united extensions. This is possible although it requires careful programming and good hardware support. Note that some of the exceptions would not occur in Kahan's more complete system, e.g., tan x for x = π/2 or log $\underline{z}$ with $\underline{z}$ a complex interval containing zero. SIGN(∞) should be zero because 0 and ∞ separate the positive and negative reals. INT should use outward rounding for intervals. Intrinsic functions are required to form an interval from two real or complex endpoints and for closing an interval or getting its interior if open intervals are admitted. Also required are two logical functions to test whether an interval is bounded or whether the intersection of two intervals is an interval. Finally, a set of service routines is necessary to compute the left endpoint, right endpoint, width, midpoint, or norm of a given interval.

10. Conclusion

FORTRAN or PASCAL are numerically oriented programming languages which could easily be extended to provide for interval arithmetic. Writing the necessary system routines, however, is a formidable task, especially for the extended version of interval arithmetic admitting unbounded, open, and outer intervals. Algorithms designed for conventional arithmetic should not be used with interval arithmetic.

REFERENCES

[1]   Alefeld, G. and Herzberger, J., *Einführung in die Intervallrechnung*
        (B.I.-Wissenschaftsverlag, Mannheim, 1974).

[2]   Bohlender, G., Kaucher, E., Klatte, R., Kulisch, U., Ullrich, C., Wolff
        v. Gudenberg, J., and Miranker, W.L., *FORTRAN for Contemporary Numerical
        Computation*, Research Report RC 8348, IBM Thomas J. Watson Research Center
        (July 1980).

[3]   Caplat, G. and Frêcon, L., *Specifications for Interval Programming
        Languages*, in: Nickel, K.L.E. (ed.) *Interval Mathematics 1980* (Academic
        Press, New York, 1980) 267 - 279.

[4]   Coonen, J., Kahan, W., Palmer, J., Pittman, T., and Stevenson, D.,
        A Proposed Standard for Binary Floating Point Arithmetic, ACM SIGNUM
        Newsletter, Special Issue, October 1979, 4 - 12.
        A revised form is published in Computer 14 (March 1981) 51 - 62.

[5]   Crary, F.D., *A Versatile Precompiler for Nonstandard Arithmetics*,
        ACM Trans. Math. Software 5 (1979) 204 - 217.

[6]   Fateman, R.J., *High-Level Language Implications of the Proposed IEEE
        Floating Point Standard*, submitted to ACM Trans. Math. Software.

[7]   Gargantini, I. and Henrici, P., *Circular Arithmetic and the Determination
        of Polynomial Zeros*, Numer. Math. 18 (1972) 305 - 320.

[8]   Guenther, G. and Marquardt, G., *A Programming System for Interval Arithmetic
        in ALGOL 68*, in: Nickel, K.L.E. (ed.) *Interval Mathematics 1980* (Academic
        Press, New York, 1980) 356 - 366.

[9]   Kahan, W.M., *A More Complete Interval Arithmetic*, Lecture notes prepared for
        a summer course at the University of Michigan (June 17 - 21, 1968).

[10]   Kahan, W.M., *Interval Arithmetic Options in the Proposed IEEE Floating Point
        Arithmetic Standard*, in: Nickel, K.L.E. (ed.) *Interval Mathematics 1980*
        (Academic Press, New York, 1980) 99 - 128.

[11]   Kaucher, E., Klatte, R., Ullrich, C., *Programmiersprachen im Griff, Band 2:
        PASCAL*, B.I.-Hochschultaschenbücher Band 796 (B.I.-Wissenschaftsverlag,
        Mannheim, 1981).

[12]   Kulisch, U., *Grundlagen des Numerischen Rechnens* (B.I.-Wissenschaftsverlag,
        Mannheim, 1976).

[13]   Kulisch, U.W. and Miranker, W.L., *Computer Arithmetic in Theory and Practice*,
        (Academic Press, New York, 1981).

[14]   Moore, R.E., *Interval Analysis* (SIAM, Philadelphia, 1979).

[15]   Rabinowitz, H., *Implementation of a More Complete Interval Arithmetic*,
        Masters' Project, EECS Dept., Univ. Calif., Berkeley (June 1979).

[16]   Wilkinson, J.H., *Rounding Errors in Algebraic Processes* (Her Majesty's
        Stationery Office, London, 1963).

[17]   Yohe, J.M., *Software for Interval Arithmetic: A Reasonably Portable Package*,
        ACM Trans. Math. Software 5 (1979) 50 - 63.

[18]   Yohe, J.M., *Floating Point Exception Handling for Interval Arithmetic*, in:
        Nickel, K.L.E. (ed.) *Interval Mathematics 1980* (Academic Press, New York,
        1980) 547 - 554.

DISCUSSION

Kahan   *For complex interval arithmetic it is preferable to use circles (represented by the points on them closest to and furthest from the origin) rather than rectangles.   This was suggested by I. Gargantini and P. Henrici ["Circular Arithmetic and the Determination of Polynomial Zeros", Numer. Math. 18 (1972), 305-320].*

Reinsch   *It is still disappointing. For example, the product of two disks with centre at 1 and radius $\sqrt{0.5}$ contains zero!*

Rice   *Suppose someone has a computation which must achieve a certain accuracy ε. Assume all input to the computation is exact.   Three possible strategies come to mind:*

A)   Use interval arithmetic with higher and higher precision until the interval solution is shorter than ε.

B)   Use multiple precision of higher and higher precision until one is satisfied that the error is less than ε.

C)   Use perturbation or sensitivity analysis until one is satisfied that the error is less than ε.

Note that (A) and (C) may require high-precision computations, and (B) obviously does.

Do you know of anyone who has studied the trade-offs between these three approaches? Which do you feel is easier to use?

Reinsch   *(A) (Interval arithmetic) requires less human effort than (C) (perturbation analysis) because the human does not have to do the error analysis. Inspecting the intervals obtained at the end of the first run, one can reliably decide whether a rerun with higher precision will be necessary and what the required precision will be.   In the Karlsruhe interval package a precision of up to 30 decimal places is used.*

(B) can only be discussed with a clear understanding of what "until one is satisfied that the error is less than ε" means.   Usually this will require some a posteriori error analysis. A typical example of this technique can be found in G. Peters and J.H. Wilkinson ["Practical Problems arising in the Solution of Polynomial Equations", J. Inst. Maths. Applics. 8 (1971), 16-35]*.

(C)   usually requires the least computer time but the most human effort.

Meissner   *Wilkinson's methods apply to a class of problems, not just one problem.*

Reinsch   *I was thinking of Wilkinson's "running" error analysis [Sections 4 and 6 of (*) above], which gives a posteriori bounds for specific problems and data, rather than the sort of rounding error analysis described in Wilkinson's book "Rounding Errors in Algebraic Processes", which gives a priori bounds for a class of problems.*

Roberts   *Can you comment on the difference between bounds and actual errors in a large computation, where a large number of small errors produce a "random walk" of the error?*

Reinsch *See the printed version of my talk.   One does not usually just convert naïvely from real to interval arithmetic;   instead one uses interval arithmetic at strategic points, e.g., when computing a residual at the end of an*

*iteration to validate the result.   Here the intervals are usually quite
realistic.*

Hull   Correlations between errors must be taken into account.

*Reinsch Correlations between errors are certainly important.   Interval
arithmetic does not assume that rounding errors are uncorrelated and therefore
its bounds are never invalid.   However, they might well become useless.*

  *When computing the eigenvalues of a symmetric matrix A by the reduction of
Givens/Householder, the tridiagonal matrix T obtained may have large errors, but
it can be shown by backward error analysis that the errors are correlated in
such a way that the eigenvalues of T are very close to the eigenvalues of A.*

*Kahan   Roundoff is not the main problem.   More serious is that interval
arithmetic does not allow for correlations, e.g. $\underline{x}^2 \neq \underline{x}.\underline{x}$ if $\underline{x}$ is a non-diagonal
interval.*

*Boyle   Can we analyse the effect of correlations by working with correlated
intervals?*

*Reinsch   No, since intervals refer to variables and correlation to pairs of
variables.*

*Dekker   We can reduce the lengths of intervals by reformulating algorithms.
For example, residuals can be calculated with interval arithmetic after an
approximation has been computed with ordinary arithmetic.*

*Fateman   We should not fall into the morass of adding new data types to
Fortran. NaNs and their trapping facilities could be used to handle intervals
or arbitrary-precision floating-point numbers.*

Kahan   The cure might be worse than the disease.

*Fateman   My suggestion at least moves the problem from the compiler to the
trap routine.*

*Reinsch   It is customary to use heap storage only when storage requirements are
not known in advance.*

Fateman   I agree that using the heap involves some overhead.

Moler   With Fortran II we just used "I" in column 1!

Meissner   How about an internal compiler switch?

*Manfred Paul   Is there any reason, other than the fact that numerical analysts
mostly use Fortran and Pascal, that you have singled out these two languages
as being easily extendible to provide for interval arithmetic?*

*Reinsch   Honestly not, except that extensions to Fortran are currently being
considered and solicited.*

*Delves   It is not true that Algol 68 has failed to consider the provision of
arithmetic (or other) extension modules.   The language was explicitly designed
to make the provision of such modules, written wholly within the language, a
straightforward matter; and the NAG Algol 68 library includes (MK3) or will
include (MK4) modules to carry out: arbitrary length integer arithmetic,
arbitrary length rational arithmetic, arbitrary length real arithmetic (with
dynamic precision control), and arbitrary length interval arithmetic.*

<u>Reinsch</u>   *The second part of this paper discussed the basic language
constructs required for an implementation of the "more complete interval
arithmetic" of Section 3 and complex intervals of Section 4, viz:*

   A)   eight data types,

   B)   coercion rules for expressions with operands of mixed data types,

   C)   syntax for interval constants,

   D)   the partial ordering of intervals,

   *E)   additional operations (intersection, convex hull) and additional
        relations (see theoretic, inclusion).*

   *All these elements must be provided by the implementor.   There are
languages (e.g., Fortran) which require extensions for this and others which can
do all this within their framework (e.g., Algol 68).   However, the discussion
applies to both cases.*

<u>Brown</u>   *We should discuss the relationship between the last three talks,
especially between the IEEE standard and Hull's proposal.   Both are valuable
because their ideas will be used in libraries even if users don't know about
them.   Also, there is little if any conflict between Hull's proposal and the
IEEE radix-independent standard being developed to generalize the IEEE binary
standard.   Finally, Hull's proposal and both of these IEEE standards are fully
compatible with my model.*

<u>Reid</u>   *What is the policy of TOMS and the ACM Algorithms editor concerning
machine parameters?*

<u>Rice</u>   *The new TOMS Algorithms editor, Richard Hanson, was unaware of the
previously adopted policy when he became Algorithms editor last fall.   We have
discussed the matter recently and he agreed that the announced policy is a good
one and he will take steps to enforce it.   He is concerned about how to
automatically detect violations of the policy and the (admittedly rare) cases
when an algorithm uses a machine parameter which is awkward to compute from the
standard parameters.   The previous TOMS Algorithms editor, Webb Miller, did
enforce the policy whenever he became aware that an algorithm used machine
constants.*

<u>Delves</u>   *It seems that every speaker today has advocated language extensions;
those suggested include linked lists, differentiation and summation operators,
multi-length real and complex, interval arithmetic, and many others are being
considered for Fortran 8X by X3J3.   Such a variety of additions inevitably leads
to a bulky language, indigestible in its entirety for both users and compiler
writers;   and X3J3 has recognised this by considering a "core plus applications
modules" approach to language definition.   But in doing this, it runs the
risk of providing both too much and too little.   Too much, if a wide variety
of applications modules each contain separate language extensions which many
main frame compilers will, for competitive reasons, try to support in their
entirety;   the result will probably be bulky and expensive compilers.   Too
little, if we consider the contents of a particular module and assume that its
extent is frozen.   For example, suppose that we include a differentiation
operator in the math module, then almost certainly some (many?) groups will wish
that partial derivatives, or fractional derivatives, had been included.   Will
these users wait for the next language update?   What they want is the ability
to <u>add</u> their own features, extending the use of existing language syntax or
developing new syntax, to meet their special requirements;   and to do this
wholly within the existing language, so that existing compilers will compile*

their own language extensions.   This is certainly achievable, given only suitable features within the "core" module;   can I plead that we (X3J3?) aim towards this, rather than towards yet another "extended" but inextensible language?

Moler   *Someone said "I don't know which language I'll be programming in by the year 2000, but it will be called 'Fortran'".*

Ford   *M. Griffiths said it!*

**SESSION 3
ARITHMETIC AND
SYMBOLIC MATHEMATICS**

Chair
W. Cowell

Discussant
W.S. Brown

THE RELATIONSHIP BETWEEN NUMERICAL COMPUTATION
AND PROGRAMMING LANGUAGES, J.K. Reid (editor)
North-Holland Publishing Company
© *IFIP, 1982*

The Near Orthogonality of Syntax, Semantics, and Diagnostics in Numerical Programming Environments

W. Kahan and Jerome T. Coonen

Mathematics Department
University of California
Berkeley, California 94720
U.S.A.

We can improve numerical programming by recognizing that three aspects of the computing environment belong to intellectually separate compartments. One is the syntax of the language, be it Ada, C, Fortran or Pascal, which gives legitimacy to various expressions without completely specifying their meaning. Another might be called "arithmetic semantics". It concerns the diverse values produced by different computers for the same expression in a given language, including the values delivered after exceptions like over/underflow. The third compartment includes diagnostic aids, like error flags and messages; these too can be specified in language-independent ways. However imperfect, this decoupling should spell out for all concerned the nature of arithmetic responsibilities to be borne by hardware designers, by compiler writers and by operating system programmers.

> "Another of the great advantages of using the axiomatic approach is that axioms offer a simple and flexible technique for leaving certain aspects of a language *undefined*, for example...accuracy of floating point... This is absolutely essential for standardization purposes..."
>
> — C. A. R. Hoare (1969)

Professor Hoare's attitude toward floating point semantics reflects the anarchy that befell commercial floating point hardware early in the 1960's [1], and worsened in the 70's. That anarchy confounded attempts to characterize all floating point arithmetics in one intellectually manageable way. Now there is hope for the 1980's. A new standard for binary floating point arithmetic has been proposed before the IEEE Computer Society, and a radix-independent sequel is in the works. Since the binary standard has been adopted by a broad range of computer manufacturers, including much of the microprocessor industry, we expect numerical programs to behave more nearly uniformly across different computers, and perhaps across different languages as well. A draft of the binary standard, along with several supporting papers, may be found in the March 1981 issue of *Computer* [2-5].

Starting in the 1960's programming language designers came to be the arbiters of most aspects of the programming environment. With control of the programmers' vocabulary, language designers could control fundamental features such as the number of numeric data types available and the extent of run time exception handling. The language even limited the numeric *values* available by constraining the literals in the source text. This is not to say that language designers acted capriciously. They were disinclined to mention any capability not available on all computers. In this respect computer architects have laid a heavy hand on the computing environment. Languages must reflect the least common denominator of available features, and so they tend to vague oversimplifications where floating point is concerned. An extreme case is the new language Ada which, by incorporating W. Stan Brown's very general model for floating point computation [6], pretends that the difference between one computer's arithmetic and another's is merely a matter of a few environmental parameters. But sometimes the programmer must know his machine's arithmetic to the last detail, especially when trying to circumvent limitations in range or precision. These details, dangling between language designers and computer architects, too often receive short shrift from both. Tying up these loose ends would improve the computing environment.

Of course the computing environment invites numerous improvements, to graphics, file handling, database management and others, as well as floating point and languages. But enhancements to which high-level languages deny access are enhancements destined to die. Those of us working on the proposed IEEE floating point standards have had to face this problem. We believe the solution is a proper division of labor, rather than grand attempts to improve too many aspects of the computing environment simultaneously; the latter way would require impractical coordination. For example, to encourage independent development of programming languages and floating point hardware, we propose that language (syntactic) issues be decoupled from arithmetic (semantic) issues to the extent possible. We present our view of the interplay between syntax, semantics, and diagnostics as parts of the computing environment, and discuss how they interface with each other. Given an adequate interface discipline, we hope that responsibility for these parts can be divided among language designers, numerical analysts, systems programmers, and others. In the past this division has been unclear. Unfortunately, when everybody is responsible, or when nobody is responsible, then everybody can be irresponsible.

Portability

We regard the programming language as just one layer of the computing environment, dissenting from a more traditional view that the language *is* the environment. What does this mean for program portability? Until very recently, portability of numerical programs was considered to be a quality of source code that could be compiled and run successfully without change on a variety of computers. The issues appeared largely syntactic. For example, programs like the PFORT verifier [7] were developed to check Fortran codes for adherence to a standard for ''portable Fortran'', their principal task being to weed out various quirks of dialect. Nowadays, we acknowledge that the portability issues go deeper than differences among Fortran dialects. They entail the (semantic) subtleties of over/underflow and rounding that, if ignored, can cause ostensibly portable programs that function beautifully on one machine to fail on another. Programming languages that lack the vocabulary required to address these issues aren't very helpful here. If we cannot ''mention'' these issues how can we resolve them?

Ideally, the variation of floating point arithmetic from one machine to another should be describable with a few parameters [8] which portable programs could determine through system-dependent environmental inquiries [9]. This scheme works satisfactorily for many programs that do not depend critically upon the finer points of the arithmetic. However, any such parameterization must be based upon an abstract model encompassing simultaneously all current arithmetic engines, some of them disconcertingly anomalous [1, 10]. To insist that this model underlie portable programming is to dump upon programmers the onus to discover and defend against all mishaps the model permits, some of them mere artifacts of generality. This in turn would burden programs with copious tests against subtle (and certainly machine-dependent) thresholds to avoid problems with idiosyncratic rounding and over/underflow phenomena. A programmer who shirks his responsibility to produce robust code obliges the user of his program, possibly another programmer, to unravel a more tangled web. Ultimately, the buck may be passed to users who find either their programs or their computers to be inexplicably unreliable. We doubt that any semantic analog of the PFORT verifier will ever be able to test for robust independence of the underlying arithmetic. Computer arithmetics are too diverse to allow every potentially useful numerical algorithm to be programmed straightforwardly in a fashion formally independent of the underlying machine.

Portability at the source code level is nice when inexpensive. When not, we are content with "transportability", whereby algorithms can be moved from one environment to another by routine text conversion, possibly with some aid from automation. An algorithm may depend critically upon the underlying arithmetic semantics and upon a system's ability to communicate error reports between subprograms. It is transportable to the extent that the dependencies can be communicated in natural language using mathematical terms, if not in Fortran. We are not advocating yet another programming language. We prefer that programmers accompany their codes with some documentation that explains, and can even be used to verify, how the program handles its interactions with the underlying system. Because computing environments are so diverse, we expect some algorithms to be transportable to only a few systems, not all; this does not undermine the notion of transportability. Essential to transportability is a manageable corpus of information about

* syntax – the programming language to be used,
* semantics – the arithmetic of the underlying computer, including the run-time libraries of functions like cos(), and
* diagnostics – the system's facilities for error reporting and handling,

preferably no more than can fit on a short bookshelf, and yet enough to cover a wide range of manufacturers' equipments.

Syntax

In this paper, *syntax* refers to the expressions in a language – which ones are legitimate and how they are parsed. Issues relevant to numerical calculations include the number of data formats available, how they combine to form arrays and structures, and the order of evaluation in unparenthesized expressions. Languages vary greatly in their provision of numeric data formats, usually called "types". Both Basic and APL have just one numeric type, which is to be used for both integer and floating point calculations; Pascal and Algol 60 have just one real type. Fortran and C have single and double types, although in C all floating expressions are of type double. PL/I programmers may specify the precision of their floating point variables, though they typically map into the single and double types supported by the underlying system. The new language Ada provides syntactic "packages" in which floating types may be defined to correspond to the host system's facilities, but its strong typing prohibits mixing of different user-defined types in expressions without explicit coercions, even if the underlying hardware types are the same.

Expression evaluation is just as varied. For example, in

$$1.0 \ + \ 3/2$$

most compilers would recognize the 3 and 2 as integers. Their ratio would be evaluated as the real 1.5 or truncated integer 1 depending upon the strength of the 1.0 to coerce their types. Different Fortran compilers have disagreed in this situation. In Ada such an expression would be illegal unless the 3 and 2 were written with decimal points to indicate that they were real literals. What about the unparenthesized expression

$$A \ * \ B \ + \ C \ \ ?$$

Most languages, like Fortran, evaluate it as if it were written $(A*B) + C$, but APL evaluates $A \times B + C$ as if it were written $A \times (B+C)$. The situation gets more complicated when relational and boolean operators are involved. In Pascal, the attempt to simplify the language by keeping the number of levels of operator precedence small led to some surprises for programmers. For example, because the conjunction $\cap$ has greater precedence than <, the expression

$$x < y \; \cap \; y < z \; ,$$

used for checking bounds on the variable y, has the bizarre interpretation

$$(x < (y \cap y)) < z$$

which is illegal because of the appearance of the real y as an operand to $\cap$.

Perhaps the widest syntactic liberties are taken by standard C compilers. Expressions of the form

$$a + b + c \; ,$$

where a, b, and c may be subexpressions, are evaluated in an order determined at compile time according to the complexity of a, b, and c. This is so *regardless of parentheses* such as

$$(a + b) + c \; .$$

Such a convention is disastrous in floating point where, say, $(a+b)$ cancels to a small residual to be added into the accumulation c. In such cases all accuracy may be lost if $(b+c)$ is evaluated first at the compiler's whim. The cautious programmer who writes

$$(x - 0.5) - 0.5$$

to defend against a machine's lack of a guard digit during subtraction will always be vulnerable, if not to a C compiler then to an optimizer that collapses the expression into the algebraically, though not numerically, equivalent form $(x - 1.0)$.

To jump the gun a bit, it is clear from the examples above that *syntax constrains semantics*. Syntax also constrains programmers who, C compilers notwithstanding, are well advised to preclude any ambiguity in expression evaluation by inserting parentheses liberally.

Semantics

We concentrate here on arithmetic semantics. That is, after an expression has been parsed — so the computer knows which operations to perform — what does its evaluation yield? Floating point semantics depends vitally on the underlying arithmetic engine. The initiated reader realizes that this is where the real headaches set in. For example, on machines such as programmable calculators where the fundamental constants π and e are available in a few strokes, we might expect

$$(\pi \times e) - (e \times \pi)$$

to evaluate to 0.0 since, *semantically*, we expect multiplication to be commutative despite roundoff. Unfortunately, even this simple statement is not universally true. Different Texas Instruments calculators yield different tiny values for the expression above; and it's not just a matter of machine size and economy, for early editions of the Cray-I supercomputer exhibited similar noncommutativity.

Another well-known example of murky semantics is the expression

$$X - (1.0 \times X)$$

which is exactly X rather than 0.0 for sufficiently tiny nonzero values X on Cray and CDC computers. On these machines $(1.0 \times X)$ flushes to 0.0 for those tiny X. On some other machines that lacked a guard digit for multiplication, the expression above was nonzero whenever X's last significant digit was odd!

Hardware-related anomalies like these seem to predominate in any serious treatment of arithmetic semantics. Such distractions are what led Professor Hoare to despair about floating point in high-level languages. We will not dig further into the lore of arithmetic anomalies. Interested readers can find an introduction in [1]. The technical report [10] studies the overall impact of anomalies and compares two approaches to improvement.

Arithmetic semantics is not restricted to simple operations. In languages like Basic that include matrix operations, assignments like

$$MAT \ X \ = \ INV(A) \ * \ B$$

are allowed. As users might expect, most implementations evaluate $(A^{-1}) * B$ (approximately), following the strict mathematical interpretation of the formula. However, more robust systems by Tektronix and Hewlett-Packard use Gaussian elimination to solve the linear system $AX = B$ for X, thereby obtaining a usually more accurate X that is guaranteed to have a residual $B - AX$ small compared with $|B| + |A| \cdot |X|$. If A is close enough to singular, the subexpression $INV(A)$ may be valid or not depending upon good or bad luck with rounding errors − on all machines except the Hewlett-Packard HP 85. All machines solve $(A + \Delta A)X = B$ with ΔA comparable to roundoff in A though possibly differing from column to column of X. The HP 85 further constrains ΔA to guarantee that $(A + \Delta A)^{-1}$ exists. Thus it has no "SINGULAR MATRIX" diagnostic. Consequently, a program using inverse iteration to compute eigenvectors always succeeds on the HP 85 but on other machines is certain to fail for some innocuous data. Is such a program, using a standard technique, portable or not? Who is to blame if it is not?

Arithmetic exceptions such as over/underflow and division by zero fit into our informal notion of semantics when they are given "values". We take this view in spite of a current trend among authors to consider exceptions under a separate heading *pragmatics*. This trend is understandable, given the variety of exception handling schemes across different hardware. Consider for example the expression $0.0/0.0$. When they are to continue calculation (i.e. without a trap) CDC, DEC PDP/VAX-11, and proposed IEEE standard machines stuff a non-numeric error symbol in the destination field. This symbol is then propagated through further operations. Most other machines just stop, forcing program termination. At least one will store the "answer" 1.0.

Dividing zero by itself is usually bad news within a program, so the diversity of disasters that arise on various machines is not too surprising. A quite different situation arises with the exponentiation operator in Y^X. Since this is part of the syntax of several languages, for example Fortran, Basic, and Ada, responsibility for its semantics has been taken by language implementors. Of the many problems that arise we will consider just one: what is the domain of Y^X when both X and Y are real variables? Consider the simple case $(-3.0)^{3.0}$, which is:

-27.0	...on very good machines,
$-26.999...9$	...on good machines,
TERMINATION	...on bad machines,
undefined	...on cop-outs,
$+27.0$	...on *very* bad machines.

Why this bizarre diversity of semantics? Although for arbitrary X the expression Y^X may have no real value when Y is negative, the particular case above is benign because X has an integer value 3.0. Thus restricting the domain of Y to nonnegative numbers is unnecessarily punitive. We recommend that, should X be a floating point Fortran variable with a nonzero integer value,

$$Y \ ** \ X \ = \ Y \ ** \ INT(X) \ .$$

This cannot hurt Fortran users, but will help the Basic programmer (and the conversion of programs from Basic) because most implementations of Basic, with just one numeric data type, cannot distinguish the real 3.0 from the integer 3 in the exponent. This recommendation costs extra only when Y is negative. On the other hand, if Y is 0.0 we distinguish $Y^{0.0}$, which is an error, from $Y^0 = 1.0$ which mathematics makes obligatory. Note that none of these issues are language issues, though until now they have been settled by language implementors. Ideally, these responsibilities should be lifted from language designers and implementors, and borne by people like the members of IFIP Working Group 2.5.

The point of this digression into the murk of pragmatics was to indicate that the current situation in exception handling is the result of a host of design flaws rather than inherent difficulties. We object to the connotation "pragmatics" carries with it of acquiescence to inevitable hazards. We prefer to capture all semantics, including the anomalies, under one heading even if this entails a different semantics for each different implementation of arithmetic. This exposes rather than compounds a bad situation.

A notably clean and complete arithmetic semantics is provided by the proposed binary floating point standard. The IEEE subcommittee responsible for the proposal set out to specify the result of *every* operation, balancing safety against utility when execution must continue after an exception. Even a cursory glance at the proposal indicates the extent to which exception handling motivated the design:

*   Signed ∞ for overflow and division by 0.0.

*   Signed 0.0 to interact with $\pm\infty$, e.g. $+1.0/-0.0 = -\infty$.

*   NaN — not a number — symbols for invalid results like 0.0/0.0 and $\sqrt{-3}$.

*   Denormalized numbers — unnormalized and with the format's minimum exponent — to better approximate underflowed values.

*   Sticky flags for all exceptions.

*   Optional user traps for alternative exception handling.

These features promote comprehensible semantics for "standard" programming systems.

Diagnostics

After syntax and semantics, the third aspect of the numerical programming environment is the set of execution time diagnostic aids. They may be roughly divided into anticipatory and retrospective aids, and according to whether they find use during debugging or during (robust) production use.

The principal anticipatory debugging aid is the breakpoint for control flow and, when the hardware permits, for data too. Some systems can monitor control or data flow according to compiler directives inserted in a program. Retrospective debugging aids include the familiar warnings and termination eulogies, as well as the more voluminous memory dumps and control tracebacks. Systems with sticky error flags can list those still standing when execution stops — in a sense they signal unrequited events.

For the production program that would be robust, and perhaps even portable, the situation is not so clear. Because most current systems provide neither exception flags (such side effects are anathema to some language designers) nor error recovery, a program — if it is not to stop ignominiously on unusual data — must include precautionary tests to avoid zero denominators and negative radicands, and tests against tiny, but carefully chosen, thresholds to ward off the effects of underflow to zero. The lack of flags can force the use of explicit error indicators in subprogram argument lists to communicate exception conditions. The languages Basic, PL/I, and Ada allow for anticipatory exception handlers (e.g. ON <condition> ... in PL/I) but do not allow the exception handler to discover anything about the exception beyond a rough category into which it has been lumped, thereby making an automatic response by the program very cumbersome.

Another variety of anticipatory diagnostic aid is available through an option in the proposed floating point standard. It is essentially an extension of the PL/I "on-condition" except that it is outside any current language syntax. This feature, which might be called trap-with-menu, allows the programmer to preselect from a small list of responses an alternative to the default response. By devising the menu carefully, we should be able to give the user sufficient flexibility without having to cope with a voluminous floating point "state" at the time of the exception.

The Syntactic—Semantic Interface

From the point of view of the numerical analyst, the semantic content of programming languages is given by the following list.

* What are the numeric types, and what is their range and precision?
* Which numeric types are assigned to anonymous variables like intermediate expressions, converted literals, arguments passed by value,...?
* Which numeric literals are allowed, and are they interpreted differently in the source code than the IO stream?
* Which basic arithmetic operations are available, and what is in the library of scientific functions?
* Is there a well-understood vocabulary reserved for the concepts and functions we need, and defended against collision with user-defined names?
* What happens when exceptions arise? How can error reports be communicated between subprograms?
* Is there a way to alter the default options (for, say, rounding or handling of underflow) by means of global flags?

These are among the knottiest issues in numerical computation. But, to a large extent, they can be freed from the more conventional language issues and thus resolved within the numerical community. Only questions about data types and the change of control flow on exceptions are necessarily tied to language syntax.

Consider a hypothetical language with only skeletal numerical features. Assume that integer types and arithmetic and character strings are "fully" supported. The language supports single and double real variables, pointers to them, and allows real variables to be embedded in arrays and structures. There is also provision for functions returning real values, and for real parameters passed either by value or reference. But the *only* operation on real types is assignment of a single value to a single variable, and of a double value to a double variable.

To be useful numerically, this hypothetical language would require a support library providing the basic arithmetic operations as well as the usual complement of elementary functions. But because each operation more complicated than a straight copying of bits would result only from an explicit function call, the programmer would in principle have complete control of the arithmetic semantics (by choosing a suitable library). As an example, consider the evaluation of the inner product of the single arrays $x[]$ and $y[]$ using a double variable for the intermediate accumulation to minimize roundoff:

```
double_precision temp_sum;
temp_sum := DOUBLE_LITERAL( "0.0" );
for i in 1..n do
    temp_sum := DOUBLE_SUM( temp_sum,
            SINGLE_TO_DOUBLE_PRODUCT( x[i], y[i]));  od
inner_product := DOUBLE_TO_SINGLE( temp_sum );
```

Even this simple example exposes many of the questions that arise in numerical programs. Would the constant 0.0 require a special notation (such as 0.0D0) to be assigned to a double variable? In a more conventional rendition of the program the inner loop would involve a statement of the form

$$temp\_sum := temp\_sum + x[i] * y[i];$$

Would the product be rounded to single precision before the accumulation into *temp_sum*, destroying the advantage of double precision?

Semantic Packages

The skeleton language above may be unambiguous, but it is clearly much too cumbersome for calculations involving complicated expressions. What we must do is bridge the gap between the handy syntactic expression $x[i] * y[i]$ and the semantically well-defined

SINGLE_TO_DOUBLE_PRODUCT($x[i]$, $y[i]$) .

We propose to do this through so-called semantic packages.

It may be a sign of progress that the new language Ada comes very close to suiting our needs. Although Ada incorporates the Brown model for arithmetic by providing a set of predefined attributes for each real type available to the programmer, this is in general insufficient for programs that would be robust. More important for us, Ada allows the overloading and redefinition of the infix operators +, −, etc. and in so doing provides the *explicit* connection between the operators and the real hardware functions they represent. The semantic packages, corresponding directly to the (syntactic) **packages** construct in Ada, could contain exact specifications of the arithmetic functions (which are actually implemented in hardware). Thus there would be a semantic package for each basic architecture, for example IBM 370, DEC PDP/VAX-11, and the proposed IEEE binary standard. Some semantic packages could be more general, encompassing several machines whose arithmetic is similar enough that a few environmental inquiries supply all the distinction that is necessary for a wide range of applications. For example, one such package might include IBM 370, Amdahl, Data General MV/8000, HP 3000, DEC PDP/VAX-11 and PDP-10, relegating TI, CDC 6000, Cray 1 to another.

Our attempt to force the gritty details of arithmetic semantics upon programmers may dismay readers who embrace the modern trend to elevate the programming environment above machine details. Such an attempt is made within Ada, by means of a small set of predefined attributes associated with each real type. We have already explained that this is not enough; sometimes the program that would be robust must respond to machine peculiarities that defy simple parameterization. The report [10] on why we need a standard contains several examples.

An effort to "package" arithmetic semantics within various programming languages may seem impossible. For example, the details of floating point, especially in the proposed IEEE standards, involve global flags to indicate errors, and modes to determine how arithmetic be done. In Fortran, such state variables may be defined as local data within the standard library functions whose job is to test and alter the flags, although the actual implementation involves collusion with the hardware flags. This is not a complete formalization, since Fortran provides no way to describe the connection between the flags and the arithmetic operations. Current trends in language design eschew error flags as side effects of the arithmetic operations (functions). Modes and flags seem to violate the principle that all causes and effects of expression evaluation should be visible within that expression. Perhaps surprisingly, Ada again provides us with the desired facility − but without excessive or expensive generality. In accordance with the Steelman requirements of the United States Department of Defense, Ada permits side effects " limited to own variables of encapsulations". This is exactly our intention in using semantic packages to describe arithmetic.

Optimization

Any treatment of floating point semantics must deal with that favorite whipping boy, the code optimizer. We considered a most extreme example above, in which C compilers would calculate floating sums like

$$(a + b) + c ,$$

without regard to the parentheses, in whatever order makes best use of the register file. This is simply a mistake in the language design.

Not all anomalies are so clear-cut. Some questions arise when, as in architectures suggested by the proposed IEEE standard, extended registers with extra precision and range beyond both single and double types are used as intermediate accumulators. Consider the typical code sequence

$$x := a * b;$$
$$y := x / c;$$

in which all variables are assumed to be of type single. If $(a*b)$ were computed in an extended register, should that value or the single value x be used in the evaluation of y? Efficiency dictates the former, saving one register load and lessening the risk of spurious over/underflow. But common sense dictates the latter, so that what the programmer sees is what the programmer gets.

A similar situation arises in inner product calculations of the type discussed above. Consider the loop

> **double_precision** *temp_sum*;
> *temp_sum* := 0.0;
> **for** i **in** 1..n **do**
>   *temp_sum* := *temp_sum* + $x[i]*y[i]$; **od**
> *inner_product* := *temp_sum*;

in which, like the earlier example, all variables are single except for the double *temp_sum*. The fully "optimized" compiler might run this loop with just two extended registers, one to compute the products $x[i]*y[i]$ and one to accumulate *temp_sum*, thereby avoiding $(n-1)$ register loads and stores by simply keeping *temp_sum* in a register. Alas, the programmer asked for a double precision intermediate, not extended, so such optimization is precluded.

The moral of these examples is that declared types must be honored. Also, the type assigned by the compiler to anonymous variables must be deducible syntactically, or, better, it should be under the programmer's control. The alleged optimizations above were disparaged because named variables were replaced surreptitiously by extended counterparts that happened to be in registers. This is not to say that extended evaluation is unhealthy; on the contrary, extended temporaries can reduce the risk of spurious over/underflow or serious rounding errors, and therefore should be used for anonymous variables. But the advantage of extended is lost if languages prevent programmers from requesting it for declared temporaries. The expression

$$temp\_sum + x[i]*y[i]$$

in the loop above would best be computed entirely in extended before the store into *temp_sum*. These facilities for extended expression evaluation are not unique to the proposed IEEE standard; the benefits of wide accumulation were realized in the earliest days of computing. The Fortran 77 standard includes some intentionally vague language about expression evaluation in order not to prohibit extended intermediates, and the Ada standard, which seems to avoid some problems by strict typing and requirements for explicit type conversions in programs, uses a so-called *universal_real* type (at least as wide as all supported real types) for the evaluation of literal expressions at compile time.

The use of an extended type for anonymous variables is prone to one class of problems. When real values or expressions may be passed by value to subprograms there may be a conflict between the implicit type of the expression and the declared type of the target formal parameter. This problem arises in current implementations of the language C, which supports both single and double types but specifies that all real expressions are of type double. Suppose that a C program contains the statement

$$y := f(a * b / c);$$

where all variables are of type float (single) and the function $f()$ is defined by

float $f(x)$
   float x;
  { ...... }

How can the type of the expression $(a * b / c)$ be double while the type of the formal parameter x is float? C resolves the discrepancy by silently countermanding the declaration of x and replacing **float** by **double**. Once again, what you see is not what you get. This use of wider intermediates, exploiting the PDP-11 floating point architecture, is exactly analogous to one use of extended registers. Though it is efficient and straightforward to implement, it is not acceptable.

Conclusion

We have cited examples to show that progress in numerical computing has been slowed by questionable decisions in the design of computing languages and systems. We have suggested a rough division into three categories, syntax, semantics and diagnostics, so that the difficult issues could be resolved by those most qualified — and most profoundly impacted. IFIP Working Group 2.5 might well take responsibility for the interfaces with semantics. Ideally their efforts will lead to fully specified environments for which reliable numerical software can be derived, possibly automatically, from algorithms expressed in a mathematical form if not already in a programming language. Programming then becomes a three phase translation involving the language (syntax) to be used, the underlying arithmetic engine (semantics), and the host system (diagnostics). We acknowledge that these categories are not completely independent, and that the boundaries between them cannot be drawn precisely, at least not yet. Nonetheless, we remain convinced that those boundaries must be drawn if we are to bring the required expertise to bear on the current morass.

Acknowledgement

This report was developed and originally typeset on a computer system funded by the U. S. Department of Energy, Contract DE-AM03-76SF00034, Project Agreement DE-AS03-79ER10358. The authors also acknowledge the financial support of the Office of Naval Research, Contract N00014-76-C-0013.

References

[1] Kahan, W., "A Survey of Error Analysis," in: *Information Processing 71*, (North-Holland, Amsterdam, 1972) 1214-1239.

[2] "A Proposed Standard for Binary Floating-Point Arithmetic," Draft 8.0 of IEEE Task P754, with an introduction by D. Stevenson, *Computer*, **14**, no. 3, March (1981) 51-62.

[3] Cody, W. J., "Analysis of Proposals for the Floating-Point Standard," *Computer*, **14**, no. 3, March (1981) 63-68.

[4] Hough, David, "Applications of the Proposed IEEE 754 Standard to Floating-Point Arithmetic," *Computer*, **14**, no. 3, March (1981) 70-74.

[5] Coonen, Jerome T., "Underflow and the Denormalized Numbers," *Computer*, **14**, no. 3, March (1981) 75-87.

[6] Brown, W. S., "A Simple But Realistic Model of Floating-Point Computation," to appear in *ACM Transactions on Mathematical Software*, 1981.

[7] Ryder, B. G., "The PFORT Verifier", *Software — Practice and Experience*, **4** (1974) 359-377.

[8] Sterbenz, P. H., *Floating-Point Computation* (Prentice-Hall, Englewood Cliffs, N.J., 1974).

[9] Brown, W. S. and S. I. Feldman, "Environment Parameters and Basic Functions for Floating-Point Computation," *ACM Transactions on Mathematical Software*, **6** (1980) 510-523.

[10] Kahan, W., "Why do we need a standard for floating point arithmetic?", Technical Report, University of California, Berkeley, CA, 94720, February (1981).

DISCUSSION

Summary by discussant (Brown)

Professor W. Kahan observed that language syntax, arithmetic semantics, and execution-time diagnostics are approximately independent features of the numerical programming environment. Furthermore, he suggested that standards for arithmetic semantics and execution-time diagnostics can and should be specified by experts in those areas in a language-dependent form.

Boyle   What precision should be used in the evaluation of an expression?

Kahan   In the absence of explicit instructions from the programmer, I favor the "scan for widest" rule. Look at all the constituents of the expression, and choose the widest precision that occurs. Then evaluate the expression using the chosen precision for all operations and for all intermediate values. In the case of an assignment statement, the left hand side should be included in the scan, and the value of the expression should be rounded, if necessary, to the precision of the left hand side before the assignment is performed.

Hull   I think the "scan for widest" rule is too rigid. There are times when I want to specify the precision of operations independently of the operands.

Kahan   I agree, and a numerically-oriented language should enable you to do so.

Hull   Unfortunately, the usual functional notation for this is quite cumbersome, and I think we should seek language facilities (e.g., precision blocks) that are better suited to the purpose. (For further discussion, see Hull's paper elsewhere in this volume).

Sedgwick   The "scan for widest" rule prohibits the analysis of subexpressions independently of each other.

Smith   What precision should be used in the evaluation of a subexpression that occurs as an argument of a function?

Kahan   Apply the "scan for widest" rule to the subexpression, and include the corresponding formal parameter of the function if its precision is available.

Reid   Because of the vast number of programmers who are accustomed to other rules, we should impose a new rule only if there is good cause. Here I believe there is. The "scan for widest" rule will often lead to better results, and will rarely if ever cause unpleasant surprises.

Waite   I believe that "syntax", "pragmatics", and "semantics", respectively, are the words normally used for the three concepts that you discussed.

Feldman   In practice, the compartmentalization that you want isn't really possible. For example, it is difficult to know at what level to put out messages for run-time exceptions.

Meissner   I agree that "language designers" (such as X3J3) should welcome guidance from numerical analysts in areas such as those described in this presentation.

When we discuss the subject of coercion of subexpressions, the semantics of integer division tends to "overwhelm" the other issues. The guidance needed from the numerical viewpoint is on those other issues - I hope it will be possible to get this guidance separately from the concerns about integer division.

Comment (E. Hehner)

Hehner   Yesterday Professor Kahan characterized a difference between himself
and Stan Brown by saying that Brown sees the world as it is, whereas he (Kahan)
sees it as it could be.   In reply to Tom Hull, however, Kahan said that Hull,
by beginning with the language, is defining a new world, whereas he (Kahan)
lives in the world of existing machines: The opposite difference.

    Like Bill Waite, I would object to Kahan's use of the word "semantics";
I too prefer to put the peculiarities of machine arithmetic under the heading
"pragmatics".   I agree that the language designer should be relieved of such
considerations, but perhaps for a different reason.   There are many ways that
an implementation of a language is incomplete.   This is not a pejorative;
all interesting languages are infinite, and all implementations are finite.   The
language designer does not say what happens when a recursion stack overflows,
or a program contains too many nested ifs, or too many statements.   He should
design a language with a clean, mathematically expressed semantics, and leave
the problems of implementation limits to others.   I put arithmetic in the same
category;   its semantics is well-known to mathematicians.   Someone has to be
concerned with the fact that machine arithmetic is not the mathematical
arithmetic, but subject to space limitations, just as someone must say what
happens when an array is too large.   But at first, the language designer should
design a clean language free of these problems, and only later should someone
(perhaps the same person under a different hat) consider them.

THE RELATIONSHIP BETWEEN NUMERICAL COMPUTATION
AND PROGRAMMING LANGUAGES, J.K. Reid (editor)
North-Holland Publishing Company
© *IFIP, 1982*

SYMBOLIC MANIPULATION LANGUAGES AND
NUMERICAL COMPUTATION: TRENDS

Richard J. Fateman[1]

Computer Science Division
Department of Electrical Engineering and Computer Sciences
University of California
Berkeley, California
U.S.A.

This is a review of several approaches to numerical computation facilities from the point of view of symbolic and algebraic manipulation systems. We indicate some of the more prominent efforts of the past and present, and suggest directions for the future. We provide some relevant details concerning developments at the University of California, Berkeley, and include extended examples of the development of simple numerical programs using algebraic manipulation.

1. Past and Present

In this paper we discuss some of the implications of the use of symbolic or algebraic manipulation systems in the realm of numeric computation.

Some of the earliest descriptions of mechanical computation, dating back to Babbage and Zuse[2] have mentioned algebraic formula manipulation as an appropriate application domain.

Certainly a few hardy souls have written single-use programs for the solution of non-numeric problems from the earliest days of working computers: Jean Sammet [10] provides a survey of most early published efforts. However it was not until the mid-1960's that the cost of computation had dropped sufficiently to make it attractive to try to provide a "high level language" for algebraic computation, and attempt to provide this tool to users outside the privileged circle of the designers themselves. Several of such efforts to provide formula manipulation were based on the assumption that conventional programming languages (in particular, Algol and Fortran), in common use, would provide an appropriate vehicle for dealing with formulas, too.

Several means for providing this melding were attempted. A number relied upon the introduction of special statements which fit the general pattern of Algol, Fortran, or PL/I, but were pre-processed into calls to subroutines for algebraic processing. There was some hope that the programmer could use the host language statements for program control; unfortunately the two languages' data structures rarely were compatible, and special recognition had to be made to convert from an expression (which happened to be a number) to a number in the host language. Typically it was not transparent to "evaluate a formula numerically" and certainly not efficient. In essence, numerical evaluation was the result of simplification of a internal tree-like expression with numeric operands and rational

---

1. Work reported herein was supported in part by the U. S. Department of Energy, Contract DE-AT03-76SF00034, Project Agreement DE-AS03-79ER10358.

2. Knuth and Trabb Pardo report in [7] p. 203 that c. 1944, Konrad Zuse wrote algorithms in his programming system for matrix calculations, series expansions, and testing whether or not a given logical formula is syntactically well formed.

operators.

The other principal approach to incorporating both numeric and algebraic types of processing in the same language, dating from the early sixties, was to start with algebraic manipulation programs written in a convenient language, usually Lisp. Then numeric floating-point capabilities were added as part of the implementation of the Lisp language. The PDP-10 MACLISP system was one of the earlier systems to have a good compiler and run-time library [12]. One difficulty in this approach is that material from scientific software libraries generally had to be re-coded before incorporation in the Lisp system.

At first glance, the combination of systems which produce symbolic expressions and algebraic languages such as Fortran would seem to be quite natural, but in fact Fortran is designed to do very little with symbolic expressions: it cannot do anything with them at run-time, and at compile-time, all it can do is convert an expression into a sequence of machine operations. This means that the appropriate interface between two systems may very well be that the algebraic manipulation system's run-time environment should have at its disposal a compiler for a numeric language. If the interchange is to be two-way, there must also be efficient routines for converting the data types in the "symbolic" realm to floating-point numbers or other numeric types. For example, there may be requirements for a polynomial with integer coefficients to be converted to an array of floating point numbers; "high precision" (linked-list) integers must be converted, if possible, into some Fortran-like form.

In the past, Fortran systems generally did not have run-time access to their own compilers, although some clever programmers were able to accomplish essentially this type of access in particular cases. One system which we are currently using (a version of Bell Laboratory's UNIX operating system) does allow this kind of access, and examples in this paper show how this can be used.

It is possible to write algebraic manipulation programs in languages other than Lisp, but this does not mean the interface problem indicated above is necessarily solved unless the run-time environment is appropriate. Among the large number of alternative languages and systems, we would like to point to one of the more interesting efforts, by van de Riet [14], using Algol and Pascal.

Another tack has been taken by several programming systems which provide rather specific assistance of a symbolic flavor, by making a more elaborate numeric compiler. Consider the compilation of an algebraic assignment statement as a transformation of a string of characters into an internal tree form, and then into a text to be assembled into machine code. Starting from the internal form, a compiler can be constructed which will also compute the derivative of an expression (another tree) and produce that assembly code as well. Thus compile-time symbol manipulation is a facility that can be provided with relative ease, so long as sufficient attention to details can be included. Clearly the compiler must know about derivatives of built-in functions, and have techniques for differentiating user-defined functions (and possibly in several variables).

Such differentiating compilers may not produce high-quality code, but the user is relieved of a certain attention to details.

Several systems (see [11]) provide convenient user interfaces for optimization problems involving numerical differentiation. Another approach based on representation of functions (and programs!) by Taylor series expansions is pursued in [6].

There are also systems which are written especially for specific problem solving domains, and translate their input into small subroutines (in assembler, Fortran, etc). These routines are then called by standard drivers to, for example, solve differential equations. The most prominent of these appear to be [2], and [9], where the solution of ordinary differential equations, or systems of them, are accomplished in a two-pass approach. The statement of the problem is translated by means of a symbolic system into an iteration based on Taylor series expansion and analytic continuation. This can be used for one or two point boundary value problems, and has in at least some cases been claimed to be superior to a general numerical scheme which must, perforce, depend on rather more limited information about the differential equation. Taylor series may not be valid in the region of interest, but probably contributing to the lack of wide-spread use is the difficulty of comparing the achievements of these programs to convention formulations: the Taylor series methods do not use "function evaluations". It also seems that these systems, while somewhat portable, have not actually been engineered and distributed for use by the occasional and casual programmer or engineer.

More general Taylor series packages have been written, e.g. in Scratchpad, Macsyma, Reduce[3], although the authors of these programs have usually not provided automatic drivers for solving systems of ODE's by Taylor series. These systems have, by and large, taken the view that interlanguage communication is a more reasonable approach than combination languages. This approach is described in the next section.

2. Interlanguage communication

Most popular interactive algebra systems in use today have provided a form of output that can be fed into Fortran compilers. That is, for people who wish to shift their attention from algebraic to numeric computation, a specialized "print-fortran" routine is provided. Thus differentiated expressions can be re-parsed by a fortran compiler and evaluated subsequently, when imbedded in a suitable template program.

This would seem to be fairly straightforward, since spewing out internal forms (typically trees) in an infix form is an elementary program for students of data structures.

Unfortunately, attention to detail, and sometimes outrageous detail, is necessary in these output forms. For example, in Fortran, 1/2 is 0, while 1.0/2.0 is 0.5. Fortran, PL/I, Pascal, etc. each have their own particular use of "=" ".EQ." and other special tokens. Should the user wish to emit conditional statements, Fortran provides some very inconvenient forms stretching over several statements, and requires not only the generation of one or more statement labels, but global knowledge of other statement numbers already in use.

If the symbolic system is producing floating point constants, the precision and range of the Fortran processor must be considered. For this reason, it is generally preferable to produce 1.0/3.0 to 0.33333333, since in the latter case, one cannot predict how many 3's to print. It may be even better to produce 1.0d0/3.0d0. Curiously, we have discovered that in the latest Fortran standard there is no FORMAT specification that can be depended upon to print 1.0d0 from a floating point number. Instead, 1.0e0 may be printed, regardless of the precision! This means that it is difficult to print a high-precision constant in such a way as to guarantee that, if it were read in again as part of a program, it would be accurately represented to double-precision.

If the range and/or precision of the Fortran processor exceeds that of the symbolic system's representation, as would be the case if a PDP-10 Lisp system were to produce CDC-6000/7000 code, occasionally unsatisfactory results can be expected.

In the case of   the most generally available algebra systems on the PDP-10 (Reduce, Macsyma) which are implemented in Lisp, none have permitted the direct use of the Fortran compiler and subsequent loading of the object module back into the algebra system. The reasons for this are variously consequences of the already large size of the algebra system, the incompatible run-time environment, and possibly the incompatible use of data-types.

It is not clear that current "standards" deliberations will be able to produce a machine-independent and quirk-free Fortran language specification relative to the above considerations. In spite of all these problems, the facilities for the support of Fortran on most systems is vastly superior, for numerical computation, to any of the other "advanced" languages, and will probably remain so, for some time.

---

3. General references for these systems can be found in the various proceedings of SIGSAM and Eurosam conferences (e.g. [8],[15]).

3. More specific information, closer to home

The implementation of Macsyma produced by a group at the University of California, Berkeley, for the VAX-11 computers, combines features of the earlier PDP-10 Macsyma with prototype interlanguage communication at an improved level.

In Macsyma itself it is possible to evaluate algebraic expressions in one or more variables numerically, either using exact rational arithmetic, machine floating point, or "arbitrary precision" floating point numbers. These evaluations are done interpretively, and thus the machine floating point computation is, compared to the equivalent Fortran quite slow.

The VAX system also allows for the generation of Fortran code, much as could be done by the PDP-10 system, and in particular, the Mactran system of M. C. Wirth. Its further enhancements by Douglas Lanam are illustrated in the examples following this text. (both systems are described in reference [15]).

The environment of Macsyma on the VAX (Vaxima, as it has been called) allows for compiling, loading and executing the program that may have just been generated by the symbolic system. Thus for a modest overhead involving the use of an external language processor, an optimized machine-language routine is produced. The resulting routines are particularly attractive in the case of a detailed plot (especially contour plots) which require many function evaluations, or the use of numerical zero-finding or differential equation solving.

Of course these routines need not be used by Vaxima itself, but can be placed in a subroutine library for used by arbitrary programs.

Adding to the versatility of the Vaxima system is the ability to generate enhanced Fortran-language programs. That is, the code need not be restricted to Fortran 77, but can include the use of several packages in a pre-processor stage. These packages provide for interval arithmetic computations, high-precision floating point [3], and a simulation of the proposed IEEE standard double and extended precision arithmetic.

Within Macsyma, there are many tools which provide for source to source transformations of symbolic expressions. These provide not only for pre-computation of common sub-expressions, but rearrangement of expressions for the reduction of round-off error, the minimization of operation counts, economization of series, rearrangement of rational or exponential approximations, etc. These are illustrated in the examples following.

4. Future trends

There are several areas for progress and problems in the development of algebraic manipulation systems in conjunction with numerical computation. These are the advent of new languages, in particular Ada, the standardization of arithmetic along the model proposed by the IEEE [4], and progress in the automation of programming tasks.

4.1. Ada

Probably the most difficult to evaluate situation with regard to programming languages in the future, is the impact of Ada. It appears that Ada forbids or at least makes inconvenient by its strong typing constraints, many of the programming techniques that have been used in the past with great success in the algebraic manipulation systems written in Lisp.

It may be that the feasibility of producing quality Ada code for applications (e.g. control programs embedded in machinery) will be enhanced by the availability of algebraic manipulation systems which in some sense understand the Ada language, much as current systems tend to have built-in Fortran sensibilities.

4.2.

The adoption of the IEEE floating-point arithmetic standard has much more sanguine prospects: The difficulties of inter-language communication of floating point values will be drastically reduced; the ability to rely on high-quality evaluation of elementary functions, and a general upgrading of software quality will undoubtedly aid in providing more useful bases on which to build systematic floating point / algebraic interfaces. The use of "not-a-number" symbols to carry symbolic entries is tantalizing, but probably of limited use. The shifting of elaborate processing requirements to trap-handling programs is no more attractive in current compiler/systems technology than treating new data-types at the compiler stage. This may, however, change in the future.

4.3. Environments

In the future, we believe the proper emphasis will be on environments for computation: on systems, rather than programming languages. By limiting the scope of individual languages, yet providing for communication on the level of files, input/output streams, subroutine parameters, shared memory, and inter-process communication, challenging problems of the future can be handled more readily; any language should have ability to interact with others and system resources.

We are seeing the beginning of this type of synergism in several areas. The range of applications is indicated in a survey article by Ng in [8] and more recent proceedings (e.g. of SYMSAC-81 [15]). In part, these applications are the result of the rapidly decreasing cost and higher availability of relatively large-address-space computers: the controlled growth of large knowledge-based systems including algebraic manipulation facilities is an area with potentially great impact on scientific computation.

4.4. Better models of computation

Associated with the attempts to provide better control of programming and software engineering tasks in general, there has arisen a significant literature which combines abstraction and reasoning with concrete program implementation. A model for computation which permits algebraic systems to grow in an orderly fashion, and perhaps allow for the efficient implementation of much simpler -- e.g. numeric -- systems, has been sought after by several groups of researchers, for at least the past 10 years.

The notion is to provide a set of mathematically clear basic routines for a suite of manipulations. These routines, whose validity can be extended by the use of additional routines combine generality, and special-case efficiency "tricks" to extent the basics routines to additional abstract mathematical domains such as rings, fields, and algebraic and transcendental extensions of these domains. The associated algorithms naturally must be stated in a form to allow smooth transitions, and yet provide appropriate levels of efficiency in data storage and manipulation. A recent reference to such work is the paper by Jenks and Trager [5].

It would be most valuable if in the future we could provide powerful packages of an algebraic computation nature for use in a scientific programming environment in as routine a manner as we can now provide scientific library routines for computing elementary functions.

References

1. J. D. Anderson, E. L. Lau, and R. W. Hellings, "Use of MACSYMA as an automatic Fortran coder," in *Proc. 1979 Macsyma Users' Conf.*, Lab. for Comp. Science, M.I.T. 583-595.

2. D. Barton, I.M. Willers and R.V.M. Zahar, "Taylor series methods for ordinary differential equations -- an evaluation," in: J. Rice (ed.), *Mathematical Software,* Acad. Press, 1971. 369-390.

3. R. P. Brent, "A Fortran multiple-precision arithmetic package," *ACM Trans. Math. Softw.* 4, 1 (March, 1978), 57-70.

4. IEEE Computer Society Microprocessor Standards Committee Task P754, "A proposed standard for binary floating-point arithmetic, draft 8.0" *Computer* 14, 3 (March, 1981) 52-63.

5. R. D. Jenks and B. M. Trager, "A language for computational algebra," in: P.S. Wang (ed.), *Proc. of the 1981 ACM Symp. on Symbolic and Algebraic Computation,* ACM, 1981.

6. G. Kedem, Automatic differentiation of computer programs, *ACM Trans. Math. Softw.* 6, 2 1980, 150-165.

7. D. E. Knuth and L. Trabb Pardo, "The early development of programming languages," in: N. Metropolis et al, (ed.), *A History of Computing in the Twentieth Century,*, Acad. Press, 1980. 197-274.

8. Edward Ng (ed.), *Symbolic and Algebraic Computation, Proc. of Eurosam 79,* Springer-Verlag Lecture Notes in Computer Science 72, 1979.

9. A. C. Norman, "A System for the solution of initial and two-point boundary value problems,"in: *Proc. ACM 25 Annual Conf.*, Boston, August 1972, 826-834.

10. Jean Sammet, "An annotated descriptor-based bibliography on the use of computers for non-numerical mathematics," *Comput. Rev.* 7, (1966), also subsequent modifications in SIGSAM Bulletin.

11. B. Speelpenning, *Compiling Fast Partial Derivatives of Functions Given by Algorithms,* Ph.D thesis, January 1980. Also Tech. Report UIUCDCS-R-80-1002 Comp. Sci. Dept., Univ. Illinois.

12. G. L. Steele, Jr., "Fast arithmetic in Maclisp," *Proc 1977 MACSYMA Users' Conference,* NASA CP 2012, Berkeley Ca., July 27-29, 1977. 215-224.

13. R.G. Tobey, "Eliminating monotonous mathematics with FORMAC," *Comm. ACM,* 9, Aug. 1966.

14. R. P. van de Riet, Formula Manipulation in Algol 60, part I, Math. Cent. Tract 17, Mathematisch Centrum, Amsterdam, 1970 (2nd edition), 1968 (1st edition). (189 pages)

15. P.S. Wang (ed.) *Proc. of the 1981 ACM Symp. on Symbolic and Algebraic Computation,* (SYMSAC-81) Snowbird, Utah, Aug 5-7, 1981.

Some Examples from Macsyma, using a VAX/UNIX System

These examples were discussed in the IFIP conference talk.

(c4) /* consider the following expression */

exp: log(-2*atan(log(r))/acos(r)^2)/(1-r);
Time = 66 msec.

$$\frac{\log(-\dfrac{2\,atan(\log(r))}{acos(r)^2})}{1-r} \tag{d4}$$

(c5) /* evaluation of this for r in (0,1] by simply evaluating the formula
produces results that are inaccurate.
In particular, evaluation at r=1 provokes several divisions by zero.
At r=0, the left endpoint, there is an essential singularity.
We can expand exp in a Taylor series about r=1,
and near this point use the Taylor series instead.
We can generate the Taylor series as follows: */

tay1: taylor(exp,r,1,3);
Time = 10750 msec.

$$\frac{1}{3} + \frac{7\,(r-1)}{45} - \frac{1229\,(r-1)^2}{5670} + \frac{8699\,(r-1)^3}{113400} + \dots \tag{d5}$$

(c6) /* This is a pretty display, but evaluating it in the implied
order as given is not very attractive.
Consider, for example the blind conversion of tay1 to Fortran syntax: */

```
fortran(tay1)$
    7.0*(r-1)/45.0+8699.0*(r-1)**3/113400.0+(-1229.0)*(r-1)**2/5670.0+
1   1.0/3.0
Time = 216 msec.
```

(c7) /*better would be to convert it via "Horner's Rule", viz: */

hornertay1:horner(tay1);
Time = 200 msec.

$$\frac{r(r(8699\,r - 50677) + 92897) - 13119}{113400} \tag{d7}$$

(c8) /* which would, in Fortran, look like */

```
fortran(hornertay1)$
    (r*(r*(8699*r-50677)+92897)-13119)/113400.0
Time = 66 msec.
```

(c9) /* Another technique, described by C. Meszteny and C. Witzgall,
["Stable Evaluation of Polynomials," Journal of Research of the
Nat'l Bur. of Stds.-B, vol. 71B no. 1 (Jan 1967) p. 11-17.]

involves a reformatting to minimize round-off error, with respect
to evaluation in a particular interval, in this case, say, 0.99 to 1.0.

This produces another result, illustrated below */

loadfile ("minform.saved") $
Time = 200 msec.

(c10) floatform:ev (tay1,expand,numer);
Totaltime = 2116 msec. Gctime = 1783 msec.

$$0.07671075837742504 \, r^3 - 0.4468871252204586 \, r^2 \tag{d10}$$
$$+ 0.8191975308641975 \, r - 0.1156878306878307$$

(c11) stable:minform (floatform,r,0.99,1.0);
Time = 1516 msec.

$$((0.07671075837742504 \, r - 0.2942327160493827) \, (r - 1.0) \tag{d11}$$
$$+ 0.1577307751322751)$$
$$\cdot (r - 0.99)$$
$$+ 0.3317560255820106$$

(c12) /* For about the same arithmetical work, we can provide a Pade
approximation based on the Taylor series, thus */

padeappx: pade (tay1,2,2);
Time = 1166 msec.

$$[\frac{11176809 \, r - 841029}{11268151 \, r^2 - 3475967 \, r + 23215156}, \tag{d12}$$
$$(-\frac{11268151 \, r^2 - 41596637 \, r + 7100386}{24661665 \, r + 45022635})]$$

(c13) /* This is a list of two approximations of requested degree.
We can choose the second, rearrange it via Horner's rule,
and view it in Fortran syntax:
*/

pade2:horner (padeappx [2]);
Time = 150 msec.

$$\frac{(41596637 - 11268151 \, r) \, r - 7100386}{24661665 \, r + 45022635} \tag{d13}$$

(c14) /*we can combine these ideas in another way
by Meszteny's technique: */

pade2: minform (num (padeappx [2]),r,0.99,1.0)/denom (padeappx [2]);
Time = 716 msec.

$$\frac{(30441167.51 - 11268151.0\,r)\,(r - 0.99) + 23036369.8349}{24661665\,r + 45022635} \tag{d14}$$

(c15) /* Let us chose one of these, for example, "stable" in the program below. One way to determine those values for which the approximation is to be used is to test if acos(r)^2 is zero. (in which case, use the approximation.)
This is implemented below.

The "vaxtran" system allows for generation of a Fortran program which is valid for r as r approaches and reaches 1 from the left. */

/* This is the command that produces, compiles, and loads the Fortran program */

vaxtran(func(r : double):double);

```
creating func
*begin
    double precision function func(r)
    double precision r

    double precision z
    z=acos(r)**2
    if ( z .eq. 0.0) then
c    func=stable
    func = ((0.07671075837742504*r-0.2942327160493827)*(r-1.0) +0.15773
1   07751322751)*(r-0.99) +0.3317560255820106

    else
c    func=subst(z,acos(r)^2,exp)
    func = log(-2*atan(log(r))/z)/(1-r)

    end if
    end
*end
[f77 func]
func.f:
   func:
1.3u 1.8s 0:10 30% 21+22k 43+6lio 178pf+0w
[ffasl func.o]
```

Time = 1533 msec.

func created (d15)

(c16) /* Now try it out. There is a funny problem here: Fortran 77 constants are only single precision, so the answer is not exactly the closest value to 1/3. This can be fixed...*/

func(1.0);
Time = 0 msec.

0.3333333341́1809116 (d16)

(c17) func(0.5);
Time = 0 msec.

$$0.2004427740157895 \tag{d17}$$

(c18) /* There are other computational forms that can be produced which
result in accuracy improvements, or run-time efficiency. It is
possible to generate a Taylor series, use Chebyshev polynomials
to express the same polynomial, and then economize by truncation. */

/* Let us define the Chebyshev polynomials: */

(T[0](x):=rat(1), T[1](x):=x, T[r](x):=2*x*T[r-1](x)-T[r-2](x))$
Time = 50 msec.

(c19) T[5](z);
Time = 583 msec.

$$16\,z^5 - 20\,z^3 + 5\,z \tag{d19}$$

(c20) /* we can convert a polynomial to Chebyshev form by dividing by
successive Chebyshev polynomials. Thus if we consider a
Taylor series for exp(z) about z=0, */

ez:taylor(%e^z,z,0,5);
Time = 116 msec.

$$1 + z + \frac{z^2}{2} + \frac{z^3}{6} + \frac{z^4}{24} + \frac{z^5}{120} + \ldots \tag{d20}$$

(c21) for i:hipow(ez,z) step -1 thru 1 do (quorem:divide(ez,T[i](z),z),
                ez:'T[i](z)*quorem[1]+quorem[2]);
Time = 1216 msec.

$$done \tag{d21}$$

(c22) ez;
Time = 0 msec.

$$\frac{t_5[z] + 10\,t_4[z] + 85\,t_3[z] + 520\,t_2[z] + 2170\,t_1[z] + 2430}{1920} \tag{d22}$$

(c23) /* a more pleasant form might be */

ez:ratexpand(ez);
Time = 200 msec.

$$\frac{t_5[z]}{1920} + \frac{t_4[z]}{192} + \frac{17\,t_3[z]}{384} + \frac{13\,t_2[z]}{48} + \frac{217\,t_1[z]}{192} + \frac{81}{64} \tag{d23}$$

(c24) /* The Chebyshev series could be truncated and the computation
reordered again: */

ez:subst(0,'t[5](z),ez);
Time = 66 msec.

$$\frac{t_4[z]}{192} + \frac{17\, t_3[z]}{384} + \frac{13\, t_2[z]}{48} + \frac{217\, t_1[z]}{192} + \frac{81}{64} \tag{d24}$$

(c25) ev(ez,t);
Time = 200 msec.

$$\frac{8\, z^4 - 8\, z^2 + 1}{192} + \frac{17\,(4\, z^3 - 3\, z)}{384} + \frac{13\,(2\, z^2 - 1)}{48} + \frac{217\, z}{192} + \frac{81}{64} \tag{d25}$$

(c26) fortran(horner(%));
   (z*(z*(z*(16*z+68)+192)+383)+384)/384.0
Time = 233 msec.

$$done \tag{d26}$$

(c27) /* This is not the best set of coefficients since they are derived
from the Taylor series coefficients rather than the integral
formulation. Unfortunately, these integrals involve Bessel functions
and cannot be computed by Macsyma (now, at least). They would
look like this, though: */

a[k]:=1/%pi * integrate(%e^z*t[k](z)/sqrt(1-z^2),z,-1,1);
Time = 0 msec.

$$a_k := \frac{1}{\pi} \int_{-1}^{1} \frac{e^z\, t_k[z]}{\sqrt{(1 - z^2)}}\, dz \tag{d27}$$

(c28) /* An interesting program to perform rational function integration
is presented below. Of course Macsyma can perform many integrations
in closed form, but in the case of rational functions with denominators
which are irreducible over the integers, it generally refuses to
produce the answer because of the need to introduce algebraic numbers.
Even if algebraic numbers were introduced, the result would generally
be unwieldy.
Furthermore, many problems have numeric coefficients for which it
would be pointless to convert to exact integers. The program below
uses an IMSL zero-finder to factor a polynomial over the
"field" of complex floating point numbers, computes the partial
fraction expansion, and provides a formula for the indefinite
integral. */

/*floating point rational integration program. P is arbitrary rational
function in x */

fpint(p,x):=
block ([num,den,div,res,keepfloat,

    lroots, distinctroots, qq,i,c],

```
      local(r,q),
/* simplify the input */
      p:rat(p), num:ratnumer(p), den:ratdenom(p),
/* separate polynomial and rational parts */
    · div:divide(num,den),
/* integrate polynomial if any, and initialize result */
      res: integrate(ratdisrep(div[1]),x),
/* set num to numerator/leading coefficient of denom */
      num: ratdisrep(div[2]/ratcoef(den,x^hipow(den,x))),
      keepfloat:true,
      r[i]:=0,
/* call the IMSL polynomial zero-finder */
      lroots:zrpoly(den,x),
/* The number of roots at c, a pt in complex plane is stored in r[c] */
      for c in lroots do  r[c]:r[c]+1,
/* distinctroots is mysteriously set to a list of the roots [c1,c2, ... ] */
      distinctroots: map(first,rest(arrayinfo(r),2)),
/* qq is a reconstruction of the denominator, factored */
      qq:1,
      for c in distinctroots do qq:qq*(x-c)^r[c],
/* q[c] is qq with the r[c] roots at c taken out */
      for c in distinctroots do q[c]:qq/(x-c)^r[c],
/* The formula below gives the (constant) residue needed in
numerators in the partial fraction expansion. Integration
of expressions of the form K/(x-c)^i is easy. */
      for c in distinctroots do for i:1 thru r[c] do
      res: res+1/(r[c]-i)!*ratsimp(subst(c,x,diff(num/q[c],x,r[c]-i)))
          *(if i=1 then log(x-c) else (x-c)^(1-i)/(1-i)),
      return(res))$
Time= 16 msec.
```

(c29) prob:1/(x^3+x+1);
Time = 66 msec.

$$\frac{1}{x^3 + x + 1}$$
(d29)

(c30) /* just display a few digits of precision
in the answer. */

floatformat(f,0,8)$
Time = 16 msec.

(c31) ans:fpint(prob,x);
Time = 1783 msec.

$$-\frac{0.43046249\,i\log(x + 1.16154140\,i - 0.34116390)}{1.16154140\,i - 1.02349171}$$

$$-\frac{0.43046249\,i\log(x - 1.16154140\,i - 0.34116390)}{1.16154140\,i + 1.02349171}$$

$$+0.41723799\log(x + 0.68232780)$$
(d31)

(c32) /* Although formally correct, this doesn't really *look* right.
The answer should not require complex logarithms.

Assume the argument of the log is always positive, and use
some magic utterances to reformat the answer. Convert to
rectangular form, then simplify (to show the imaginary part is 0.). */

```
block([keepfloat],
    keepfloat:true,
    assume(x> -0.6),
    ans2:rat(rectform(ans)));
Time= 1633 msec.
```

$$0.41723799\log(x+0.68232780) \tag{d32}$$
$$-0.20861899\log((x-0.34116390)^2+1.34917842)$$
$$-0.36764907\,atan2(1.16154140,x-0.34116390)$$

(c33) /* Now use full (double) precision for the Fortran statement. */

```
floatformat(g,0,16)$
Time= 33 msec.
```

```
(c34) fortran('ans=ans2)$
    ans = 0.4172379879262188*log(x+0.6823278038280193)-0.2086189939631
1   094*log((x-0.3411639019140097)**2+1.349178423907576)-0.36764907
2   38633923*atan2(1.161541399997252,x-0.3411639019140097)
Time= 283 msec.
```

(c36) /* Whether evaluation of this formula is preferable to quadrature is
a matter for consideration, of course. */

DISCUSSION

Summary (W.S. Brown)

Prof. Richard J. Fateman reviewed a variety of approaches to the problem
of combining symbolic and numeric computation, and suggested that inter-
language communication may be more effective than combination languages.
In particular, he noted that a user of MACSYMA on the DEC VAX-11/780
can invoke the FORTRAN compiler as a subroutine to compile a MACSYMA·
expression into a FORTRAN function, and can then invoke the resulting function to evaluate the expression numerically.

Comment (J. Rice)

Fateman's history of the symbolic and numerical systems is through the
path of symbolic algebra systems. In the context of the discussions at this
conference, one can view this history as having symbolic algebra as the
"hardware" and now one is defining "languages" (systems) which use this
hardware for general purpose scientific computation. The other line of development starts with the language objective itself and then explores the needed
facilities. A number of efforts of this type were made in the 1960's; e.g., the
Culler-Fried system, the Klerer-May system, and NAPSS. All of these efforts
failed (primarily because of the lack of sufficient computing resources), but it is
instructive to study the designs made and to see how various combination
symbolic/numerical facilities were implemented. The book *Interactive Systems
for Experimental Applied Mathematics* by Klerer and Rheinfelds (Academic Press,
1968) contains information about most of these efforts.

Comment (L. Meissner)

To embed FORTRAN in an applications-oriented environment may be better than to "attach" the applications to FORTRAN. This could be a way to relieve the pressure on X3J3 to expand the Fortran language to accommodate various applications.

Comment (S. I. Feldman)

The more capabilities one requires of a computing environment the more difficult it is to move from one environment to another.

Question (S. I. Feldman)

How many real applications have there been involving the ability of MACSYMA to invoke the FORTRAN compiler?

Answer (R. J. Fateman)

Recently there have been a number of serious problems in finite differences.

Comment (M. Hennell)

It would be desirable to avoid the need for a special interface whenever one system wants to invoke another.

Reply (T. Lake)

There is no need for such interfaces if one can always escape to :ommand level.

Comment (T. Lake)

Perhaps we are getting close to applications involving the symbolic manipulation of FORTRAN expressions; e.g., for proving the correctness of a program or analyzing its numerical behavior.

Question (L. Meissner)

Can FORTRAN be a tool in a larger problem solving environment that also includes symbolic computation, graphics, and so forth? If so, this may be better than adding such facilities to FORTRAN.

Question (C. Lawson)

Is MACSYMA now available to those who have a VAX-11/780?

Answer (R. J. Fateman)

Technically it is ready, and I want it to be used as widely as possible. However, MIT is planning to license it, and there may be a few legal problems remaining.

THE RELATIONSHIP BETWEEN NUMERICAL COMPUTATION
AND PROGRAMMING LANGUAGES, J.K. Reid (editor)
North-Holland Publishing Company
©*IFIP, 1982*

Programming Languages for Symbolic Algebra
and Numerical Analysis

W. Morven Gentleman

Computer Science Department
University of Waterloo
Waterloo, Ontario, Canada

Symbolic algebraic computation can be a valuable aid
when used with numerical mathematics to derive
numerical programs and to analyze numerical programs.
To be effective, there must be a correspondence between
the expression class the symbolic algebra system
manipulates and those the numerical programming
language uses to express formulae which must be
evaluated. The operations of the symbolic algebra
system must correspond to operations the numerical
analyst wants to perform on the expressions in the
numerical program code. Inability, or even awkwardness,
in fitting the two languages together this way, is
probably a large part of why symbolic algebra has been
less useful to numerical analysts than might have been
hoped.

1. Introduction

The working numerical analyst does large amounts of simple
algebraic computation. Typically the computations are not enormous, but they are tedious, and the consequence of even trivial
sign errors can be severe. The numerical analyst is thus an excellent candidate for assistance in the form of symbolic algebra
manipulation. Two recent surveys[1,3] cite various ways
numerical mathematics and symbolic mathematics can interact in
solving problems, and provide extensive lists of references to
such uses. On the other hand, few numerical analysts have had
access to such tools, yet amongst those who have, the results
have been more mixed than might be expected. This paper examines
the reasons behind this.

2. Applications

The algebraic computations the numerical analyst does occur
in several quite different phases of his work and call for quite
different kinds of support. Broadly speaking, they break into
formulation of the problem, derivation of algorithms and formulae, and analysis of programs. Let us examine each of these
more closely.

The most common, but least well defined, are the algebraic
computations done while formulating the problem. Typically
problems are posed for the numerical analyst in terms of a
geometric model, or a physical situation, or sometimes a
mathematical situation which must be approximated. The numerical

model to be solved must be derived from this. Typical things which must be chosen are the coordinate system to be used, the discretization taken, basis functions for expansions, how derivatives or integrals will be estimated, approximations to be used for "known" functions, how error should be assessed, etc. One thing often done at this stage is to use classical approximating methods to obtain analytic approximate solutions to general forms of a problem. Appropriate use of symbolic computation can enormously reduce the dimensionality of the numeric problem as when classical methods reduce differential or integral equations to fitting undetermined coefficients in a symbolic representation of the solution. It is not unusual for several alternatives to be tried before the numerical model to be used to represent the original problem is chosen. It is also not unusual for many facts about the algebraic structure of the problem to be investigated, which requires algebraic computation not directly related to solving the original problem.

Once the numerical problem to be solved has been formulated, the algorithms to solve it must be derived. Sometimes these are independent of the particular problem to be solved, applying instead to a whole class, but sometimes they are very specific to the problem. One way in which algebraic calculation can be very useful at this stage is in deriving versions of formulae to be used for intermediate calculations taking explicit advantages of what is known about the specific problem. Additional approximation can be avoided, by using analytical derivatives instead of numerical derivatives, if the analytical derivatives can be obtained from symbolic representations of the functions. Numerical instabilities can often be avoided, for instance by symbolic evaluation of recurrence relations which facilitate explicit and exact cancellation between like terms appearing in both numerator and denominator of rational expressions. Large rounding errors present in direct computation of quantities can sometimes be avoided by deriving expressions for incremental change. Cost can sometimes be reduced by a preliminary stage of performing part of the computation symbolically, followed by substituting into the resulting expressions rather than repeating the entire computation for each set of numerical values. (This must be done thoughtfully, however. It is not just that the cost of the symbolic computation will outweigh the cost of the numeric computation saved unless they are in some critical inner loop. Naive evaluation of the symbolic expressions can actually be more expensive than the purely numerical approach if the latter effectively takes advantage of common subexpressions which the former ignores.)

The analysis of programs introduces some special requirements. Error analysis, in the sense of studying the truncation error induced by approximations to the analytical model, can be studied in terms of algebraic expressions. The same is true for sensitivity analysis, that is, assessing the effects of changes in the input data. The other analyses one wants to make, however, must work with something much closer to the actual code that will be executed in obtaining the numerical solution. Cost analyses, for example, are much more meaningful when the overheads are properly accounted for, which can include details of how expressions are written as well as details of loop structure.

Careful analysis of the effectiveness of special architectural features such as vector hardware, more general pipelining, chaining multiple functional units, array parallelism, caches or other memory hierarchies requires knowledge of the exact code as well as requiring automated aids representing the specific hardware. Correctness analysis is most usefully done after the last point at which human intervention can introduce further errors into the code. Error analysis, in the sense of studying the fixed or floating point rounding errors, requires a complete and detailed specification of what arithmetic operations will be performed and in what order.

3. Problems

3.1. Function space

To be useful, the function space from which the values manipulated by the symbolic algebra system are drawn must include the expression class used by the numerical calculation system. This may sound trite, but it is not. The richer the function space over which the symbolic algebra system must work, the less it can do. Even for the functions of one variable, the class of functions generated by +, -, x, /, exp, and abs is so large that recognition of expressions whose value is identically zero is uncomputable! Many symbolic algebra systems have thus limited themselves to manipulating rational algebraic expressions, or perhaps Poisson series, in order to be efficient and definitive. Unfortunately this eliminates these systems from practical use by the numerical analyst. The algebra the numerical analyst wants to do involves a much richer class of functions. Actually, by recognizing that certain subexpressions can effectively be treated as independent indeterminates, it is often possible to do the desired computations in such a limited system-but it is so inconvenient that people will resort to hand calculation instead.

A more serious problem is that most symbolic algebra systems work over the field generated by integers and indeterminates with the rational operations, extended by various functions. In particular, complex numbers are not part of the underlying field, even though many mathematical theories of relevance to the numerical analyst are simplest when explained in terms of functions of complex variables. The standard answer to this is that the complex field is trivially obtained from the rational field by extending the indeterminates with the symbol "i", then applying the "side condition" i**2 = -1 by reducing every expression to its ideal generated by i**2 + 1. While mathematically valid, this approach is computationally unsatisfactory. It only solves the problem for polynomials, leaving one to have to "rationalize the denominator" manually for rational algebraics (of which more later) and probably leaving arguments to functions unreduced. The intermediate expression swell caused by having the lowest level arithmetic operations unaware of the side condition may well be unacceptable. (This same criticism applies to side conditions which implement approximations such as series or asymptotic expansions: unless the lowest level arithmetic operations are aware of them the intermediate expression swell may be unacceptable.)

An equally serious problem is that existing symbolic algebra systems work over scalars. This permits small matrices all of whose elements are represented explicitly, but not the general matrix algebra which is common in numerical analysis. Much of the algebraic computation the numerical analyst does is with vectors whose dimensionality is irrelevant, or with matrices whose dimensionality is represented symbolically, or with block matrices whose elements, being themselves matrices, are from a non-commutative division ring. These things cannot even be expressed in current symbolic algebra systems. Nor can common matrix operations, such as transposing, applying standard factorizations, etc. These shortcomings are not just felt at the stage of formulating the problem. If the numeric computation is to be expressed in some language like APL or the various vector and array processor extensions to Fortran, which do understand the concept of arrays, the derivation of algorithms and expressions and the analysis of algorithms may well require vector, matrix, or tensor calculation.

Yet a third serious problem is that symbolic algebra systems tend to work with functions that have convenient mathematical properties and are easy to represent, such as analytic functions. Numerical analysts often need to work with functions that have convenient numeric properties instead. Splines are a case in point, as are continued fractions. Symbolic algebra systems do not handle them well.

A fourth serious problem arises because of the operations that the numerical analyst often wishes to perform on expressions. Frequently these operations involve limiting behaviour: infinite sums, integrals, limits of sequences, etc. For completeness, not only must the input expressions, the "general term", be in the class of expressions which the symbolic algebra system manipulates, but the limit must too. Often this is not possible, as these have no closed form in any useful class of expressions. The numerical analyst solves this in hand calculations by approximation: Taylor expansion, Fourier expansion, asymptotic expansion, dominating series, etc. The effect allows him to stay in the same function class (an expansion in Bessel functions, for instance, may introduce Bessel functions of another kind but probably won't introduce any new kinds of functions). Recognition that the terms in an expansion are those of a known function can allow series to collapse. Unfortunately, these operations usually are not directly available in symbolic algebra systems.

3.2. Representation

When algorithms, expressions, or even just subscript forms and loop control variable limits are derived by a symbolic algebra system for a numerical computation program, it is critical that the representation produced by the symbolic algebra system can be directly included in the code for the numerical program. Any manual intervention introduces the possibility of error, and avoiding this possibility is one of the attractions of using the symbolic step. Since the input languages for numeric computation programs are hardly suited to insightful examination of expression form (partly because they do not look like conventional two dimensional mathematical notation), and since such examination is

often crucial to making the right design choices in the code,
many symbolic algebra systems offer a choice of output form:
readable or Fortran. Even so, variable names, subscript
representation, numeric coefficients, etc. may need changing.

But the problem is more complex. The symbolic algebra
system represents expressions internally in a form convenient for
symbolic calculation. For instance, canonical representations
are often used. These same representations are by no means the
preferred choice for numerical evaluation, hence more control has
to be possible over the way expressions are produced. For instance, recursive representation of polynomials in many variables
as polynomials whose coefficients are polynomials in one fewer
variable may not, in some circumstances, even be a good choice
for evaluation. These problems get worse when there are multiple
expressions, possibly sharing common subexpressions. One solution has been to write programs that restructure output from symbolic algebra systems in order to produce source to be included
in numerical codes.

The reverse set of problems occur when one wants to use the
symbolic algebra system to analyze a numerical computation algorithm or program, expressed as a specific piece of code. Here
it is necessary to read the specific code, and from it to form
expressions in the class that the symbolic algebra system can
manipulate. The difficulty of this depends on the kind of
analysis to be performed.

For analysis of correctness, the primary requirement is
merely that the expressions which form the right side of assignment statements in the numerical program be expressible as values
in the symbolic system. These expressions may then be symbolically evaluated. Sometimes the expressions which form loop
control must similarly be expressed and symbolically evaluated
too. In my experience, this is an interesting difference between
correctness of numerical mathematics software and correctness of
typical software to which program verification methods have been
applied. In the latter, the bulk of the effort is in identifying
invariant properties preserved by loops and multiple alternative
constructs, proving that these invariants indeed are preserved,
then using them to prove higher order properties of the code.
For numerical mathematics software, on the other hand, proving
that an expression, or a straight line sequence of expressions,
really produces useful approximations to the values intended, is
frequently the bulk of the work. This usually necessitates
knowing the derivation of the algorithm. Inferring from the code
that J.C.P. Miller's backward recurrence from arbitrary initial
values actually produces accurate Bessel function values seems
most unlikely. Less extreme, inferring the function to which a
polynomial or rational approximation with floating point coefficients is intended to be an approximation also seems unlikely.
These examples illustrate, however, how the expressions can often
be readily checked, once the derivation is known.

For sensitivity analysis, the requirements are quite
similar. Sensitivity analysis is fairly straightforward when the
quantities dependent on the potentially changed data are continuously differentiable. It is merely necessary to evaluate

those quantities symbolically, then take partial derivatives with
respect to the input data. As with analysis of correctness, there
is some complication in sorting out what parts of the source code
for the numerical program are relevant to the symbolic evaluation, and it would be nice to handle loops and multiple alternative statements, although they are of lesser importance.
Libraries and separate compilation form a barrier in that properties of separately compiled functions can not be assessed. A
practical problem is how to look at the expressions which are
sensitivity multipliers, and rearrange terms or apply approximations until they are simple enough to be interpreted.

Rounding error analysis has a different character. Existing
symbolic algebra systems have great difficulty in carrying out
the manipulation of inequalities which is at the heart of
traditional rounding error analysis. Instead, the analyses performed are in the spirit of Webb Miller's approach: a new
variable is introduced for each possible rounding error, the error in the quantity of interest is calculated, then this error is
maximized by constrained optimization over the possible values of
the errors. This approach is only successful for small, fixed
dimension expressions, although sometimes the results for more
general problems, for example with arbitrarily dimensioned arrays, can be inferred from a few special cases.

Analysis of cost is quite different again. Here the control
flow structure is the only significant part of the computation.
It is quite straightforward to scan a basic block in the
numerical program to count the number of multiplications, additions, subscript references, function calls and other operations
known to have substantial cost. The complication comes in examining the control flow to count, symbolically, how often each
basic block is executed. In simple cases this is possible to
determine algebraically from subscript bound expressions, but in
others it must depend on guessed statistics with respect to conditional branches. Modelling higher level operations, such as
vector or matrix operations which are primitives at the language
level, can be more challenging if it requires that dimensions and
similar parameters must be calculated.

3.3. Language and compiler limitations

Another problem facing the designer of the symbolic algebra
system is that the compilers for numerical computation languages
are built under the assumption that source code for them is
produced by humans, not machines. In the past, this has been an
inconvenience mostly in that symbolic algebra systems regularly
generate expressions that overflow Fortran compiler expression
tree space. Today, with languages designed to enforce restrictive structured programming control structure dogma, there can be
other odd inconveniences such as having to introduce Boolean flag
variables in loops. Another complication occurs in producing expressions in a form that will trigger optimizers in the compiler
for the numerical computation language. When languages for
numerical computation lack the ability to have compile time
initialized constant data structures, as for instance with
PASCAL, it can be impossible for the symbolic algebra system to
generate as output a numerical code that efficiently evaluates

one or more expressions - the analytic form of a gradient or Hessian might be a case where this could be important.

Data structuring differences between the symbolic algebra system and the language for numerical computing can also cause problems. Symbolic algebra systems, because they are more modern as well as because they need it, typically have data structuring possibilities such as records, generalized arrays (whose elements are possibly dissimilar arrays), lists, etc. These things can be simulated in Fortran, but being able to do so adds useless complexity to the symbolic algebra system.

3.4. Intellectual overload

It is worth admitting that most people feel that they have too much to do to invest in learning yet one more programming language. Any unnecessary differences between the symbolic and numeric languages, from rules for forming identifiers to syntax of loop control to options on the invoking command, will represent an adequate impediment to keep people who could benefit by using symbolic and numeric computation together from doing so.

4. Examples

It is not easy to find simple examples to illustrate all of the above points, but some problems I have recently worked on convey the flavour of the frustrations which can occur even in an unsophisticated context.

4.1. Error analysis of circle drawing

The problem was to draw circular arcs on the screen of a graphics terminal, using the microprocessor in the terminal to compute the points. This meant that only fixed point add, subtract, and multiply were available. Divide had to be avoided as being too slow, although shifts were available. The standard trigonometric functions were not available, and would have had to be avoided in any case as being too slow.

The approach taken was to represent a point on the circle in terms of Cartesian coordinates of its displacement from the centre of the circle, then to apply a 2 by 2 linear transformation matrix to obtain the next point on the circle. This is done repeatedly, plotting each point computed, until the required arc has been drawn. The numerical analysis problem is what to choose as the linear transformation. Of course the exact answer is well known, but it involves trigonometric functions which we have said we don't want to use.

$$\begin{vmatrix} x' \\ y' \end{vmatrix} = \begin{vmatrix} \cos(eps) & -\sin(eps) \\ \sin(eps) & \cos(eps) \end{vmatrix} \begin{vmatrix} x \\ y \end{vmatrix}$$

The standard solution in the graphics community [2, p.26], involves an unusual approximation.

$$\begin{vmatrix} x' \\ y' \end{vmatrix} = \begin{vmatrix} 1 & -eps \\ eps & 1 - eps**2 \end{vmatrix} \begin{vmatrix} x \\ y \end{vmatrix}$$

Comparison with some alternative approximations seemed worthwhile. The obvious choices are to use the Taylor series expan-

sions for cos(eps) and sin(eps), taken to first or second
order. A less obvious choice is to use the first order Taylor
series, but to correct the approximations for both cos(eps) and
sin(eps) by dividing by sqrt(1 + eps**2) so the length of the
vector is preserved. A final choice, in the spirit of Comrie
throwback, might be to modify the coefficients of the linear and
quadratic terms to approximately allow for the effect of higher
order terms.

The important comparison is not the error after a single
step, but rather the error after a substantial angle theta has
been traversed, where this requires N steps. Of course N is an
indeterminate, not a specific integer. We need to compute the
Nth power of the linear transformation, where the incremental
angle eps corresponding to a single application of the linear
transformation is theta/N. In general it is not obvious how to
compute symbolically the Nth power of a matrix in any practical
and useful way. However, since our matrices are only 2 by 2, the
eigenvalue decomposition can be found explicitly, and we can use
the factorization of the Nth power matrix as the same eigenvector
matrices pre- and post- multiplying the Nth power of the diagonal
matrix of eigenvalues.

This exact solution is not of much use, however, as it
typically does not simplify, and consequently one is left looking
at large and incomprehensible formulae. To have something useful,
one must approximate, and an obvious thing to try is the asymptotic expansion in the reciprocal of N. Not surprisingly, the
leading terms in these expansions can be recognized as cos(theta
) and sin(theta), as appropriate, and the higher order terms
can be studied to see the dependence of the error on N and theta.

Two complications become apparent when this program is attempted. First, the eigenvalues and eigenvectors are complex,
that is, have nonzero real and imaginary parts. As mentioned earlier, most symbolic algebra systems handle this by a side condition applied after expressions are computed. The result is substantial intermediate expression swell, which made the computation time excessive, and nearly made the computation impossible
on the system I was using. The other complication is that the
expressions for the elements of the eigenvector of the approximation used by the graphics community contain radicals, that is,
square roots of algebraic expressions. (Fortuitously, this does
not happen for the other approximations.) The symbolic algebra
systems available to me do not provide for radicals. Again this
can be handled, in principle, by introducing new indeterminates
for the radicals and applying side conditions through ideals. In
fact, it is essential to be able to rationalize the denominators
of expressions to simplify them, even just to recognize when the
expression is identically zero. The awkwardness of doing this
eventually lead me to abandon the whole computation.

The other thing which lead me to abandon the computation was
the results obtained for the cases I could do. Although some
things were evident, including confirmation of observations from
numerical experiment, the main thing that was obvious was that
Cartesian coordinate assessment of the error was not sensible.
Perhaps this should have been apparent from the beginning, but it

was not until I actually saw the error expressions that I
realized I wanted to think of them a different way, and in particular that the error in this computation should be assessed in
polar coordinates even though the computation is done in Cartesian. Making this change was easy, and some results were more
comprehensible. But what happened mostly was, as Stoutemeyer observed[4], that it is hopelessly awkward to satisfactorily simplify trigonometric expressions.

4.2. Global properties in nonlinear optimization

A problem in trying to understand the global behaviour of
the residual sum of squares, S, as a function of the parameters,
B[j] and L, arose in connection with trying to develop efficient,
globally convergent algorithms for the nonlinear least squares
problem
$$y[i] = \text{sum}(\ (d[i]+L)*x[i,j]*B[j], \ j=1..q)$$
$$+ \ r[i]$$
where $x[i,j]$ and $d[i]$ are given and $y[i]$ is observed for each
observation i. In particular, if the value of S corresponding to
the best choice of B[j] for fixed L is plotted as a function of
L, empirically it was observed to always have one relative maximum, one relative minimum, and to approach a constant asymp-
totically for large $|L|$. This structure could be exploited if it
could be proved.

The asymptotic behaviour is readily shown, so what is required is to examine the stationary points. The derivation of the
gradient and Jacobian by hand is straightforward but tedious and
error prone. Mathematically it would be most convenient to express this problem in matrix notation, but no symbolic algebra
system available to me can do this. One system available to me,
MAPLE, is at least able to express the sum of squares: the model
above is expressed in its notation. Unfortunately, being in an
early stage of development, differentiation and other operations
through the summation notation is as yet unimplemented in MAPLE.
(Even so, observe how much work would be required for a program
to scan a Fortran subroutine that computed the residual sum of
squares numerically, and mechanically derive from it the MAPLE
representation of S.) Consequently, some of the preliminary algebra must be done by hand. To do this, I fixed q at a specific
value, and used the model equation above as a definition for the
residual r[i] in the ith observation. I squared this, then used
pattern matching substitution to replace the sums and cros-
sproducts of the known quantities with symbols representing their
sums over all observations. This then gave a symbolic expression
for S which could be differentiated to find the gradient and
Jacobian.

The derivative of S with respect to L is linear in L, so unless the coefficient of L vanishes, this derivative being zero
very simply determines L in terms of the B[j], and this value of
L can then be substituted in the equations for the rest of the
gradient being zero. The coefficient of L is a quadratic form in
the B[j], and consequently can only vanish if all the B[j] are
zero - in which case the derivative of S with respect to L
vanishes for all L.

The derivative of S with respect to B[k] is linear in all the B[j], including j=k. The coefficients of the B[j] are quadratic in L, and the terms independent of any B[j] are a term linear in L and a constant term. If q=1, and B[1]=0 as in the last paragraph, there is one value of L such that the derivative of S with respect to B[1] will vanish. It is readily seen that this stationary point is a saddle point of S considered as a function of L and B[1], and corresponds to the maximum observed in S considered as a function of L alone. If q>1, the coefficients of the terms independent of B[j] will be different in each derivative of S with respect to B[k], so it would be entirely fortuitous if when all B[j]=0, L could be chosen so all these derivatives vanished simultaneously. Thus the only remaining stationary points to consider are those where L is determined in terms of the B[j] by the derivative of S with respect to L vanishing.

If q=1, substituting for the L determined by the derivative of S with respect to L vanishing produces a linear equation for B[1]. Thus this second stationary point is the only other one that exists, and it is readily shown to be a relative minimum, confirming the qualitative behaviour observed empirically. If q>1, however, L is the ratio of two quadratic forms in the B[j], and substituting in the equations for the other derivatives of S vanishing produces nothing nice. The question is thus unresolved as to whether the observed behaviour can be proved in general.

4.3. Cost analysis of nested dissection

In analyzing the storage requirements of parallel implementation of a nested dissection code, there are four different kinds of regions that must be considered, depending on where a region is with respect to the boundary of the original problem. The total storage requirements is the sum of the number of regions of each kind, times the requirement for a region of that kind. If the original problem is a square grid, 2^s-1 points on each side, then at the kth stage this sum is explicitly

$$\text{STORE} = \{2(2^{s-k}-2)(2^{s-k}-1)\}\{43/4 \times 2^{2k}-4 \times 2^k+1\}$$

$$+ \{4(2^{s-k}-1)\}\{30/4 \times 2^{2k}-9 \times 2^k+3\}$$

$$+ \{2(2^{s-k}-2)\}\{21/4 \times 2^{2k}-15/2 \times 2^k+3\}$$

$$+ \{4\}\{13/4 \times 2^{2k}-7 \times 2^k+3\}$$

Analyses of other aspects of this program leads to similar expressions. Manual simplification of such expressions is tedious but straightforward. Surprisingly, such expressions are outside the class of expressions most symbolic manipulation systems can simplify, unless new indeterminates are manually substituted for 2^s and 2^k.

5. Conclusions

Although symbolic algebra systems can be of considerable help to the numerical analyst, use of current systems to work with programs in current numerical programming languages reveals shortcomings that limit the effectiveness of this tool. Many of the problems are due to deficiencies in the current generation of symbolic algebra systems, but some are due to unnecessary conflicts between the symbolic algebra system and the numerical programming language, and require changes in the numerical programming languages to cure.

6. References

1. W.S.Brown and A.C.Hearn, "Applications of Symbolic Algebraic Computation", Comp. Phys. Comm., Vol 19, (1979), pp207-215.
2. W.M.Newman and R.F.Sproull, Principles of Interactive Computer Graphics, Second Edition, McGraw Hill Book Company, 1979.
3. E.W.Ng, "Symbolic-Numeric Interface: A Review", Symbolic and Algebraic Computation, Lecture Notes in Computer Science 72, Springer-Verlag, Berlin 1979, pp330-345.
4. D.R.Stoutemeyer, "sin(x)**2 + cos(x)**2 = 1", Proceedings of the 1977 MACSYMA Users Conference, NASA CP-2012, Washington, D.C.

DISCUSSION

Summary by discussant (Brown)

Professor W. Morven Gentleman discussed applications of symbolic computation that are of interest to numerical analysts. Using a variety of examples, he described both successes and disappointments in this area. The focus of his talk was on the problem of fitting the symbolic and numeric parts of an application together.

Boyle   In symbolic computation there is a recognized need for data structures and list processing. What role do these play in numeric computation?

Gentleman   They are often needed (e.g., for certain sparse matrix algorithms), and the facilities ought to be the same in both contexts.

Kahan   Some problems are parameterized by order. Typically, the first few orders can be done by hand, and a symbol manipulator can only add one or two more orders.

Gentleman   I agree, but I rarely encounter this phenomenon. Normally I use symbolic computation to avoid errors in problems that could in principle be done without a computer.

SESSION 4
DATA STRUCTURES

Chair
M. Paul

Discussant
P. Kemp

THE RELATIONSHIP BETWEEN NUMERICAL COMPUTATION
AND PROGRAMMING LANGUAGES, J.K. Reid (editor)
North-Holland Publishing Company
©IFIP, 1982

DATA STRUCTURES FOR NUMERICAL QUADRATURE

L. M. Delves

Department of Computational and Statistical Science
University of Liverpool, Liverpool, England

1. INTRODUCTION

A very large number of numerical integration routines have been successfully
written in a very large number of programming languages from the structureless
(e.g. BASIC) to the highly structured (PL/I, PASCAL, Algol 68). It is clear that
access to a rich variety of data structures is not necessary for numerical
quadrature problems ("you can do anything in FORTRAN" is a true enough statement).
It is equally true that, here as in other fields, the facilities available within
a given language do influence what is provided in practice: if something can be
provided easily, it is likely to be provided early rather than late. This is borne
out by experience with the NAG quadrature chapter, D01, where despite a late start
(because of a late start?) and lack of manpower, the facilities in the Algol 68
version were for several years rather better than in the FORTRAN version; this
paper is based on experience in developing the Algol 68 quadrature chapter.

We ask here therefore, not what are the minimal necessary structures, but rather
the more interesting question: what data structures are natural in the field of
quadrature. Quadrature routines are of two main types: automatic and non-automatic.
The latter evaluates a specified quadrature rule Q_N chosen by the user

$$If = \int_R f(\underline{x}) d\underline{x} \approx Q_N f = \sum_{i=1}^{N} w_{Ni} \, f(x_{Ni}) \tag{1}$$

and returns $Q_N f$, perhaps with an error estimate; the former tries to choose N, and
possibly also the class of rules $\{Q_N\}$, in order to achieve a specified accuracy as
economically as possible. We consider here the more interesting automatic routines;
and these again are of two main types:

1) routines which attempt to treat the region as a whole,
    using a sequence of higher and higher order rules.

2) Routines which use a fixed order rule and subdivide
    the region until the accuracy criterion is met.

Natural development, and the survival of the fittest, has meant that in practice
routines of the former type usually subdivide the region when the going gets tough;
while the latter often try a (perhaps limited) sequence of rules within each chosen
subdivision, to try to avoid the need for a grid refinement. We therefore generally
have two parts to the code:

1) A strategy for subdividing the region if it seems necessary.

2) A strategy for approximating the integral, and estimating
    the error, in a given subregion.

These two parts are reasonably disjoint, and we consider them separately.

145

2.   SUBDIVISION

At an intermediate stage in the calculations, a typical routine has split the region R into a number of subregions R_i and is attempting to find an approximate result of the form

$$R = \sum_{i=1}^{P} R_i$$

$$I = \int_R f(\underline{x})\ dx = \sum_{i=1}^{P} \int_{R_i} f(\underline{x})\ d\underline{x} = \sum_{i=1}^{P} (Q_i f + \varepsilon_i) \qquad (2)$$

For some subregions, it will have computed $Q_i f$ and estimated ε_i for some rule Q_i; others will not have been treated yet. This situation is illustrated in one dimension in figure 1.

(1)	(2)	(3)	(4)	(5)	(6)	(7)	(8)
✓	✓	x	x	✓	x	✓	x

Figure 1   The state of the play during the calculations

x : region not yet treated

✓ : approximate integral evaluated

The basic requirement is to handle this collection of subregions and sub-results neatly; and the simple data structures needed to do so are well known:

subregion: = <u>structure</u> {<u>boolean</u> evaluated, <u>real</u> approxintegral, errorestimate,

<u>substructure</u> {description of subregion}}

region: = {linked list of subregions}

The form of the linked list depends on the details of the subdivision strategy. The simplest such strategy subdivides a given subregion into two, and then places one of those on the top of the pending list and deals with the other immediately; if no further subdivision is required, it proceeds to the subregion on top of the pending pile. The linked list in this case is of linear last-in-first-out form; that is, a simple stack, as illustrated in figure 2.

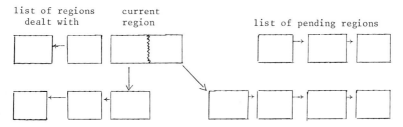

Figure 2.   The simplest subdivision strategy

More generally, a single subdivision stage may produce many new subregions (consider halving the mesh size in three dimensions); and the next subregion to be treated may be chosen globally (as the largest remaining subregion; or that with the largest contribution to the error, for example). Then a stack provides only a rather clumsy mechanism, and

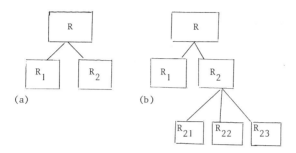

Figure 3.   A more general subdivision strategy

(a)   The region subdivided into two.

(b)   A further subdivision yields a total of four subregions.

the region is more naturally represented as the tree structure shown in figure 3, yielding a form for an individual node in the tree:

link:= <u>structure</u> {subregion, pointer to father, pointer to descendants}.

3.   TREATMENT OF INDIVIDUAL SUBREGIONS

The attention paid to individual subregions also depends on the subdivision strategy. A simple fixed-order routine may (for example) estimate the sub-integral I_i using single-panel Trapezoidal and Simpson rules, and subdivide if these fail to agree well enough; a second routine may evaluate I_i using a sequence of Gauss rules of increasing order, subdividing only as a last resort if the convergence achieved is unsatisfactory; while a third may try to construct a suitable sequence of rules based on its accumulated knowledge of the behaviour of the integral over the subregion. The latter two cases are the more interesting here; they imply a need to provide a variety of rule families for the routine to choose from (Newton-Cotes; Gauss-Legendre; Gauss-Laguerre; etc), and we therefore look at the characteristics of such families which need to be catered for.

These include the following:

(a)   <u>Obvious</u>

Rules of the type under consideration evaluate the integral

$$\int_R f(x)\ w(x)\ dx = \sum_{i=1}^{N} w_{Ni}\ f(\underline{x}_{Ni}) \tag{3}$$

Since they are often hard to compute, we need to <u>store</u> the points and weights $\underline{w}_N$, $\underline{x}_N$ for a range of N; of region R; and of weight function w(x).

(b)   <u>Pretty obvious</u>

Every family of rules has a set of associated parameters which extend its usefulness, usually via an appropriate mapping of the variables. For example, in one dimension, we use the following maps almost without thought:

Gauss Legendre: $\int_{-1}^{1} f(x)\ dx \quad \rightarrow \quad \int_{a}^{b} F(z)\ dz$

$$z = \tfrac{1}{2}[(b-a)x + (b+a)];\ F(z) = f(x)$$

Gauss-Laguerre:

$$\int_{0}^{\infty} e^{-x}\ f(x)\ dx \quad \rightarrow \quad \int_{a}^{\infty} e^{-\alpha z} F(z)\ dz$$

$$z = x/\alpha + a\ ;\ F(z) = f(x)$$

Modified Gauss-Laguerre:

$$\int_{0}^{\infty} e^{-x}\ f(x)\ dx \quad \rightarrow \quad \int_{0}^{\infty} F(z)\ dz$$

$$z = x\ ;\ F(z) = e^{-x}\ f(x)$$

Other useful rules derive from nonlinear maps. For example:

$z = x^{\alpha}$ To treat an algebraic singularity at $x = 0$

$z = 2\alpha/(x + 1) - \alpha$, to map $[-1, 1]$ onto $[0, \infty)$

As a result, it is helpful to associate a mapping with every rule. Moreover, some rules will share (unmapped) points and weights; these should therefore be stored elsewhere, and merely pointed to by the particular rule.

In most cases, the mapping procedures will be provided in advance; but it is possible for a quadrature routine to wish to accept procedures from the user, or even to construct them "on the fly", on the basis of information gathered about the integrand behaviour; this possibility causes its own problems.

(c)   <u>Obvious when you think about it</u>

Other, minor features of quadrature rules needing to be catered for in practice include:

   (i)   <u>Symmetry</u>: Any symmetries displayed by the region, the weight function $w(x)$, and the rule itself, lead to corresponding symmetries in the points x_{Ni} and weights w_{Ni} in (3), and a reduction in the storage requirements. In one dimension, the only relevant symmetry is that with respect to reflection about the mid point of the interval and this can be described by a single boolean.

   (ii) <u>Edge effects</u>: If a basic rule is used to form a compound rule, points on the edge of the region can be merged with neighbouring points. The situation is again simplest in one dimension, for which a rule is either open or closed.

Restricting discussion for simplicity to one dimensional rules, we arrive at the following description of a family of (one dimensional) rules:

family of rules = <u>structure</u> {<u>boolean</u> open or closed, symmetry,
        <u>pointer</u> to sets of points and weights,
        <u>procedure</u> (or <u>pointer</u> to procedure) to
        carry out mapping}

$$\text{set of points and weights} := \left\{ \begin{array}{l} \text{array} \\ \text{or} \\ \text{linked list} \end{array} \right\} \text{ of N-point rules}$$

N-point rule:= <u>structure</u> $\left\{ N; \; \underline{x}_N; \; \underline{w}_N \right\}$

4. MULTI DIMENSIONAL QUADRATURE

Although some of the detail above is specific to one dimensional problems, the discussion given remains valid in general for multi-dimensional routines written for a specific dimensionality. Also of interest are routines for a given region (sphere, cube, simplex) in a variable number of dimensions; and here, the facility of prime utility is the often-maligned <u>recursive procedure</u>. Lack of recursion makes even two dimensional product quadrature awkward in FORTRAN 66 or 77 ("let's keep two copies of the (one dimensional) routine in the library with different names"); its presence makes multi-dimensional routines trivial. Moreover, this is a <u>legitimate</u> (that is to say: <u>efficient</u>) use of recursion, since the recursion level needed is low (the dimensionality of the integral) and the amount of work carried out at each level relatively <u>high</u>, so that the overheads due to the recursive call itself, are (or should be) low.

5. IS IT ALL WORTH IT?

These structures look perhaps elaborate for such a simple purpose; are they over elaborate? The answer has to be: in the context of an automatic routine: not really. The subregion management problem discussed in section 2 always has to be solved, and linked lists provide a natural mechanism for doing so. Whether the discussion of section 3 is relevant depends on how ambitious the routine is; and most routines do in fact restrict themselves to a choice of one or a few rules, which are hard wired into the coding. An extreme example of this is given by coding of the form (and who has not seen it)

```
EST1  = GAUSS 3PT (FUNC, A, B)
EST2  = GAUSS 5PT (FUNC, A, B)
EREST = ABS(EST1 - EST2)
```

Here, GAUSS 3PT and GAUSS 5PT are probably (rather primitive) library routines, written to make the Gauss-Legendre rule readily available. Something better than this is needed if they are to be available for other purposes: multi-dimensional quadrature, or as "best" collocation points, or for the solution of integral equations; each of these applications either have to collect their integration rules from a central source, or keep duplicate copies in their own private format. The structures of section (3) give a natural way of storing the rules in a form suitable for general access, and are those used in [1].

6. HOW WELL DO CURRENT LANGUAGES COPE, THEN?

The structures map onto different languages in different ways, and with different degrees of difficulty

(a) <u>FORTRAN</u> is as usual do-it-yourself. Linked lists are normally implemented via integer pointers and delimiters, with points and weights all packed into a single (or two) arrays. The easiest way to store different families of rules with public access is via named COMMON blocks (unfortunately forbidden in many libraries because of overlay problems); but there is no way of attaching a mapping procedure to a rule. Recursion is not allowed.

(b) PASCAL. Supports explicit structures and linked lists in a natural way; but structures may not contain procedures nor pointers to procedures (procedures are not types); and variable-length arrays are not supported, so that points and weights have to be packed together in arrays in a way even more restrictive than FORTRAN.

(c) ALGOL 68. Provides all of the concepts introduced; but the scoping rules limit in practice the ability to construct new procedures dynamically (WG2.1 please note!)

(d) ADA. Provides structures and pointers; but as with PASCAL, procedures may not be put in structures. Nor may they even be passed as parameters except with difficulty, making every quadrature routine hard to write (or have I missed something? ADA experts please respond).

REFERENCES

1. NAG Algol 68 MK3 Library manual. NAG Central Office, 7 Banbury Road, Oxford, England.

DISCUSSION

Feldman   Why do you emphasise trees rather than other structures?

Delves   Because it is more efficient numerically.to treat large regions first, rather than to take regions in the order that they come a tree is a natural structure to use. The important point is that a variety of structuring facilities are needed in languages - the application programmer should be able to choose which he uses, and not be dictated to by the language designer.

Reinsch   A further reason for needing a structure which helps in dealing with large regions is that there is usually an upper limit on the number of function evaluations allowed, and a reasonable estimate for the integral is required when this allowance is exhausted.

Feldman   C provides the facilities you need.

Rice   Another structure which makes the treatment of regions in the right order easy is the queue, although this structure does have other, completely different weaknesses.

Kahan   While agreeing totally with the need for these data structures, it is important to be constructive in discussions with language designers. It is unreasonable to say, in effect "this is what we want, now you must provide it." Rather we must also explain why we need the features, because then the language designers may be able to provide a superior solution which we had not thought of.

Gentleman   While agreeing with most of your examples, I am curious about your use of quadrature to illustrate the need for recursion. The reason is not that recursion is inefficient, which it need not be, but that twice in the evolution of adaptive quadrature rules, thinking in recursive terms has led to significant oversights of the mathematical implications of the rules.

The first occurred in Mckeeman's original adaptive Simpson algorithm. Given the way Algol compilers implement recursion, this technique implies that the quadrature proceeds left to right across the interval. It was pointed out later, by Malcolm and Simpson, that processing the panels in a different order can produce the integral approximation at much lower cost.

The second is connected with your suggestion that multi-dimensional quadrature could be done recursively by dimension, the inner integral done adaptively serving as integrand to the next outer integral. Lyness has pointed out that this approach, while apparently plausible, can fail catastrophically because the inner integration achieves its economy by discontinuity in use of rules.

Thus, although it is true that recursion is an appropriate way to abstract an algorithm by hiding unnecessary details, care must be taken over what the abstraction hides. As in the two cases cited, the recursive conception of the algorithm tends to focus on local aspects of it, accentuating local implications, where frequently other formulations are more appropriate in dealing with the global aspects.

The issue is not the efficiency of implementation, but the way of thinking about the a'₁orithm, and the resulting human understanding of the implications.

Rice   I believe that the weakness of the recursive approach is not an artefact of a particular quadrature rule but that this appoach is inherently inefficient for higher dimensional integrals. However the more efficient methods will require even more complex data structures than are necessary for one dimensional integrals. The recursion approach is, however, a very useful quick way to attempt higher dimensional integrals when more efficient methods are not available.

Delves   One can indeed do better than the product of one-dimensional rules for multi-dimensional integrals. My reason for introducing recursion was to point out that, since there are circumstances in which it is useful, it should be available in a programming language.

Hammarling   In reply to your queries on the facilities offered by ADA, procedures may not be passed to parameters in a procedure call, but they can be passed as generic parameters to a generic procedure or package. Thus the lack of procedure parameters does not mean that every quadrature routine is hard to write.

*THE RELATIONSHIP BETWEEN NUMERICAL COMPUTATION
AND PROGRAMMING LANGUAGES, J.K. Reid (editor)*
North-Holland Publishing Company
©*IFIP, 1982*

DATA STRUCTURES FOR SPARSE MATRICES

J.K. Reid
A.E.R.E. Harwell

We describe the data structures that are widely used for
implementing sparse matrix algorithms and discuss language
features that would ease their implementation without loss
of performance.

1. INTRODUCTION

Sparse matrix algorithms depend for their efficiency on data
structures that store only the relatively few matrix coefficients that
are non-zero (or become non-zero during processing) and permit the
necessary operations to be performed efficiently. An appropriate
choice of data structure can be made only in conjunction with a careful
consideration of exactly what operations are required. Such fine
detail can be vital to the final efficient execution and leads me to
doubt the wisdom of "hard-wiring" complex data structures into
languages. Rather we should concentrate on having convenient building
blocks.

Most current sparse matrix software is written in Fortran using
"do-it-yourself" data structures. For a detailed explanation of what
can be done we refer the reader to the recent book of George and
Liu (1981). Indeed my own experience is of this mode of working.
Here we will lead the reader through a whistle-stop tour of data
structures used for sparse matrices and some of the operations applied
to them, introducing as we go those language elements that seem
necessary. In this short paper we necessarily omit some details.
Our language is informal and is based on elements of Algol 60, Simula,
Algol W and Pascal. We summarize the language elements that we
conclude to be needed in section 8. The reader should regard the
syntax of my language as "purely illustrative" after the fashion of
the X3J3 Fortran committee.

2. SPARSE VECTORS

For simplicity we begin by considering the storage of sparse
vectors. One motivation is in order to store the rows (or columns)
of a sparse matrix as a set of sparse vectors, but the right-hand
side of a sparse set of linear equations is often sparse so it is
worthwhile to consider the problem in its own right.

A sparse vector may be held in a full vector of storage. This is
wasteful of storage but is often used temporarily because of the speed
with which specific components may be accessed.

The obvious (and widely used) alternative is to pack the non-zero
values into a real array and their indices into a parallel integer
array. However the intimate connection between the real and the
integer suggests that we should rather have a single array whose
components are "records" containing a real and an integer. Such a
record might be declared thus

record nz [r: real, j: integer]

and an array of such records might be declared thus

153

$$\text{array x(len):} \quad nz$$

where len is an integer giving its length.   The kth non-zero could
then be referenced as a whole as

$$x(k)$$

or its two parts could be referenced as

$$x(k).r \quad \text{and} \quad x(k).j$$

Now let us consider what operations we may wish to apply to a
sparse vector.   Taking its inner product with a full vector f is
straightforward:

```
prod:=0;
for k:=1 to len do
      begin
          xj:=x(k).r;
          j:=x(k).j;
          prod:=prod+xj*f(j)
      end;
```

However, taking the inner product between two such vectors is not so
simple.   The obvious nested loop

```
prod:=0;
for k:=1 to lenx do
   for l:=1 to leny do
       if x(k).j=y(l).j then
           prod:=prod+x(k).r*y(l).r;
```

we regard as two inefficient to countenance.   One solution is to
order the non-zeros in each vector by components, so that the inner
product can be calculated by scanning the two packed vectors in phase.
A better solution is to keep a full vector, say f, filled with
zeros, use it temporarily to store one of the sparse vectors, then
restore it to zero:

```
prod:=0;
for k:=1 to leny do  f(y(k).j)=y(k).r;
for k:=1 to lenx do  prod:=prod+x(k).r*f(x(k).j);
for k:=1 to leny do  f(y(k).j):=0;
```

Similarly, when adding a multiple of one sparse vector to another, we
may either scan the two in phase if we know that each is ordered or we
may use a temporary copy in a full array.

3. GENERAL SPARSE MATRICES

We now consider the storage of sparse matrices that have no
special features (e.g. bandedness).   Unlike the vector case, a full
array is impractical even for temporary use because it requires too
much space.

Perhaps the most convenient way to specify a sparse matrix A is as its set of non-zeros with their row and column numbers. Here a suitable record is

$$\text{record nza[r: real, i,j: integer]}$$

and the whole sparse matrix, with nnz non-zeros could be held in

$$\text{array A(nnz): nza}$$

This form can be used directly for forming y:=Ax if x and y are full vectors:

```
for i:=1 to n do y(i):=0;
for k:=1 to nnz do
    begin  i:=A(k).i ;
           j:=A(k).j ;
           aij=A(k).r;
           y(i):=y(i)+aij*x(j);
    end;
```

but is uneconomical for almost anything else one might want to do. The problem is that access is often wanted to a single row or column or a small set of rows and columns and the whole set of records would need to be scanned to obtain this.

Two solutions to this problem are in wide use. The first involves holding the matrix as a set of sparse vectors, one for each row (or column). A single temporary full vector permits us to add a multiple of one row to another efficiently, as sketched at the end of section 2, and this is the basic operation of Gaussian elimination. Once the elimination is complete and we have triangular factors stored by rows, forward and backward substitution for solving a set of equations is easy if full vector storage is used for the original right-hand vector b and the solution x since at each stage we essentially need the inner product of a packed sparse vector with a full vector.

The language problem here is that we need an array of arrays of varying lengths. In fact this can be achieved by working with an array of pointers which point to the individual rows. One may think of the name of a conventional array as a fixed pointer to the actual array. Here we need an array of pointers

$$\text{array row(n): pointer}$$

and a declaration for row i of the form

$$\text{array row(i) (len(i)): nz}$$

where len(i) is the length of row i and nz is the record we introduced in §2. The name of the array is row(i). This would permit us to refer to the kth non-zero in row i as

```
                        row(i)(k)
```

and its column number **as**

```
                    row(i)(k) .j
```

When a fill-in takes place it is usual to allocate a fresh array for the new
row and temporarily waste the storage used for the old version. Ideally then
our language should permit us to dynamically create and destroy arrays.   The
syntax we have already used permits this.   Creating the new

```
              array row(i)(len(i)):   nz
```

would imply destroying the old array referenced by row(i). We would rely on the
system to manage the space efficiently.

Another structure involves working with records including
integer pointers to other non-zeros in the same row or column.   These
pointers may be additions to the record or may replace the row and/or
column numbers. For instance the Harwell code MA17E holds the sparsity
pattern of the lower triangular part of a symmetric matrix in the
equivalent of an array of records (i,j, next-in-row, next-in-column)
while the Harwell code MA18A, for unsymmetric matrices, works with
the equivalent of records (r, next-in-row, next-in-column).

Storage management for this second structure is simple because
we have a single array of records. Fill-ins can be put in any free
record and the pointers adjusted to correspond.   If we are not
keeping both factors in the array, then records will be freed by
eliminations and the pointers can be used to chain them together in
a "free list" ready for reuse for later fill-ins. Disadvantages,
however, are that access to a row (or column) will require access to
scattered locations in storage, likely to be more expensive with
modern hierarchical systems (e.g. fast buffers, paged virtual
memories), and that the pointers need longer integers than the row and
column numbers they replace (2-byte integers usually suffice for row
and column numbers).

4. VARIABLE-BAND MATRICES

Gaussian elimination without interchanges causes no fill-in ahead of the
first non-zero in a row or the first non-zero in a column. We may therefore
hold the rows of the lower triangular part as full vectors spanning from the
first non-zero of the row to the diagonal.   The columns of the upper-triangular
part may be held similarly, but by far the most common case is where A is
symmetric and here the same vectors may be regarded as representing the columns
of the upper triangular part.

The data structure here is static, but the language we described in §3 is
suitable with the rows now of ordinary reals but of varying lengths.

5. FINITE-ELEMENT MATRICES

A finite-element matrix has the form

$$A = \sum_k B^{(k)} \tag{5.1}$$

where $B^{(k)}$, the contribution from the kth element, is non-zero in only a small number of rows and columns. $B^{(k)}$ can therefore be stored as a small full array plus an integer array labelling where it fits into A. It is possible to form A as an overall sparse matrix, but an alternative is to keep the form (5.1). If a variable is contained in only one $B^{(k)}$ then it can be eliminated by operations only within $B^{(k)}$, and the resulting overall reduced matrix will be of the form (5.1) with this one $B^{(k)}$ altered. This process is usually known as "static condensation". Having performed all possible eliminations of this kind, we may add together all $B^{(k)}$ involving a particular variable into a new "generalized element" involving all the variables of its constituents. Now the variable may be eliminated by static condensation within the new element. The process may be continued until we have a single generalized element.

The simplest form of this algorithm is Irons' (1970) frontal method, where each generalized element is formed by adding in the next $B^{(k)}$ in the original sequence. The most general case can be represented as a tree whose leaf nodes are associated with the original $B^{(k)}$ and whose internal nodes are associated with generalized elements formed by the addition of elements and generalized elements followed by the elimination of internal variables. If this tree is searched in a depth-first order, then the generated matrices will be needed in a last-in/first-out order and so can conveniently be stored on a stack.

This structure occurs naturally for finite-element problems but it can also be used for other symmetric matrices. Whenever a variable is eliminated we generate a new "element" $B^{(k)}$ whose rows and columns are those active in the step and we keep it rather than immediately add it into the overall sparse matrix. It is particularly advantageous when analysing a sparse structure without keeping the reals because the total length of the index lists for the $B^{(k)}$ does not increase (i.e. there is no possibility of storage running out because of fill).

The language elements we have described permit all these data structures to be constructed. The tree could be held as a set of records containing pointers to other tree nodes, to real arrays holding matrices $B^{(k)}$ and to integer arrays of indices. The stack that is created temporarily could be held as an array of pairs of pointers referencing arrays of reals and indices.

6. PARTITIONED MATRICES

If the matrix A can be permuted to block lower triangular
form

$$
\begin{pmatrix}
A_{11} & & & \\
A_{21} & A_{22} & & \\
A_{31} & A_{32} & A_{33} & \\
\cdot & \cdot & \cdot & \cdot & \cdot & \cdot \\
\cdot & \cdot & \cdot & \cdot & A_{rr}
\end{pmatrix}
\tag{6.1}
$$

then sets of equations Ax=b can be solved by block forward
substitution, by solving successively

$$
A_{ii}\,x_i = b_i - \sum_{j<i} A_{ij}\,x_j, \quad i=1,2,\ldots,r. \tag{6.2}
$$

The diagonal blocks need a data structure that permits fill-in
whereas the off-diagonal blocks require only access by rows (or block
rows) for the matrix-vector products in (6.2).

Similar considerations apply to blocked irreducible matrices
if implicit storage of the off-diagonal blocks of the factorized
matrix is used. For example George and Liu (1981), working with
irreducible symmetric matrices, store the diagonal blocks as a variable
band matrix and the rows of the block lower triangular part of A
as a set of packed vectors.

Again our language elements appear suitable. We would require
real arrays for the rows of the diagonal blocks and arrays of
(real, integer) records for the off-diagonal parts, and two arrays of
pointers.

7. HYPERMATRICES

An approach that has been used with success by Argyis and
his collaborators is to subdivide matrices into blocks and store only
those blocks that have at least one non-zero. Access to the blocks
is obtained through a full matrix of pointers, with null pointers
where the blocks are zero. The technique can be nested, with the
first level pointers leading to matrices of second level pointers, and
so on, with only the last level pointing to real numbers. The
actual block sizes are usually chosen to suit hardware characteristics.

For this application we need arrays of pointers referencing
arrays of pointers, etc. eventually reaching an ordinary real array.
Our syntax permits this, if a little clumsily. An individual non-
zero might be

$$A(m,n)(k,\ell)(i,j)$$

8. SUMMARY

To summarize, we believe that the efficiency of present data structures, hand-coded in Fortran, could be maintained with an improvement in clarity if the following were available

    a) simple records that group a few scalar variables

    b) pointers to components of data structures

    c) arrays of pointers and arrays of records

    d) dynamic storage allocation.

9. ACKNOWLEDGEMENTS

I would like to thank John Lloyd for the discussions we had at the IMA sparse matrix conference at Reading in July 1980 and to thank Iain Duff for reading my first draft and suggesting substantial improvements to the presentation.

10. REFERENCES

George, A. and Liu, J.W. (1981). Computer solution of large sparse positive definite systems. Prentice-Hall, New Jersey.

Irons, B.M. (1970). A frontal solution program for finite element analysis. Int. J. Numer. Meth. Engng. $\underline{2}$, 5-32.

DISCUSSION

Delves I would like to make two comments. The first is that the required sparse matrix structures can be provided in a natural way within Algol 68. Picking out two for illustration, possible realisations are:

(i)   Sparse vectors

```
mode nonzero = struct (real r, int j);
[1:lenx] nonzero x; [1:leny] nonzero y   #sparse vectors# ;
```

With these declarations, the code to compute the scalar product of two vectors becomes

```
[1:n] real f; real prod := 0.0;
for k to leny do f[j of y[k] ]:= r of y[k] od;
for k to lenx do prod +:= r of x[k] * f[j of x[k] ]od;
for k to leny do f[j of y[k] ]:= 0.0 od;
```

(ii)   Sparse matrices stored as rows of sparse vectors

```
mode sparsevector = ref [ ] nonzero #merely a shorthand#;
[1:n] sparsevector array #declares a sparse matrix#;
```

To allocate storage to array, given n[i] nonzeros in row i:

```
for i to n do
array[i] := {heap or loc} [1:n[i] ]nonzero od;
```

To access kth nonzero in row i:

value := array[i][k];

The other examples follow similarly; the code segments are given here to show how natural (or unnatural!) they would look in Algol 68.

The second comment follows from the first. The realisations above are not unique; others can be constructed to carry out the job. Moreover, as you point out, different data structures may be more convenient, or more efficient, in other sparse matrix problems. Now what happens if a need for a sparse matrix type (say) is accepted by X3J3 and frozen into FORTRAN 8x with a particular structure? The answer: yet another element of the language to put users off, and a structure which will not always provide what is really needed. Far better to provide the facilities for users to set up their own data types, and operators upon them, which they can tailor to their own needs.

van Wijngaarden   It is not necessary to unpack the elements into a full vector if the structure has three fields, value, index and a reference to the next non-zero, i.e. a linked list.

Kahan   This technique only works if the elements are sorted.

van Wijngaarden   With a linked list it is quite feasible to maintain the sorted order.

Reid   In practice the use of temporary full storage doubles the speed. In every example we have looked at, this increased speed is important.

G. Paul   The technique used in NASTRAN is to maintain a vector whose size is determined by the estimated fill-in. This vector is then gradually filled as new non-zeros appear in the course of the computation. This technique is efficient in a scalar environment, but in a vector processing environment a number of problems arise. For example storage alignment could well cause loss of efficiency. In a vector processing environment one might well keep track of non-zero elements using a bit vector, and not use structures at all.

Smith   There will be an example in my paper.

Lake   When a language with the required features is designed and the syntax completely specified, will it be effectively ALGOL-68, or could it be something smaller?

Reid   That must be left to the language designers to answer. In this session we are trying to point out that certain features are needed, and why.

Lake   Have you considered the relationship of these facilities to machines like the Cyber 200's which have architectural features supporting sparse matrices?

Reid   We only considered conventional machines. For vector machines it might be worth looking at different problems, for example finite element calculations, and investigating whether further data structures are required.

Fosdick   When a language with these features is designed, who would use it - would it be intended for scientific end users or for software developers like NAG?

Reid   I would expect the advanced features to be used only by software developers.

Fosdick   But the end user wants to write simply A*B to multiply matrices, rather than having to handle details via looping constructs.

Reid   This could be provided by a simple extension - the end user need not see the implementation details.

Meissner   One of X3J3's problems is that we perceive requests for features at all levels to be included in a single language. Fortran is asked to include features so that users can write A*B (where A and B are sparse arrays) plus many other facilities. We have the difficulty of deciding how to incorporate these diverse (and often conflicting) requirements into a single language - the core-plus-modules approach has evolved in order to address this problem.

Ford   A useful analogy is the English language. It is used both by generalists and specialists but does not need a core-plus-modules structure. Why cannot Fortran be the same?

Gentleman   Simply asking for dynamic storage allocation is not enough. Sparse matrix computations use dynamic storage in a well understood and well disciplined way, doing relatively little calculation on each entity obtained dynamically. As a consequence, efficient storage managers can be designed and built for this application, but even so storage management can be a significant part of the cost. The use of a single, general, language-provided storage manager such as the Algol-68 heap can imply wholly unacceptable overheads. Consequently we must be able to choose an appropriate storage manager to use. One way to do this would be to select an application-dependent storage manager from a library of managers written in the host language. Unfortunately, it is a consequence of many strongly typed languages that the storage manager must be provided as part of the runtime support of the language system, because it is impossible to express a storage manager in the language itself.

Feldman   If one drops the requirement for strong typing, it is possible to take this approach.

Reid   Any useable language ought to be capable of producing code as efficient as my Fortran examples.

Gentleman   Unfortunately a general dynamic storage manager has to be able to handle the worst cases, and is therefore inefficient.

Dekker   One might also look at the problem of transposition of a sparse matrix, which appears to require the building of a completely new structure for the transposed matrix. This can often be avoided by replacing column operations by certain other row operations. For example, the product of a transposed matrix and a vector can be calculated not only as a collection of inner products involving columns of the original matrix but also as linear combinations of its rows.

*THE RELATIONSHIP BETWEEN NUMERICAL COMPUTATION
AND PROGRAMMING LANGUAGES, J.K. Reid (editor)
North-Holland Publishing Company
© IFIP, 1982*

ARRAY PROCESSING FEATURES IN THE NEXT FORTRAN

Brian T. Smith[1]

Applied Mathematics Division
Argonne National Laboratory
Argonne, Illinois 60439

This paper summarizes the array processing facilities being
proposed for the next Fortran standard by the Fortran Standard's
committee X3J3.   The facilities include: automatic arrays;
element-by-element array operations; array expressions; new
conditional control constructs for logical arrays; new statements
for packing and unpacking arrays; virtual arrays; and a large
collection of new intrinsics for arrays. The paper concludes with
a comparison of the Fortran facilities with the ALGOL68-TORRIX68
system for computations in linear vector spaces.

1. INTRODUCTION

The purpose of this paper is two-fold: to summarize the new array
processing features[22] being proposed by the X3J3 Fortran Standards Subcommittee
of the American National Standards Institute (ANSI) for the next ANSI Fortran
Standard; to briefly compare these new Fortran facilities with those of the
ALGOL68-TORRIX68 package for the application area of linear vector spaces.

The facilities being introduced into Fortran are currently being developed
and described by the Fortran Subcommittee X3J3. Their work is not complete at
this time, but the committee expects to finish a draft of the proposed standard
within two years. It, therefore, may be considered premature to discuss and
compare the proposed facilities with an existing language, but it is precisely at
this time that the committee needs feedback from Fortran users on the appropriateness of its proposals.

Thus, part of the first purpose of this paper is to create feedback for the
X3J3 Fortran Committee about the array processing facilities. The committee has
completed its survey, tutorial, and proposal phases for the next standard, and is
now gathering together into a coherent language the proposals that have been
passed up to now. The committee believes that the array processing features
proposed to date represent a useful and nearly complete array processing
facility. As such, the committee is attempting to describe and place the
proposed features into the new "core+modules" architecture of the next
standard[4]. Comments from the Fortran user community concerning the
appropriateness of the facilities and their place in the new architecture are
welcomed at this time.

A complete survey of array processing facilities in current and proposed
languages would of course yield a very large document of somewhat boring details
of each language. Such a survey might range from the simplest implementation of
matrix packages, including procedure packages and libraries such as
EISPACK[6,15], IMSL[7], LINPACK[5], NAG[11], and PORT[14], to languages such as
ADA[10], ALGOL68[17,19], APL[8], LWG Fortran[9,21], PL/1[1,3], and VECTRAN[12]
with modest array facilities, to vector-matrix packages such as the ALGOL68-
TORRIX68[18] system, an extensive abstract linear algebra package built on top of
ALGOL68. Because one of our purposes is to introduce the proposed Fortran
facilities, we will limit the comparison to what can be considered to be the

ultimate in a linear algebra package, namely the TORRIX package. Such a limitation will bring out the major points in the discussion.

The remainder of this paper briefly reviews the facilities in the proposed Fortran [22, chapter 3] and compares these features with the ALGOL68-TORRIX68 system.

2. PROPOSED ARRAY PROCESSING FEATURES IN THE NEXT FORTRAN

2.1 Introduction

The standards committee is committed to what has become known as the "core+modules" architecture for the next Fortran language. To quote from the committee's working document [22, chapter 14], "the core is a complete and consistent language comprising a set of language features, conforming to a set of established criteria, sufficiently rich for the implementation of most applications." Modules represent both the obsolete parts of the language and the growth areas for the language, either extension or application modules.

The more extensive array processing facilities will probably reside in the growth modules, and the more basic array facilities will reside in the core module. Since the committee has not decided how they will partition the array features into the language modules, we will ignore the distinction between the core and other modules for the time being.

As a note of caution to the reader, the syntax used to describe the array features must be considered as illustrative. In some cases, the committee has not decided on the most appropriate syntax, and in any event, the documents produced by the committee must be considered draft proposals for the next Fortran standard up to the time they are approved by the ANSI committee X3.

The current array processing facilities have drawn from array processing facilities in other languages. For instance, there are many similarities of the proposed facilities to features in languages such as ADA[10], APL[8], PL/1[1,3], and extended Fortran languages such as VECTRAN[12] and the Fortran extensions defined by the LWG[9,21].

The remainder of this section briefly summarizes the new constructs and facilities for array processing that have been added to Fortran. The reader is referred to the X3J3 document [22] for the precise details. In lieu of detailed descriptions, examples are given to illustrate the expected uses. The topics in order are automatic arrays, array specifications, operations on arrays, new conditional statements for logical arrays, packing and unpacking of arrays, virtual arrays, elemental procedures and the new array intrinsics, and array constants. The section ends with programs written in the new Fortran which perform the familiar LU factorization of a square matrix. All of the examples use the new program form and new control structures[22, Chapters 1 and 5].

2.2 Automatic Arrays

In addition to the various ways of specifying arrays in Fortran 77[2], the proposed array processing facility permits the specification of an array with a dimension bound that is non-constant and whose name is not a dummy argument. Such specifications are permitted only in subprograms where the expressions, representing the range declarators, can be evaluated upon entry to the subprogram. On entry, the array is allocated space, and upon execution of a RETURN or END statement the space is released to the operating system. Such arrays are called automatic arrays. They cannot be saved (appear in a SAVE statement) or initialized by a DATA statement. For example, the array SCRATCH in the following subprogram is an automatic array:

```
SUBROUTINE  EXAMPLE(N, A)
REAL  A(N, N), SCRATCH(N)
    :
    :
RETURN
END
```

2.3 Specifications of Arrays and Array Sections

An array is defined to be a non-empty sequence of data of the same type and often has a name (an array that is an expression does not have a name). The type may be integer, real, complex, logical, or character. (The committee is considering new data types, such as bit and string types, that could be added to the list here, but the interaction of new facilities with the array facility has not been thoroughly considered by the committee as yet.) A named array is specified using the usual Fortran type specification statements within the language. Using the new array facility, an entire array may be referenced in many contexts not permitted by the previous standard. Parts (sections or slices) of an array may also be specified in a subset of these contexts.

Arrays have shape, which is defined as the number of dimensions and size of each dimension. An array's shape is denote by $(d_1, d_2, \ldots, d_n)$ where d_i is the size of the i-th dimension. Consistent with Fortran 77[2], the dimension bounds may be negative, zero, or positive. Two arrays are said to conform if their shapes are the same. Note that this does not require that the two arrays have the same dimension bounds (smallest and largest permitted subscripts) along each dimension.

The shape of an array may be changed when referenced, provided the referenced elements are not beyond the original dimension bounds declared in the specification statement. Such an array reference is called a section. It behaves in all contexts just like the full array except that only a portion of it is referenced or changed. For example, the entities A(2,*), A(2,4:6), A(-*,2), and A(3,IP) using the specification statement

INTEGER IP(4), A(-1:5,10)

are valid instances of array sections; the first is the array of shape (10) consisting of the second row of A; the second is an array of shape (3) consisting of three elements in the second row of A; the third is an array of shape (7) consisting of the second column of A, with the elements in reverse order (the symbol – before the symbol * indicates the reverse selection of the elements); and the last one is an array of shape (4) consisting of four of the elements of the third row of A, the elements being the A(3,IP(I)) for I from 1 to 4, provided the four integers IP(I) are between –1 and 5. (Proposals for shifted array sections, for example, A(*+1,2) yielding an array of shape (6) starting from the element A(0,2), are currently being considered by the committee.)

The use of the last section A(3,IP) is restricted in that it cannot be passed as an actual argument. Otherwise, array sections can be used anywhere an array can be used in executable statements.

A section does not inherit the dimension bounds of its parent. Instead, the range of each dimension is from 1 to the size of the particular dimension. Thus, A(*,2) has dimension bounds 1:7.

There is one other way to reshape an array, using the IDENTIFY statement. It provides a new name for the reshaped array and will be described in Section 2.7.

A scalar is considered to be an array in two ways, depending on the context. First, as a degenerate section, it is a degenerate array whose shape has zero dimensions. With this interpretation, a scalar expression involving a subscripted array element can be considered as a conformable expression. Second, as a "constant" array, it is an array, all of whose elements are equal to the scalar, and whose shape is determined by the conformity rules of the context in which it used. For example, its shape may be determined by the "other" operand of a binary operator as in the expression A+1.0, or by the "assigned-to" entity in an assignment statement, as in A=2.0, or by the rules for valid argument lists to an intrinsic function. This process of a scalar taking the shape determined by its environment is sometimes called <u>broadcasting</u> [20, p. 17]. We will see an example of broadcasting in Section 2.4

An array name can appear in most contexts in which a scalar can appear in Fortran 77. Thus, it may appear in an expression, in the i/o lists of input/output statements, DATA initialization statements, and so on. However, array sections may not appear in certain non-executable statements where array names may appear, such as DATA initialization statements, SAVE statements, and so on.

2.4 Operations on Arrays

All arithmetic, logical, character, and string operators extend element-by-element to arrays. Thus, the expression $A_1 * A_2$, for example, where A_1 and A_2 conform, creates an array of the same shape as A_1 whose i-th element is the product of the i-th element of A_1 and the i-th element of A_2. (The element order is the usual column major order, but the processor need not compute the product in that order.) The result of any expression is an array whose shape is that of the operands and whose lower subscript bound in each dimension is one and whose upper subscript bound is the size of the array along the particular dimension.

Similarly, the Fortran intrinsics extend to handle arrays of appropriate types. Thus, for example, SQRT(A) where A is real or complex is an array of the same shape as A whose i-th element is the square root of the i-th element of A. Again, the element order is column major order, but the processor need not evaluate the results in that order.

Thus, expressions can be formed from the usual Fortran operators, Fortran intrinsics, entire arrays, array sections, and scalars provided all of the arrays are conformable. Degenerate empty arrays are not permitted in such expressions, even when all arrays within the expression are empty. An empty array would naturally occur when the size along one of the dimensions of the array is zero. For example, the assignment statement in the following loop would be invalid when I reaches N:

```
DO (J = I,N)
   A(J, I+1:N) = A(J, I+1:N) - MULTIPLIER * A(I, I+1:N)
REPEAT
```

Otherwise, the loop on J must be terminated at N-1, which might be inappropriate if the loop contained other statements to be executed for J equal to N. It would be better to interpret such statements as null statements. This is an oversight on the part of the committee and should be corrected.

The semantics for an assignment statement $\underline{v} = \underline{e}$ are that the expression $\underline{e}$ is evaluated first, then any conversion of type from that of $\underline{e}$ to that of $\underline{v}$ is performed, and finally the result is assigned to $\underline{v}$. This definition therefore permits the expression $\underline{e}$ to reference all or parts of the array $\underline{v}$ without surprising effects. For example, the assignment statement

$$A = A / A(1,1)$$

scales the entire array by the first element so that if A(2,2) equals A(1,1) in the original A, then the final A has both A(1,1) and A(2,2) equal to 1. Note that in this statement the scalar A(1,1) is broadcast into a "constant" array of the same shape as the dividend; cf. Section 2.3 above.

The relational operators .EQ., .NE., .LT., .LE., .GT., and .GE. extend to array operands. That is, the expression A_1 .relop. A_2, where A_1 and A_2 are conformable, yields a logical array of the same shape as the operands such that the i-th element of the result is the value of the scalar logical expression

$$A_1(\text{i-th element}) \text{ .relop. } A_2(\text{i-th element}) \quad .$$

For example, A .NE. 0.0 where A is a 10 x 10 array yields a 10 x 10 logical array whose (I,J)-th element is the value of the scalar logical expression A(I,J) .NE. 0.0 .

The above example also illustrates the use of a scalar as a constant array. Here, the scalar 0.0 is broadcast to an array of the shape of A, thereby conforming with A.

2.5 New Control Structures for Logical Arrays

Two new control structures are being added to the language that permit array assignment statements (in addition to the new statements PACK and UNPACK described in the Section 2.6) to be controlled or masked by a logical array. The new control structures use the keywords WHERE, OTHERWISE, and END WHERE, and play analogous roles for arrays as the current statements with keywords IF, ELSE, and END IF play for scalars. Note that the conditional expression in an IF, or ELSEIF statement does <u>not</u> extend to a logical array expression.

First, the logical WHERE statement, analogous to the logical IF statement, permits a single assignment statement (as well as the the PACK and UNPACK statements described in Section 2.6) to follow the right parenthesis enclosing the logical array expression. The result of the logical expression must be conformable with the "assigned-to" array in the array assignment statement. Thus, its syntax is

WHERE(<u>lae</u>) <u>v</u> = <u>e</u>

As for the semantics, the assignment <u>v</u> = <u>e</u> is limited by the mask <u>lae</u>. That is, wherever <u>lae</u> is true, the corresponding element of <u>v</u> is assigned the corresponding element of <u>e</u>, and wherever <u>lae</u> is false, no assignment to <u>v</u> is made. For example, the statements

B = A
WHERE(A .NE. 0.0) B = B/A

yield a two-dimensional incidence array B for A, such that B has the value 1.0 wherever A is non-zero and zero elsewhere.

A natural extension of the logical WHERE statement is the block WHERE construct. Its form is

WHERE (<u>lae</u>)

    A block of assignment statements $\underline{a}_w = \underline{e}_w$, or
        PACK statements PACK($\underline{v}_w$, $\underline{a}_w$, $\underline{ct}$), or
        UNPACK statements UNPACK($\underline{a}_w$, $\underline{v}_w$, $\underline{ct}$)
    called the WHERE block.

[OTHERWISE

    A block of assignment statements $\underline{a}_o = \underline{e}_o$, or
        PACK statements PACK($\underline{v}_o$, $\underline{a}_o$, $\underline{ct}$), or
        UNPACK statements UNPACK($\underline{a}_o$, $\underline{v}_o$, $\underline{ct}$)
    called the OTHERWISE block.]

END WHERE

where <u>lae</u> is a scalar or logical array expression that is conformable to the arrays $\underline{a}_w$ and $\underline{a}_o$ in the WHERE and OTHERWISE blocks. (Throughout the remainder of this paper, language constructs that are optional are enclosed in square brackets [...].) The semantics are straightforward but may have surprising results when the same array is referenced in both the WHERE and OTHERWISE blocks. Note that the OTHERWISE block is optional.

As for the semantics, the statements in the WHERE block are executed in order followed by the statements in the OTHERWISE block. The elements of the logical expression $\underline{lae}$ that are true specify that the corresponding elements of the entities $\underline{a}_w$ in the WHERE block are to be defined with the corresponding elements of $\underline{e}_w$ (or are to be defined with certain elements defined by the PACK and UNPACK statements). In addition, the elements of the logical expression $\underline{lae}$ that are false specify that the corresponding elements of the entities $\underline{a}_o$ in the OTHERWISE block are to be defined with the corresponding elements of $\underline{e}_o$ (or are to be defined with certain elements defined by the PACK and UNPACK statements). The semantics for the PACK and UNPACK statements in a WHERE or OTHERWISE blocks are deferred until Section 2.6.

Because statements in both blocks can be executed, these semantics may yield surprising results when the arrays $\underline{a}_w$ or $\underline{a}_o$ appear in the expressions $\underline{e}_o$ or $\underline{e}_w$ respectively. Of course, this difficulty does not occur with the IF- and ELSE-blocks because, for each execution of the IF-THEN statement, only one of these blocks is executed.

The semantics become further complicated when the expressions $\underline{e}_w$ or $\underline{e}_o$ contain references to array transformational functions that are either Fortran intrinsics or user-defined function subprograms with array arguments returning arrays. The semantics in such a case are that any transformational functions referenced in the WHERE or OTHERWISE blocks are fully evaluated, including all arguments, without reference to the mask $\underline{lae}$. For example, if F is a transformational function with one array argument, the statement in the WHERE block in the program segment

```
WHERE( A .NE. 0.0 )
    C = F( B/A )
END WHERE
```

is not protected from a division-by-zero exception when some element of A is zero. Of course, these semantics for the evaluation of the transformational function apply to the logical WHERE statement as well.

The committee is treating logical arrays in these contexts as bit array masks. Fortran 77 of course requires logical arrays in a storage sequence (such as in an EQUIVALENCE or COMMON statement) to consist of an ordered sequence of numeric storage units. Since the next standard is relegating all storage association rules for entities of different types to the compatibility module, a processor could implement, in order to improve efficiency, logical arrays as bit arrays for all but the compatibility module. Indeed, the committee is attempting to avoid any language construct or rule that would prohibit such a compact and potentially efficient implementation of logical arrays.

2.6 Packing and Unpacking Arrays

Two new statements are being proposed to permit the packing and unpacking of arrays into and from one-dimensional arrays. For reshaping an array $\underline{a}$ into a one-dimensional array $\underline{v}$, the statement

$$PACK(\underline{v}, \underline{a} [, \underline{ct}])$$

where the size of $\underline{v}$ is at least the size of $\underline{a}$, both $\underline{a}$ and $\underline{v}$ are of the same type, and $\underline{ct}$ is an optional integer variable, copies the elements of $\underline{a}$ in array storage order to successive elements of $\underline{v}$. If the PACK statement is within the scope of a logical WHERE statement, a WHERE block, or OTHERWISE block, the elements taken from $\underline{a}$ are only those selected by the mask $\underline{lae}$ for the logical WHERE statement and the WHERE block, and the mask .NOT. $\underline{lae}$ for the OTHERWISE block. Of course, $\underline{lae}$ and $\underline{a}$ must be conformable. The size restriction in these latter cases requires that the number of elements selected by the mask does not exceed the

size of v. In all cases, the integer ct, if present, returns a count of the
number of elements selected.

Similarly, the UNPACK statement

UNPACK(a, v [, ct])

where a is an array, v is a one-dimensional array, and both a and v are of the
same type, assigns consecutive elements of v to elements of a in array storage
order. If the UNPACK statement is within the scope of a logical WHERE statement,
a WHERE block, or OTHERWISE block, the appropriate mask (as defined above for the
PACK statement) specifies the elements of a that are assigned consecutive
elements of v. The size restriction in these latter cases requires that the
number of elements selected by the mask (or the size of a if there is no mask)
must not exceed the size of v. Of course, the logical expression lae and the
array a must be conformable. The integer ct returns the number of elements
assigned to a.

The PACK, and UNPACK statements, used in conjunction with the WHERE
construct, seem to be sufficient primitives to copy any array from one shape to a
quite different shape through one-dimensional arrays. Other uses may involve
computation with sparse unstructured arrays of medium size. To illustrate this
latter use, consider adding two sparse two-dimensional arrays for which only the
non-zero elements are explicitly represented and two-dimensional incidence arrays
indicating the locations of these non-zero elements are available. The following
program segment performs the sum; an example illustrating the multiplication of
two sparse vectors appears in [13].

```
SUBROUTINE   NZ_SUM(N, NZ_A, INC_A, NZ_B, INC_B, NZ_SUM, INC_SUM)

INTEGER   N
REAL      NZ_A(*), NZ_B(*), NZ_SUM(*)
LOGICAL   INC_A(N, N), INC_B(N, N), INC_SUM(N, N)

! This subroutine adds the sparse two-dimensional arrays A and B, whose
! representations are a pair of one-dimensional arrays NZ and INC.  NZ
! consists of the non-zero elements of the two-dimensional array.  INC is a
! mask whose j-th true element in storage order indicates the location in the
! two-dimensional array of the j-th non-zero element with value NZ(j).

! Local variables.  They are all automatic arrays.

! SIZE is an intrinsic function that returns the size of an array.

REAL      SUM_A( SIZE(NZ_A)+SIZE(NZ_B) ), SUM_B( SIZE(NZ_A)+SIZE(NZ_B) )
LOGICAL   A_IN_SUM( SIZE(NZ_A)+SIZE(NZ_B) ), B_IN_SUM( SIZE(NZ_A)+SIZE(NZ_B) )

! Determine the incidence array of non-zeros elements for the sum.

INC_SUM = INC_A .OR. INC_B

! Form a one-dimensional incidence array A_IN_SUM locating the non-zero ele-
! ments of A in the one-dimensional array of sums NZ_SUM.

WHERE( INC_SUM ) PACK( A_IN_SUM, INC_A)

! Place the non-zero elements of A at the appropriate positions of the array
! of sums NZ_SUM, indicated by the incidence array A_IN_SUM.  The intrinsic
! MERGE  merges two arrays, taking an element from the first argument where
! the third argument is true, and taking an element from the second argument
! where it is false.

SUM_A = MERGE(NZ_A, 0.0, A_IN_SUM)

! Repeat for B forming SUM_B

SUM_B = MERGE(NZ_B, 0.0, B_IN_SUM)

! Form the sum.

NZ_SUM = SUM_A + SUM_B

RETURN
END
```

The use of intrinsic functions in specification statements has not been accepted as yet; see Section 2.10.

In this example, INC_A (and similarly for INC_B) is a square logical array of order N. Everywhere INC_A is false, say at position (i,j), the corresponding element of A is zero, that is, A(i,j)=0. Everywhere INC_A is true, the corresponding element of A is non-zero, and its value can be found in NZ_A. That is, if the (i,j) position is the n-th position of INC_A in column major order that is true, then the value of A(i,j) is NZ_A(n). Using this data representation, the logical array INC_SUM locates the potential non-zero elements in the sum A+B. The one-dimensional array A_IN_SUM locates the positions of the non-zero elements of A in the array NZ_SUM of potential non-zero elements of A+B. The MERGE intrinsic forms a one-dimensional array SUM_A whose non-zero elements are those of A, with zeros in the places where A is zero and B is not. Similarly, the array SUM_B contains the non-zero elements of B. The sum of these two arrays then represents the non-zero elements in the sum A+B. Of course, for completeness, the array of sums needs to be checked for newly created zero elements in the sum, and the incidence array INC_SUM appropriately modified. This would be accomplished by the additional statements:

```
TEMP = NZ_SUM
WHERE( INC_SUM )  UNPACK(INC_SUM, TEMP .NE. 0.0 )
WHERE( TEMP .NE. 0.0 )  PACK(NZ_SUM, TEMP)
```

where TEMP is an array of the same shape as NZ_SUM.

2.7 Virtual Arrays -- The IDENTIFY Statement

The Fortran committee is committed to provide a replacement for the EQUIVALENCE statement. In addition, the array facility needs some way to form sections of arrays that are not parallel to the coordinate axis, such as diagonals. As its first attempt to provide such a feature, the committee has adopted a modified form of the IDENTIFY statement from VECTRAN[12]. It is new executable statement that permits a linear mapping of a virtual array onto a host array of the same type.

The form of the IDENTIFY statement (the syntax here is still being worked on) is

$$\text{IDENTIFY } \langle \underline{rd}[,\underline{rd}...]\rangle \quad \underline{v}(\underline{i}[,\underline{i}...]) = \underline{h}(\underline{ie}[,\underline{ie}...])$$

where the rd's are range declarators of the form $[\underline{r}_1:]\underline{r}_2$, the entities i are dummy integer variables whose scope is the IDENTIFY statement, v is the virtual array name whose explicit or implicit type must be the same as that of h, the entity h is an array, and the ie's are integer expressions that are linear in the dummy variables i.

The statement is used for extracting portions of arrays and to give them new names. For example, the diagonal of a square array, treated as a one-dimensional array, can readily be expressed as

$$\text{IDENTIFY } \langle 1:N\rangle \quad \text{DIAG\_A(I)} = A(I,I)$$

Other safe and effective uses of this facility would be to extract rows from an array, giving them names that simplify and clarify the computation. For example, consider the following code segment:

```
IDENTIFY <I:N> PIVOT_ROW(J) = A(I,J)
DO (K = I+1, N)
   IDENTIFY <I:N>  ROWK(J) = A(K,J)
   ROWK = ROWK - MULTIPLIER * PIVOT_ROW
REPEAT
```

Here, the IDENTIFY statement is used to define virtual row vectors over the matrix A. Many references to ROWK can be made in the subsequent code without using the cumbersome section selectors on A every time the K-th row of A is referenced.

However, as it is defined the statement can be misused to create rather obscure code. To avoid many difficulties, most related to ambiguous semantics, there are several restrictions on its use that are not given here. The model for its definition is that, upon executing such a statement, a dope vector for the virtual array name is computed, that contains a base address and strides for each dimension much like what is needed for implementing arrays in the rest of the language. Whereas other language constructs restrict the values that can be placed in the dope vector, this statement does not. For example, the mapping need not be one-to-one, that is, two different subscript selectors for the virtual array may yield the same host location. Consequently, if such a virtual array is defined the order of evaluation, which is subject to the implementation, will determine the data assigned to the host array. Also, with this model in mind it is clear that the variables such as I and K in the row-extraction example above are not parameters that are reevaluated on every reference to the virtual array, but are evaluated at the time of execution of the IDENTIFY statement. The IDENTIFY statement has several features that make it superior to the EQUIVALENCE statement that it replaces: the virtual and host array names must be of the same type; the mapping is clearly one way, that is, it is clear which is virtual array and which is the host array; it does not permit aliasing with scalars.

2.8 Elemental Procedures and Intrinsic Functions

The new facilities permit a new specification statement in the calling program, declaring that a scalar user-defined function subprogram may be used with array arguments. A reference to such a function with all its arguments conformable arrays causes the scalar function to be evaluated element-by-element returning a conformable array; that is, the scalar function is called once for every position in the conformable arrays. For example,

```
ELEMENTAL EXTERNAL   F
REAL   F(10), C(10)
REAL   A(10), B(-10:-1)

C = F(A, B)
   .
   .
   .
END

REAL FUNCTION   F(X, Y)
REAL X, Y
   .
   .
   .
F = ...
RETURN
END
```

causes the function F to be called 10 times with arguments A(I) and B(I-11), and the value of F(A(I),B(I-11)) is assigned to C(I) for each I=1,...,10.

The Fortran 77 intrinsic functions such as SQRT and SIN as well as the new intrinsics are classified into three types: elemental intrinsics, transformational intrinsics that reduce the dimensionality of one of the array arguments, and other transformational intrinsics.

The elemental intrinsics consist of the new intrinsic MERGE plus all of the Fortran 77 intrinsics extended to have array arguments. The Fortran 77 intrinsics are extended in the same way as described above for elemental user-defined functions. The new elemental intrinsic MERGE forms the merge of two conformable arrays under control of a third conformable logical array. Where an element of the logical array is true, the corresponding element of the first argument is taken; otherwise the corresponding element of the second argument is taken.

The reduction transformational intrinsics reduce the dimensionality by taking the sum or product (disjunction or conjunction for logical arrays), counting true values, or selecting algebraic minimums or maximums of the entire array argument or optionally along some specified dimension of the array argument. Each of these operations can further be controlled by a mask. For example, where A is a two-dimensional array of order 10, SUM(A) returns the sum of all the elements, SUM(A,1) returns a one-dimensional array of size 10 that contains the column sums of the array A, and MASK_SUM(A, A .GT. 0) returns a scalar that is the sum of all the positive elements of A.

The remaining transformational intrinsics perform various utility and linear algebra operations on arrays. They include, for linear algebra applications, dot product (where the first argument is conjugated if it is of type COMPLEX), matrix transpose, matrix multiplication, formation of a diagonal matrix (a two-dimensional array whose only non-zero elements are on the diagonal), and for general array utilities, spreading (increase rank along a dimension), replicating (increase the size with copies along a dimension), shifting, projecting (to a single element, or along a dimension), locating the first and last true value in a logical array or along a dimension in a logical array, creating logical one-dimensional arrays with repeated alternating sequences of logical values, and the determination of dimensionality, dimension bounds (lower and upper bounds in a dimension), array sizes, and so on. One missing utility is a subscript locator for a particular element such as the maximum or minimum. Needed with such an intrinsic is some way of referencing the element from, say, a one-dimensional array of the subscripts.

There is no swap operation available in the array facilities. Although its usefulness is not limited to arrays, the swap operation is a common array operation that requires three statements plus a scratch array to implement. It could readily be implemented by a generic intrinsic function for swapping two entities, such as scalars, arrays, or structures, of the same type.

2.9 Array Constants

As currently proposed, the only language construct that creates array constants is an extension of the DATA initialization statement that most Fortran compilers use now. This is inadequate, particularly when the language does not consider data initialized entities to be constants. The argument against such an array constant facility is that its use is generally limited to arrays of small size. To make it otherwise would require extensive syntax. I disagree that it is necessary to add such extensive syntactical constructs and believe that the facility given in ADA[10] is modest but adequate even for large arrays.

2.10 Intrinsics in Specification Statements

Currently, the proposed Fortran does not permit the use of any of the Fortran intrinsics in specification statements. The committee has discussed this topic and the need for such a facility in the context of several other facilities, but up to now, has not passed a proposal specifying that the intrinsics can be used in specification statements. At a minimum, the permitted intrinsics for the array processing facility should include the SIZE intrinsic, the dimension bound intrinsics EXTENT, LBOUND, and UBOUND, and the minimum and maximum intrinsics MIN and MAX for scalars.

2.11 An Example -- An LU Factorization of a Matrix

The following program segments illustrate the new Fortran array processing facilities to perform an LU decomposition[16] of a square matrix. The methods and examples used have been chosen to illustrate the features in the language rather than to suggest that any particular version would be the most efficient or most suitable way to code this operation on a particular processor.

The first example uses the WHERE statement, logical masks, and the SPREAD intrinsic to formulate Gaussian elimination in terms of two-dimensional array operations. Such a formulation is potentially useful for array processors, and is given here to illustrate that algorithms can be formulated in this way with the array facility.

```
SUBROUTINE  LU_FACTORIZATION(N, A, P)

INTEGER  N, P(N)
REAL     A(N, N)

! This subprogram computes the factorization PLU of a matrix A using the
! the usual Gaussian elimination algorithm with partial pivoting, where P
! is a permutation matrix represented as a one-dimensional of integers, L
! is a unit lower triangular matrix returned in the lower triangle of A,
! and U is an upper triangular matrix returned in the upper triangle of A.

! Local variables.  They are both automatic arrays.

REAL  T(N)
LOGICAL  ELIMINATE(N, N)

ELIMINATE = .TRUE.

! SEQ is a transformation intrinsic that returns a one-dimensional integer
! array consisting of an arithmetic progression of integers, starting at
! the first argument, up to the second argument, in steps specified by the
! third argument, if present, and otherwise in steps of 1.

P = SEQ(1, N)

DO (I = 1, N-1)

    ! Determine the pivot row PIVOT.
       .
       .
       .
    IF( PIVOT .EQ. 0.0 )  PIVOT = 1.0
    A(I+1:N, I) = A(I+1:N, I) / PIVOT

    ! Perform the matrix elimination.

    ELIMINATE(*,I) = .FALSE.
    ELIMINATE(I,*) = .FALSE.

    ! SPREAD is a transformational intrinsic that, given an array of
    ! dimension d as the first argument, returns an array of dimension d+1,
    ! by spreading n copies of the given array along the dimension
    ! specified by the second argument, where n is the third argument.

    WHERE( ELIMINATE )  A = A - SPREAD(A(*,I),2,N) * &
                            SPREAD(A(I,*),1,N)

REPEAT

RETURN
END
```

For vector or scalar processors, it may be more efficient to use more explicit code for the elimination step. The revised code for the DO-REPEAT loop without comments would be:

```
DO (I = 1, N-1)
       .
       .
    IF( PIVOT .EQ. 0.0 )  PIVOT = 1.0
    A(I+1:N, I) = A(I+1:N, I) / PIVOT

    IDENTIFY <I+1:N>  PIVOT_ROW(K) = A(I,K)

    DO (J = I+1, N)

        IDENTIFY  <I+1:N>  ROWJ(K) = A(J,K)

        ROWJ = ROWJ - A(J,I) * PIVOT_ROW

    REPEAT

REPEAT
```

A more efficient version designed to better match the Fortran storage order
for arrays (column major order) would be to use columns instead of rows. The
elimination step for such an algorithm would be:

```
IDENTIFY <I+1:N>  PIVOT_COLUMN(K) = A(K,I)
DO (J = I+1, N)
  IDENTIFY  <I+1:N>  COLUMNJ(K) = A(K,J)
  COLUMNJ = COLUMNJ - A(I,J) * PIVOT_COLUMN
REPEAT
```

3. A COMPARISON WITH ALGOL68-TORRIX68

The ALGOL68-TORRIX68 facility is described in detail in [18]; due to the
limited space in this publication, it is not reviewed here.

In general, there is no doubt that the combination of ALGOL68 and TORRIX68
yields a superior package for computing with abstract linear vector spaces. It
is not only that the TORRIX68 system is tailored to the specific application area
of abstract linear vector spaces, for Fortran could standardize an application
area module that mimicked the syntax and functionality of the TORRIX68 system.
The source of the superiority of this package is the power and extensibility of
the root language ALGOL68.

In terms of array processing facilities, why is the TORRIX68 package
superior?   First, the package is motivated by a formalized abstract mathematical
theory, namely that of linear vector spaces.   Its rules are derived from
properties of linear vector spaces, whereas the Fortran facility is an ad hoc
collection of primitives that represents a useful collection of array facilities.
For example, TORRIX68 introduces the concept of the total array; that is, a
particular vector of n elements is considered embedded in a very large
dimensional vector space in which those components that are not specified are
zero.   In such a space, the element-by-element product for vectors of unequal
length is well-defined, for everywhere the subscript spaces of the two vectors do
not overlap, the product is zero. This concept is applied to all operations in
the TORRIX68 repertoire, and in general simplifies the semantics for them.

Second, the core language for TORRIX68, ALGOL68, permits the definition of
new types and new operators on existing and new types. Thus, TORRIX68 is
applicable to arbitrary vector spaces for which the designated scalar type forms
a ring or a field. The user need only define the new scalar type and designate
that type to TORRIX to immediately have access to all the TORRIX68 operations for
linear vector spaces.   The ability to define operators permits the run-time
TORRIX system to verify the rules of applicability for the operators.   If
desirable, such checks can be defeated by the user defining his own operator of
the same name without the consistency checks.   Finally, the syntax is more
natural and convenient for the programmer. For example, it is much nicer to
write A x B to multiple two matrices A and B than to write the Fortran intrinsic
MATMUL(A,B).

Third, the core language for TORRIX68 has strongly-typed pointers, that is,
pointers that specify the type of the entity to which they point. Consequently,
the TORRIX package can distinguish between references to arrays which are just
descriptors and arrays with data stored in them.   The descriptors can be
manipulated in controlled ways. The analogue for Fortran arrays, though somewhat
of a weak feature in comparison, is the virtual array. With pointers,
unstructured arrays can be manipulated with more flexibility; for example, an
array of pointers can be used to point to an array of diagonals of a matrix, each
diagonal of a different size. The corresponding construct in Fortran would use
the IDENTIFY statement as follows:

```
IDENTIFY <N, N>  DIAGONALS(I,J) = A(I+J,I)
```

Here the super-diagonals of A are columns of the two-dimensional arrays, but the columns of DIAGONAL are all of the same length whereas the data in them is of decreasing length. The ALGOL68-TORRIX68 package with pointers creates an array of pointers to vectors of decreasing length.

As a fourth point, the TORRIX package is not part of a standard, but is a nearly portable package of ALGOL68 code. Thus, except for the implementation of four operations, the TORRIX package can be readily moved to other ALGOL68 processors. A program written to the specifications of an application module is only portable to those processors that have implemented the application module. There is no requirement that the processor implementing an application area support module for Fortran be written in Fortran.

The obvious question now is why bother with the Fortran array processing facilities at all. The reason is, of course, somewhat independent of all the superior features of ALGOL68. Programs written in the new Fortran will be compatible with programs written to the previous Fortran standard. The Fortran committee has attempted to define the array processing facilities (as well as the other extensions) so that, with a core+modules architecture for the language, processors can be written that permit new Fortran programs and old Fortran programs to coexist.

ALGOL68 started over from ALGOL 60, never pretending to be compatible with ALGOL 60, and therefore, its authors were free to design a compact, consistent, orthogonal, and well-defined language. TORRIX68 extends this approach by designing a package motivated by the modern theory of abstract linear vector spaces. On the other hand, the next Fortran will be a language that is compatible with its past definition, but is attempting to yield a cleaner, more compact, consistent core language than were the previous Fortrans.

Another issue that becomes apparent in comparing the array facilities in the two languages is the language designers' concept of its user community. ALGOL68-TORRIX68 appeals to the computer scientist, interested in the elegance, compactness and sound abstract and mathematical grounding of the facilities. The user base is most willing to and is required to abandon previous approaches and techniques to perform programming tasks. Users are required to think in the ALGOL68 way in order to effectively use the language. The approach of starting afresh yields a more consistent, more elegant language. Fortran, being a commercially used standard against which much software has been written, must unfortunately be compatible with its past. Fortran's users are not willing and not able to abandon old programs and old programming techniques.

Of course, Fortran could be extended to include packages, pointers, user-defined types and operators, and so on. In fact, several of these items have been considered by the committee and have been rejected as too drastic a change for Fortran at this time. In the meantime, the current collection of primitives appears to be and has been demonstrated in VECTRAN to be a useful set of facilities to manipulate arrays.

4. SUMMARY

We have summarized the array processing features proposed for the next Fortran and briefly compared these facilities with the ALGOL68-TORRIX68 system for programming operations on elements of linear vector spaces. We have observed that the ALGOL68-TORRIX68 system is superior in many respects; many of the features in this system are appropriate long range goals for Fortran, but they seem too innovative for Fortran at this time.

5. ACKNOWLEDGMENTS

I would like to thank Jim Cody for some helpful criticisms of early drafts of this paper.

6. REFERENCES

[1] Abrahams, Paul, The PL/I Programming Language, Courant Mathematics and Computing Laboratory, Mathematics and Computing Report COO-3077-151, March 1978.

[2] American National Standards Institute, American National Standard Programming Language FORTRAN, ANSI X3.9-1978, New York, 1978.

[3] American National Standards Institute, American National Standard Programming Language PL/I, ANSI X3.53-1976, New York, 1976.

[4] Brainerd, Walt, and Adams, Jeanne, Fortran for the 1980's, Information Processing 80, IFIP, 1980.

[5] Dongarra, J. J., Moler, C. B., Bunch, J. R., and Stewart, G. W., LINPACK Users' Guide, SIAM, Philadelphia, 1979.

[6] Garbow, B. S., Boyle, J. M., Dongarra, J. J., and Moler, C. B., Matrix Eigensystem Routines, EISPACK Guide Extension, Lecture Notes in Computer Science 51, Springer-Verlag, 1977.

[7] International Mathematical and Statistical Libraries Reference Manual, Edition 7, International Mathematical and Statistical Libraries Inc., Houston, TX, 1979.

[8] Iverson, K. E., A Programming Language, John Wiley and Sons, New York, 1962.

[9] Language Working Group, Fortran Language Requirements, Fourth Report of the Language Working Group of the Advanced Computing Committee, 1979, Available from P. C. Messina, Applied Mathematics Division, Argonne National Laboratory.

[10] Ledgard, H., ADA, An Introduction, and the Ada Reference Manual(July 1980), Springer-Verlag New York Inc., 1981.

[11] Numerical Algorithms Group Mark VIII Manual, Numerical Algorithms Group Ltd., Oxford, U. K., February 1981.

[12] Paul, G., and Wilson, M. W., The VECTRAN Language, IBM Palo Alto Scientific Center, Palo Alto, CA, 1975.

[13] Paul, G., Studies in Vector Processor Architecture and Scientific Applications, Oral Presentation at the Applied Mathematics Division, Argonne National Laboratory, 1981.

[14] PORT Mathematical Subroutine Library Manual, Bell Telephone Laboratories Inc., Murray Hill, NJ, January 1976.

[15] Smith, B. T., Boyle, J. M., Garbow, B. S., Ikebe, Y., Klema, V. C., and Moler, C. B., Matrix Eigensystem Routines, EISPACK Guide, Lecture Notes in Computer Science 6, Springer-Verlag, 1974.

[16] Stewart, G. W., Introduction to Matrix Computations, Academic Press, New York, 1973.

[17] Tanenbaum, Andrew S., A Tutorial on Algol 68, Computing Surveys, Volume 8, No. 2, pp. 155-190, June 1976.

[18] van der Meulen, S. G., and Veldhorst, M., TORRIX, A Programming System for Operations on Vectors and Matrices over Arbitrary Fields and of Variable Size, Volume 1, Mathematical Centre Tracts 86, Mathematisch Centrum, Amsterdam, 1978.

[19] van Wijngaarden, A., Mailloux, B. J., Peck, J. E. L., Koster, C. H. A., Sintzoff, M., Lindsey, C. H., Meertens, L. G. L. T., and Fisker, R. G., Revised Report on the Algorithmic Language ALGOL 68, eds, Springer-Verlag, New York 1976.

[20] Wetherell, C. W., Array Processing for Fortran, Lawrence Livermore Laboratory Computer Documentation UCID-30175, August 1979.

[21] Wetherell, C. W., LWG FORTRAN Manual, Draft document, July 1981.

[22] X3J3 Fortran Committee, Standing Document /S6.79, June 1981, obtainable from the Secretary of X3J3, L. P. Meissner, Lawrence Berkeley Laboratory.

DISCUSSION

Hehner I am very concerned about the ad hoc way these features are being added to Fortran. Remember that the area is fairly well explored, having been in languages like APL for many years. In this design there seems to be no consideration of language design principles such as uniformity, generality and orthogonality. Automatic arrays are proposed, rather than automatic variables in general. Arrays are only allowed to be of certain types. The WHERE construct allows some kinds of statement, but not all, in certain places. I am amazed that it is thought necessary to explain that in an assignment statement, the expression is evaluated first, then the value stored (although, as pointed out, this is not necessarily true in PL/1, where A = A/A(1) is equivalent to A(1)=1). There is no attempt to give a mathematical semantics - this would be a useful design guide.

Meissner You perhaps do not understand the position of X3J3. Fortran is not an elegant language, but it is heavily used and its users want more features. It is very important that these should be added in an upwards compatible fashion. Moreover it is impossible to give the full details of the proposals in a short presentation.

Kahan Users are sure to confuse arrays (which have little structure) with matrices (which obey certain mathematical rules). Also, will you get the efficiency Reid requires using these structures?

Smith The trouble with array processing features is that different groups of users expect different things. Some see them merely as a convenient means of computing on sets of data. Others, such as mathematicians, see them as a means of manipulating objects which obey certain mathematical rules. On the efficiency question, I have merely chosen a few expository examples; on these we may not achieve efficient computation, with others we certainly will.

G. Paul These array facilities are very similar to APL. Remember also, when comparing efficiency on scalar machines with Reid's methods, that his methods do not extend trivially to vector machines.

Kahan Doesn't that imply that hardware designers should really re-think what they are offering?

G. Paul   Probably, yes. It is also important to remember that these proposals for extensions to Fortran are not fully consolidated; there is still a great deal of detail to be ironed out.

Feldman   Many of the comments seem to be directed at the syntax rather than functionality. When the syntax is properly defined and agreed, it could well all work.

Boyle   Is it intended to provide means of access to hardware features provided by new machine architectures?

Smith   Many comments from users on these proposals have newer machine architectures in mind. When designing these new language features, though, we must take care to ensure that reasonably efficient implementation is possible on scalar machines. It is also worth pointing out that array processing features will almost certainly go into the extension module of Fortran 8X.

---

[1]This work was supported by the Applied Mathematical Sciences Research Program (KC-04-02) of the Office of Energy Research of the U.S. Department of Energy under Contract W-31-109-Eng-38.

THE RELATIONSHIP BETWEEN NUMERICAL COMPUTATION
AND PROGRAMMING LANGUAGES, J.K. Reid (editor)
North-Holland Publishing Company
© *IFIP, 1982*

EXAMPLES OF ARRAY PROCESSING IN THE NEXT FORTRAN

Alan Wilson
International Computers Limited
London, England

Many operations performed on the FORTRAN array structure occur regularly, right across the spectrum of applications from reservoir modelling to image processing. The next FORTRAN will provide a comprehensive set of these frequently performed operations in the form of standard built-in functions. Practical examples of the use of several of the new FORTRAN functional operations are given.

1. Introduction

The array intrinsic functions described in section 2.8 of Brian Smith's paper [1] fall into five groups:

5 functions for the measurement of arrays

14 functions for computation on arrays

2 functions for manipulation of arrays

3 functions for accessing arrays

6 functions for the construction of arrays

The names of these 30 functions (and a breakdown of the uses of the computational functions) are:

measurement:		RANK, SIZE, EXTENT, LBOUND, UBOUND
computation:		
	counting:	COUNT
	arithmetic:	SUM, PRODUCT, MASK_SUM, MASK_PRODUCT
	logical:	ANY, ALL
	extremal:	MAXVAL, MINVAL, MASK_MAXVAL, MASK_MINVAL
	algebraic:	MATMUL, DOTPRODUCT, TRANSPOSE
manipulation:		CSHIFT, EOSHIFT
access:		FIRSTLOC, LASTLOC, PROJECT
construction:		SEQ, ALT, DIAGONAL, SPREAD, REPLICATE, MERGE

2. The basic aim of array processing

The very general concepts of measurement, computation, manipulation, access and construction correctly suggest that the array intrinsic functions are intended to provide operations on arrays that are basic to the whole range of applications which use the Fortran array as their fundamental data structure. Only the small selection of algebraic computational functions MATMUL, DOTPRODUCT TRANSPOSE is at all specialized.

In other words these functions arise naturally from the basic aim of array processing, which is to operate on the arrays themselves so far as possible rather than on their individual elements.

Thus it would not achieve the aim of array processing if it was necessary to fall back on the use of loops and conditional code at every turn in order to compute such basic things as the scalar sum of the elements of a vector (= SUM(V)) or to determine the vector of the smallest positive numbers in each row of a matrix A (= MASK_MINVAL(A,A.GT.0.0,2)).

3.   FORmula TRANslation as array processing

Here are some simple examples of array processing FORmula TRANslation.   We assume the following array declarations

          REAL X(N), Y(N), A(M,N)

          COMPLEX C(N,N)

1. $\sum\limits_{j=1}^{N} \prod\limits_{i=1}^{M} a_{ij}$ = SUM(PRODUCT(A,1))

          (dimension = 1 means the product is down the columns).

2. $\sum\limits_{x_i>0.1} x_i$ = MASK_SUM(X,X.GT.0.1).

3. $\sum\limits_{i=1}^{N} (x_i-\bar{x})^2$ = SUM(X-SUM(X)/N)**2).

4. $\nu = \max\limits_{1\leq i\leq n} \left\{ \dfrac{\sum\limits_{j} |c_{ij}|x_j}{x_j} \right\}$, where x_i, i=1,2,...,n, are all positive.

          NEW = MAXVAL(MATMUL(ABS(C),X)/X).

4.   The radii of Gerschgorin's circles

The vector of radii

$$r_i = \sum\limits_{j\neq i} |c_{ij}|, \qquad i=1,2,...,n,$$

of Gerschgorin's circles associated with the complex matrix C is computed as follows:

COMPLEX C(N,N)
LOGICAL OMIT(N,N)
OMIT = .NOT.DIAGONAL(.TRUE.,N)
R = MASK_SUM(ABS(C),OMIT,2).

5.   An example from statistics: Chi-squared

The statistic

$$\chi^2 = \sum\limits_{i,j} \frac{(t_{ij} - e_{ij})^2}{e_{ij}}$$

where

$$e_{ij} = \left(\sum\limits_{k} t_{ik} \right) \left(\sum\limits_{k} t_{kj} \right) \Big/ \left(\sum\limits_{k,l} t_{kl} \right)$$

is computed as

```
REAL T(M,N),R(M),C(N),E(M,N)
R = SUM(T,2)
C = SUM(T,1)
E = SPREAD(R,2,N)*SPREAD(C,1,M)/SUM(T)
CHI_SQ = SUM((T-E)**2/E)
```

6.   Queries (without loops or conditional code)

The intrinsic functions allow quite complicated questions about tabular data to be answered, without use of loops or conditional code. Consider for example the questions asked below about a simple tabulation of, say, test scores.

Suppose the rectangular table $T(M,N)$ contains the test scores of M students who have taken N different tests.   (TABLE is an integer matrix with entries in the range 0 to 100).

Questions

1.   What are the top scores for each student?   (The list of each student's highest score).

2.   How many scores in the table are above average? (average over the whole table).

3.   Increase those above average scores by 10%!

4.   What was the lowest score in the above average group?

5.   Was there a student all of whose scores were above average?

Answers

1.   MAXVAL(T,2)

     ABOVE = T .GT.(SUM(T)/SIZE(T))

2.   COUNT(ABOVE)

3.   WHERE(ABOVE) T = 1.1*T

4.   MASK_MINVAL(T,ABOVE)

5.   ANY(ALL(ABOVE,2))

7.   Vector solution of a tridiagonal system

The following algorithm for the solution of a tridiagonal system $Tx = y$ appears in Thomas Jordan's document [2], where it is described in conventional FORTRAN.   The version below operates on 5 vectors of length N which represent the solution X, the righthand side Y and the 3 diagonals L,D,U (lower, diagonal, upper).   It is a log_base_2 algorithm that successively drives L and U out of the picture and ends with a purely diagonal backsolution.   The EOSHIFT (D,1,K) function returns D shifted left (K>0) or right (K<0) along dimension 1, with zero fill.

```
REAL X(N),Y(N),L(N),D(N),U(N)
K = 1
DO
L = L/D
U = U/D
Y = Y/D
D = 1 - L*EOSHIFT(U,1,-K)-U*EOSHIFT(L,1,K)
Y = Y - L*EOSHIFT(Y,1,-K)-U*EOSHIFT(Y,1,K)
L =   - L*EOSHIFT(L,1,-K)
U =   - U*EOSHIFT(U,1,K)
K = 2*K
IF(K.GT.N) EXIT
REPEAT
X = Y/D
```

8. REFERENCES

[1] Smith, B.T., Array Processing Features in the Next Fortran, this volume.

[2] Jordan, T.L., Gaussian elimination for dense systems on STAR and a new parallel algorithm for diagonally dominant tridiagonal systems, Los Alamos Scientific Laboratory LA5803, 1975.

DISCUSSION

Adams A worry to members of X3J3 here is an apparent firm rejection of the array processing proposals. Is this really the view of numerical analysts? Such a rejection would mean a decrease in functionality. The array processing proposals offer a great deal of generality in processing data.

Moler The most vociferous objections have come from the language designers, not the numerical analysts.

Hennell Many of the objections arise from the potential consequences of inadvertant misuse of the facilities; this is a problem faced by every programming language.

Roberts My concern is not with the form of the proposals but with when the facilities will become available to users. Present estimates of the value of 'X' in Fortran 8X, together with the experience with Fortran 77 suggest that it may be the early 1990's before these array processing features have been standardised, implemented in compilers and then in corporated into working codes. By this time array and vector machines will have been in use for more than 10 years.

As a first approach, would it not be preferable to keep the language small (and unchanged) and incoporate the new features into subroutines by adding one extra argument to the functions? Standardisation and implementation of these subroutine calls could be achieved more quickly and new facilities could be added year by year as required. A preprocessor could be used to convert a single assignment statement into a subroutine call if desired.

In my view many language enhancements could be (and probably should have been) achieved in the way, including input-output, mathematical functions, system routines and graphics. Program portability could be substantially enhanced if a standards body such as X3J3 could take on the task of standardising library subroutine calls which usually amount to several hundreds at the major installations.

Facilities proved by use could later be incorporated into the language inself at periodic updates.

Kahan   This approach would seem to be the right one. I would like to add a strong comment on syntax. Since the notation used in mathematics play an important role in conditioning the way we think, it must be as natural as possible. In a programming language, the syntax takes the role of notation in mathematics. It is vital, therefore, as an aid to thinking clearly, that the syntax should be natural.

Schlechtendahl   To some extent I disagree with these comments. The standardisation of vector machine type operations on the Fortran level is urgently needed, since otherwise calls to the manufacturers supplied software will be used. This will restrict portability of vector programs.

The expression language approach proposed by X3J3 is preferable to a subroutine approach because it avoids the introduction of many intermediate results and reduces the number of program statements required. While the language constructs do not appear as well designed as they might if they were extensions of Pascal or Algol 68, this is a consequence of using Fortran as a starting point. For Fortran, taking the goal of having a useable language extension as soon as possible, they appear to be an acceptable first step.

Delves   It isn't the functionality that I believe is being objected to, nor the definition of the intrinsic functions, but the way that the array operations are being shoveled into the language in an ad-hoc, restrictive manner, with the restrictions also ad-hoc and not the sort of restrictions which should be necessary. That is what is being objected to.

Kahan   If those functions were part of a library of programs (if they weren't intrinsic), then we would say: "ok, use them, see how you like them". And then somebody else will have to try to change the language later to put some of the functionality into the language in a natural way. But this wasn't putting a library into the language. Who knows if it is the right library? By using the names SUM,PRODUCT,MAX,MIN for intrinsic functions you are appropriating words that others may wish to use with different meanings. For example Rice cannot have his SUM operator.

Sedgwick   I found the keyword ELSEWHERE much easier to understand than OTHERWISE.   And programming language people already have preconceived meanings for what OTHERWISE should mean, and the ELSEWHERE would fit in much better with the WHERE construct.   I still have some reservations about the whole WHERE construct however.

Rice   I would be pleased to see vector and matrix arithmetic and to stop there, but I also see other requirements like in linear programming.

Meissner   X3J3 array (and other) proposals are not rigidly specified as yet. Inputs from this group can be useful in the refinement process that X3J3 will be undertaking over the next few months and years.

**SESSION 5
PARALLELISM;
LANGUAGE EVOLUTION**

Chair
J.A. Nelder

Discussant
G. Paul

THE RELATIONSHIP BETWEEN NUMERICAL COMPUTATION
AND PROGRAMMING LANGUAGES, J.K. Reid (editor)
North-Holland Publishing Company
©*IFIP, 1982*

EXCEPTION HANDLING IN ARRAY LANGUAGES

T W Lake

International Computers Ltd.
Bracknell, Berks,
England

The availability of powerful array computers has raised a
demand for array languages expressing parallel primitives.
These have exception handling requirements additional to
those of serial languages. The introduction of conditional
execution of parallel operations, which allow a
simplification of program structure, places further demands
on exception handling. The integration and implementation
of these facilities is discussed.

1   INTRODUCTION

The following is an account and development of a view of exception handling
that arose during the development of the parallel language DAP FORTRAN [1] for
the ICL Distributed Array Processor (DAP), in which the author participated.
These thoughts on exception handling had two strands - one emphasised the
special requirements of library routines as independently satisfied in the
design of ADA - the other dwelt on the special requirements of array languages.
These requirements, if accepted, have implications for high-level array
languages and for array machine architectures. The author has not attempted a
review of array languages, partly because the requirement for array languages
has come to the fore only recently with the availability of powerful array
machines [2] - ILLIAC-IV, CDC Cyber-200 series, DAP and CRAY-1. The only
related work known to the author is the special attempt made to deal with
exception handling in the ILLIAC-IV language GLYPNIR.

2   REGULAR COMPUTATION

All of the array machines mentioned in section 1 achieve high performance
through regularity of computation. While the pipeline processors (CRAY-1 [4]
and CYBER-200s [5]) concentrate on regularity in time the array processors
(ILLIAC-IV [6] and DAP [7]) have regularity in space. Many large computational
problems are quite regular and some of the algorithms for solving them are
regular too. There is a great demand for portable languages capable of
expressing this regularity clearly. Such languages would greatly facilitate the
production of software for solving regular problems even on serial machines.

The ANSI X3J3 committee is considering an array module for inclusion in the
next ANSI FORTRAN standard. One aim of such a standard must be to cover a wide
variety of machine architectures in which machines for regular computation
should find a place.

3   ARRAY LANGUAGES

The recent interest has brought about a broader view of array processing in
which the rectangular array of FORTRAN, APL etc. is but a special case. Feldman

[3] describes an array as "a collection of data of identical type; each element is identified by a member of some index set." (The index set distinguishes arrays from tables). Where arrays have the same index set they will be described as 'of the same size'. He describes two classes of array operation - (i) operations that combine all the elements of an array such as SUM , MAX, COUNT and (ii) operations that apply a scalar n-ary operator element by element to n arrays of the same size. An example of the second class is the operation of vector addition, component by component. Whole array assignment can be included in this class. These operations are to be considered as fundamental and indivisible primitives of array processing. To these classes of operations can be added (iii) motions such as shift, transpose, permute and sort, (iv) subsettings such as rectangular array sectioning and (v) array transformations such as linearisation to a vector in index order.

An important element by element operator not explicitly mentioned by Feldman is derived from the ternary conditional operator which we may write (with operands A,B and L) as

$$\text{IF} \quad L \quad \text{THEN} \quad A \quad \text{ELSE} \quad B \quad \text{FI}$$

which merges values from two arrays A and B of the same type and size according to the value of a boolean array L of the same size. For each index value the result value is the A or B value with that index according as the L value with that index is true or false. This can be written as the array function MERGE(A,B,L). DAP FORTRAN also allows the masked assignment

$$X(L) = A$$

with the same effect as

$$X = \text{MERGE}(A,X,L)$$

The ANSI FORTRAN proposal is for a WHERE statement or block, allowing conditional assignment, as in

```
WHERE ( L )
    X = A
OTHERWISE
    X = B
ENDWHERE
```

(the last line to be pronounced 'anywhere'). Despite the risk of confusion the author prefers ELSEWHERE to OTHERWISE. If we allow the order of execution of the two branches to be undefined this is equivalent to

$$X = \text{MERGE}(A, B, L)$$

with the proviso that the third parameter to MERGE is evaluated first. The activity is defined as the value of L for the first branch, .NOT.L for the second. A single statement form analogous to the FORTRAN LOGICAL IF is also proposed.

4   CONDITIONAL COMPUTATION

The MERGE and WHERE constructs are particularly important for machines good at regular computation since they benefit by avoiding control jumps. Even serial machines are poor at making frequent decisions; array machines can often substitute control by logical arrays for control of program flow. This can make the flow of control of a program much simpler. It is not the only technique available since sometimes the explicit scatter or gather of data values is the most effective.

The array machines mentioned above all support conditional computation in some way, CRAY-1 has a vector merge, Cyber-200s have vector operations where result storage is controlled by a bit string and also sparse vector operations whose operands have only the non-zero elements stored with their positions indicated by the one bits in bit strings corresponding to the full operand , ILLIAC-IV's processing elements could be selectively disabled and the DAP has an elementary conditional store operation controlled by a processing element register. Of course conditional computation is very natural to parallel machines.

It will be seen immediately that the introduction of conditional computation must involve changes to exception handling because we want to avoid interruptions by exceptions where the result is irrelevant, that is, corresponding to an activity value false. This aim is assumed in the following sections.

Since we are dealing with parallel primitives exceptions can arise in several components of the result of a single operation. A parallel language should provide the means to determine which components are in error when an exception is signalled. The provision of exception maps is discussed in section 9.

5   EXAMPLE USE OF WHERE

To illustrate the use of WHERE to avoid interruptions by unwanted exceptions consider the example of a routine intended to solve in parallel many quadratic equations taking care of all the cases of zero coefficients which could cause division by zero. The routine is intended to return in each case the number of real roots($0,1$ or 2) and as many result values as available. We use a wishful thinking FORTRAN with dynamic storage allocation, array valued functions and nested WHERE blocks in which each branch subsets the activity of its environment. The method is to compute the root of largest magnitude of $Ax**2 + Bx + C$ first and then use that to derive the smaller.

```
SUBROUTINE SOLVE(A,B,C,ROOTS,R1,R2,N)
INTEGER  N, ROOTS(N)
REAL A(N), B(N), C(N), R1(N), R2(N), D(N)
WHERE ( A .EQ. 0)
    WHERE ( B .EQ. 0)
        ROOTS = 0
    ELSEWHERE
        ROOTS = 1
        R1 = -C/B
    ENDWHERE
ELSEWHERE
    D = B*B - 4*A*C
    WHERE (D .LT. 0)
        ROOTS = 0
    ELSEWHERE
        ROOTS = 2
        R1 = =(B + SIGN(SQRT(D), B)) / (2*A)
        WHERE (R1 .EQ. 0)
            R2 = 0
        ELSEWHERE
            R2 = C/(A*R1)
        ENDWHERE
    ENDWHERE
ENDWHERE
RETURN
END
```

If WHERE control were to inhibit interrupts as described above then any
interrupts from this routine would indicate genuine difficulty with the
solution rather than the interference of different cases.

6 POST MORTEM INVESTIGATION

Although array computation does not necessarily involve large amounts of data
this is often the case and the presentation of relevant portions of the program
state on interrupt can be very cumbersome. The interrupted or halted program
should stop in a predictable and if possible restartable state. It should be
possible to save the program state and later examine it using the names given
in the source program. It is useful to be able to alter the program state in a
controlled manner and then restart it. Graphical displays help in reviewing
large volumes of data. An interactive development environment with these
facilities speeds the commissioning of software and the analysis of algorithms.
This level of support is often neglected.

7 LIBRARY ROUTINES

Before further analysis of array exception handling it is useful to identify
the special requirements of library routines which provide the most stringent
test of serial exception handling. These are generally separately compiled
routines intended for general distribution and use by a demanding community
including the mischievous and careless. Requirements of this kind were clearly
a determinant of the exception handing in ADA.

1 Library routines must retain control until ready to relinquish it.
2 Library routines must be able to control their interrupting and
   trapping environments be able to reset the caller's environment and be
   able to set another before calling a new routine. (In ADA the
   environment is tied to the block structure.)
3 Library routines should be able to indicate exceptions of their own
   just like standard arithmetic exceptions.

These requirements apply equally to array languages.

8 AN ANATOMY OF EXCEPTION HANDLING

	Exception registered as data	Exception registered by change of control
Direct Action	VAL IEEE arithmetic (a)result NaN etc. Pipeline processor (a)result NaN etc.	ADA (b) jump to exception handler IEEE arithmetic (b)take trap DAP FORTRAN (b)take interrupt
Indirect Action	ADA (a)raise exception DAP FORTRAN (a)set reporting variable	Pipeline processor (b)delayed interrupt

Table 1:   Classification of exception handling

Table 1 is an attempt at a classification of exception handling. For a given machine the position on the diagram may depend on the language used since the language run-time system may translate the machine response to a different high level response; for example, it may call a user-nominated interrupt handler.

The major division is between systems that report on an exception through the primary or secondary result of an operation and those that change the flow of control in response to an exception.

The change of control flow may be orderly, resulting in a predictable, resumable state or it may be chaotic. Pipeline processors are particularly unsuited to reporting exceptions through control changes since control flexibility has been sacrificed to regularity of control and it is hard for them to stop at the right (or any) program state. Some systems appear on two places in the diagram according to the setting of a user-visible switch.

The VAL language [8] expresses the primitive operations of a dataflow machine. As control flow is data driven exceptions must be expressed entirely by data values. To this end every elementary type has, in addition to its standard range of values, at least an INVALID value and possibly others, e.g. UNKNOWN, POSITIVE-OVERFLOW, UNDERFLOW etc. with the usual operators extended to these values. There is also an INVALID-DATA test. Unfortunately there is no agreement or standard for these extended ranges and operators and hardware implementations may differ.

The current array pipeline processors have exceptional values, referred to in the table as NaNs (Not a Number), defined for floating point types with a user inhibitable interrupt being taken on their appearance either as result or operand in arithmetic operations. The machine response is usually little disguised by the high level software since the program state after interrupt is generally unrecoverable.

The draft IEEE arithmetic standard [9] specifies exceptional values (NaNs) for data and also several user inhibitable traps.

ADA specifies system and user defined exceptions which are raised implicitly by exceptional result or explicitly through the RAISE statement. On occurence of an exception control escapes outwards until a block with an appropriate exception handler is reached when the exception is lowered and the handler executed. In contrast to the IEEE floating point standard trap which is activated by forced call there is no possibility of return to the scene of the crime. This is awkward for array languages where an exception can apply to a subset of the components of a value.

DAP FORTRAN maintains a map indicating in which components exceptions have occurred and has programmer inhibitable interrupts. (DAP FORTRAN variables are currently of fixed dimension 64 or 64 by 64). The user may nominate the variable to be used to hold this map and so can access it. A routine may remember the interrupt switch state and exception map variables on entry and restore them on return.

9   CHOOSING THE EXCEPTION HANDLING FOR AN ARRAY LANGUAGE

In comparison with scalar languages we have one advantage and some complications. The control overheads are proportionately lower in array languages. But an exception now requires some sort of exception map to indicate in which components exceptions have occurred. Exceptions subject to WHERE or MERGE constructs may require different treatment.

The exception map, if explicit, can have several forms. Its components could be simple truth values or integers (identifying particular exceptions) or declared named exceptions (as in ADA).

The control structures of exception handling schemes reviewed in section 8 are not always convenient, the IEEE trap having to be supplied for each different treatment required and the ADA exception handler not allowing computation to continue for elements unaffected by the exception.

An attractive suggestion is to have a language without traps or interrupts and having instead exceptions registered in exceptional data values. In this case the exception map would be associated with the variable. A correct implementation must have an INVALID representation for every elementary data type. This is a large requirement and constrains the choice of stored representation of data. If LOGICAL data is to support an INVALID value its storage must be doubled and the simplicity of bit strings and word logical operations lost. Conditional statements would require a third branch or an interrupt for an INVALID result.

The IEEE standard, not covering LOGICAL values, relies on a trap to detect invalid (unordered) comparisons or on the computation of both the comparison and its inverse.

If a totally data oriented exception handling scheme is rejected then explicit exception maps and interrupts must appear. An interrupt mechanism has at least the advantage of highlighting rogue program states without the introduction of explicit tests into the program.

10   EXCEPTION HANDLING WITH CONDITIONAL EXECUTION

Exceptions arise in the evaluation of expressions. Conditional evaluation of an expression can be specified with a MERGE function as defined in section 3. Since many expressions in FORTRAN occur on the right side of assignment statements it is convenient in that language to have a WHERE statement to apply a LOGICAL control to storage on the left and evaluation on the right. Constructs such as actual parameters and subscripts would be shielded from WHERE control and must involve a MERGE if they are to be executed conditionally.

There are two syntactic levels at which interrupts could be implemented - at complete array operations or at complete array assignments. With interrupts at operation level we need to predict which exceptions would propagate through the expression to active positions and which to inactive ones in order to report only effective exceptions. When the expression contains array motions such as shifts or summing operators this is more complicated. But the correct activity can still be applied by appropriately transforming the activity as applied to the function and applying it to the operand before applying the function. The appropriate transformation is a motion inverse to the data motion signified by the function. All of the array moving and rearranging functions proposed for the ANSI FORTRAN array extension have simple inverses of this kind. For example the inverse motion to a circular shift is the opposite circular shift on the same axis, the inverse to a summation along an array axis is the repetition (SPREAD) of the result-shaped value along that axis to the extent of the value summed. Conditional execution constructs can be nested straightforwardly. When we have a MERGE subject to a WHERE :-

        WHERE   (L)   X = MERGE(A,B, M)

M must be evaluated subject to L, A subject to L .AND. M, B subject to L .AND. .NOT. M to allow interrupts only where results cannot be validly given.

With the second option of interrupts to be taken at the completion of an assignment, subscript or actual parameter we have a coarser report of exceptions but an easier implementation. An accumulated exception map must be carried through the expression on evaluation simulating the extended range data type without affecting storage representation except for compiler assigned temporaries. For implementations that have to test for invalid data this option requires less control jumps than the first and can make use of exceptional data values if available on the supporting machine.

11 EXCEPTION HANDLING IN GLYPNIR

ILLIAC-IV designers seem to have thought similarly about exception handling and data motions. GLYPNIR was an intermediate level, ALGOL syntax language for ILLIAC-IV. The ILLIAC processing elements were controlled by a switch in each processing element settable by broadcast of a 64 bit word to the processing elements from the Control Unit. A problem of the language was that it was not clear whether this switch (implicitly named MODE in the language) was intended to be implicit or explicit. It was set and restored by the conditional loop constructs but was also user-visible and could be set and unset without reference to position in the control structure. And this was necessary because processing element activity was too closely linked to loop control by the absence of a scalar loop.

GLYPNIR allowed a value to be moved between processing elements before being stored and ,in order to avoid interrupts from irrelevant results (that is values that would be moved to disabled processing elements), moved the MODE value through the inverse motion before evaluating the expression to be moved and then stored.

12 DETECTING EXCEPTIONS WITHOUT INTERRUPTS

For library routines or programs wanting to detect and act on exceptions, either system or user generated, we want to accumulate exception maps without taking traps and without breaking the flow of control since we are processing many values in parallel. We require a language construct to express this. As with the WHERE and MERGE constructs we can provide FORTRAN-like constructs for either a statement or a block of conformable assignment statements or for an expression. For an assignment statement an EXCEPTIONS statement might nominate a variable to receive the exception map resulting from (conditional) evaluation of the right hand side. The size of such maps, like the sizes of results, may not be predictable by a compiler since it may depend, for example, on parameter sizes.

Instead of assigning the map of exceptions to the receiving variable it is useful to accumulate successive maps into the nominated variable with the logical OR (for LOGICAL type exception maps) so that the resulting value can summarise the exceptions of several steps or loop elaborations.

These facilities result in much clearer and more convenient program structures for programs intending to act on exception information.

13 FACILITIES FOR USER WRITTEN FUNCTIONS

The scheme proposed separates exception information from data values at the language level whatever the underlying data representation. The calculation of exception states is carried out as part of expression evaluation. Facilities are required to allow user written functions to take part in expressions fully. The facilities required depend on which of the interrupt schemes is adopted, operation level or assignment level.

Where interrupts are notified at operation level functions need a means to
ascertain the activity to be applied to their result and a statement to cause
an exception giving cause indication and exception map e.g.

        ERROR   integer-expression, logical-array-expression

This is sufficient to allow user written functions to simulate standard system-
supplied functions.

Where interrupts are at statement level functions need a means of specifying
the exception map on return e.g.

        RETURN   INVALID(logical-array-expression)

14   IMPLEMENTATION OF THESE PROPOSALS ON EXISTING MACHINES

The constructs suggested should be implementable on a wide variety of machines
to give portable exception handling in array languages, in particular for the
proposed ANSI FORTRAN. In many cases machine level interrupts would have to be
inhibited and the high level interrupt or trap activated on detection by
specific test of invalid or exceptional data. The cost of this favours less
frequent interrupt points and so the second option of interrupt on assignment
or expression completion is favoured.

On the ICL DAP efficient bit handling and low control costs allow either scheme
to be implemented. Current DAP FORTRAN contains a partial implementation of
the operation level interrupt scheme.

15   LESSONS FOR COMPUTER DESIGN

The following conclusions are drawn by the author:-
   1 It is necessary to think through the high level language facilities
     to be provided before fixing machine architecture if the language
     level is not well above the machine level.
   2 The utility of interrupts is much enhanced if they leave a
     predictable program state.
   3 Where LOGICAL control of array operations is provided it is necessary
     to inhibit interrupts by the same mechanism that inhibits storage
     of results.
   4 Efficient tests for exceptional values or automatic production of an
     auxiliary exception map ought to be supported by array machines.
   5 Efficient LOGICAL array operations with low overheads are required
     for effective implementation of array languages.

16   ACKNOWLEDGEMENTS AND DISCLAIMER

        The views expressed in this article are the author's and not
        necessarily those of ICL. The author expresses his gratitude to his
        colleagues and friends at ICL who sustained much discussion of
        exception handling, particularly to Roger Stokes, whose points sank
        in, albeit slowly.

17 REFERENCES

[1] DAP: FORTRAN PROGRAMMING, TP 6918, International Computers Ltd. (ICL,1981)

[2] Jesshope, C.R. and Hockney, R.W. (eds.), State of the art report on supercomputers. (Infotech International, Maidenhead, 1979).

[3] Feldman S., Array Operations, Evidence to the ANSI X3J3 committee (1980).

[4] Russell R.M., The CRAY-1 Computer System, Comm. ACM, 21 (1978) 63-72.

[5] Cyber 200 Model 203 Computer System Hardware Reference Manual, CDC 60256010, Control Data Corp.

[6] ILLIAC-IV Systems Characteristics and Programming Manual, NASA (Feb 1973)

[7] Flanders P.M., Hunt D.J., Reddaway S.F. and Parkinson D., Efficient High Speed Computing with the Distributed Array Processor, in: High Speed Computer and Algorithm organisation. (Academic Press, 1977).

[8] Ackermann W.B. and Dennis J.B.,VAL - A Value Oriented Algorithmic Language, MIT/LCS/TR-128, Lab for Comp. Sci., MIT.

[9] A Proposed Standard for Binary Floating-Point Arithmetic, Draft 8.0 of IEEE Task P754, Computer, 14 (1981).

DISCUSSION

Meissner   You have talked about machine facilities for disabling the store. Are facilities for disabling execution required?

Lake   These large machines differ considerably. We must distinguish between what is done in a hardware instruction and what we can provide in high-level languages.

Kahan   What is the difference between the WHERE construct and the IF - THEN - ELSE construct? In your example I could use IF - THEN - ELSE and specify execution for every instance of the index set.

Meissner   As long as everything is element-wise there is very little difference. Put a matrix transpose in one of those branches and you get something very different.

Lake   When we have array-valued functions in a branch then there are differences. I think we want atomic, indivisible array operators in array languages.

Gentleman   If the PACK/UNPACK operators that were described by Smith yesterday are used in the context of the WHERE then the temporary expressions are not necessarily conformable with the stored result. How can you handle that case?

Meissner   What is the inverse to a matrix product?

Lake   It is the motions for which we require inverses and it seems to me that there is no problem.

Kahan   Let me see if I can paraphrase the semantics of the EXCEPTIONS statement. "Recording the exceptions in E, assign X the value such and such." And then you are saying that the overflow map M is an implicit variable and E is updated by OR'ing E and M.

Meissner   So you want the underlying system to have to know that it has to map the motion of M before it updates E.

Lake   Yes.

Hull   Your example indicates how your scheme can do two things. One is to avoid any trouble that might be caused by an exception in those cases where you know that an exception does not matter. The other is to identify those places where exceptions have arisen. But do you provide the user with an opportunity to determine what is to happen in case an exception does arise? Of course I have in mind something along the lines I suggested in my talk, where the user can describe what kind of response he wants with ON statements.

Lake   I would like to link the suppression of an active interrupt to whether you are in an EXCEPTIONS statement. If you are going to accumulate this exception information then it is your responsibility to test it.

Hennell   When we designed the ALGOL-68 NAG library error mechanism we spent a lot of time pondering this question of whether to allow the user to return to the scene of the crime and change the evidence. After an awful lot of time someone asked that devastating fundamental question: "Does anyone know of a situation where we actually want to do this?" And a working party was set up and could not come up with anything.

Lake   We calculate integer powers of integers in parallel by accumulating binary powers and may get overflow at various stages. We accumulate the exception map internally and return it when the calculation has finished for all elements.

Kahan   The things you have been concerned about have weighed heavily upon the people who have been deliberating the proposed IEEE standard and the reason that we could not do what Hull suggested was that the trapping mechanism would be too general. An array processor should not trap out and for just one element go through an elaborate subroutine. So that meant that the equivalent of the PL/1 ON CONDITION for the specification of a complicated trap was just out of the question; that is why the NaN was created. If you say X = 'hideous expression' then you only want to know which elements of X have been contaminated and you only have to have that as the information crosses the assignment. Now there are two ways to cope. One is to have some type of invalid result like a NaN or infinity which, when it crosses the assignment, is recorded into your map E. The other way, which is even in some respects nicer, is to say that within the vector registers of your machine you have an extended range and precision so that the exception does not often occur until you cross the assignment. It may even not occur, since the final result may be within range. You do not have to do all this involved mapping stuff - it becomes irrelevant.

Meissner   If it is sufficient to record exceptions "across the assignment operator", then why is there a need to detect exceptions in the middle of expression evaluation?

Lake   Where you want to take a trap you could take it immediately after the offending operation. Recording exceptions only on assignment may entail the loss of information as to the cause of the exceptions.

*THE RELATIONSHIP BETWEEN NUMERICAL COMPUTATION
AND PROGRAMMING LANGUAGES, J.K. Reid (editor)
North-Holland Publishing Company*
© *IFIP, 1982*

A FRAMEWORK FOR LANGUAGE EVOLUTION

Jerrold L. Wagener
Amoco Computing Research
P.O. Box 591
Tulsa, OK 74102

I. INTRODUCTION

The 25 years since the advent of the first Fortran compiler have seen an
explosive growth of knowledge about computer languages, and the corresponding
design of thousands of high-level languages.   Many of these designs, including
new versions of Fortran itself, have extended and/or improved Fortran's
facilities for performing numeric/scientific computations.

There is no reason at this time to expect a decrease in the growth rate of
language technology.   Indeed, scientific applications of computing, and related
computing technology, are increasing as never before.   Language development and
evolution will therefore continue; it is the purpose of this paper to sketch
one model for language evolution that is predicated upon the inevitability of a
highly dynamic language environment for scientific computation.

This model is the one currently being considered for adoption by the American
National Standards Institute Technical Committee X3J3 for standard Fortran.   The
description here, due solely to individual interpretation by the author, is
believed to be close to the structure being developed by X3J3.   However, the
reader should bear in mind that the actual X3J3 model is subject to meeting-by-
meeting variation, will almost certainly differ in detail from that described
here, and may ultimately take quite a different form.

The model accommodates two, somewhat (though not entirely) orthogonal, dynamic
aspects of language technology.   The first involves the evolution of specific
language features, irrespective of application areas using the language.   The
second involves the language's ability to adapt effectively to the changing needs of
specific application areas.

The elements of this structure may be depicted as follows:

Common "Extensions"

The "Core Language"   Application-Specific Facilities

"Obsolete" Features

The left-hand column of this diagram, consisting of "extensions", the "core",
and "obsolete" features, comprise the actual language facilities.   The core
language, detailed below for Fortran, contains those language features necessary
for contemporary solution expression of typical problems in the principal
application areas for which the language is intended.   The core should not contain
highly specialized or experimental features, those being classified instead as
common extensions.   The nature of the Fortran core is more completely described
by the set of "core criteria" described below.

As a language evolves, inevitably certain features become vestigial, or
even "harmful" to the viability of the language.   It may not be feasible to
actually remove such features from the language, however, and at the same time
protect its existing software inventory.   Classifying these features as "obsolete"
serves to keep them in the language yet remind users that other features exist
in the language that provide equivalent functionality in a "better" way.   During
subsequent stages of evolution, certain obsolete features may be dropped from the
language.   The use of obsolete features in new software would normally result in
lower-quality software, in some sense, than if such features were not used.   The

use of obsolete features in application-specific facilities (see above diagram)
should be especially discouraged.

The horizontal dimension in the above diagram - the core language and
application-specific facilities - represents the adaptation of the language to
specific and dynamically changing application needs. Application-specific
facilities should normally be based upon the features of the core language. Such
facilities might include procedure libraries, macro libraries, and/or data
object libraries, or composed of elements that can be readily transformed
("preprocessed") into core language features.

In the next two sections this general framework will be illustrated by
its proposed application to Fortran.

II.   THE FORTRAN CORE

The composition of the Fortran core language is governed by the following
seven core criteria:

1.   GENERAL PURPOSE

Core Fortran should be a complete language, suitable for general purpose
application programs.   This core must especially maintain Fortran's
capability for general purpose scientific programming applications.

2.   PORTABLE

A principal goal of the core is (program and people) portability;
the core should contain a minimum of non-portable aspects.

3.   SAFE

Core Fortran should be effective for the development of reliable
software.

4.   EFFICIENT

Features that preclude either compilation or execution-efficiency with
conventional contemporary computing technology should not be included in
the core.

5.   CONCISE AND CONSISTENT

The core should be a small language, whose syntactic and semantic
elements exhibit regular and consistent patterns.

6.   CONTEMPORARY

The core should be characterized by language features that are broadly
accepted as currently the best means of achieving the desired
functionalities.

7.   COMPATIBLE

The union of the core and the obsolete features should not entail
unresolved conflicts;   this union should provide the complete functionality
of Fortran 77 and successfully process Fortran 77 programs.

It is recognized that these criteria cannot all be perfectly met by any specific core design. The criteria should therefore be used as guidelines in constructing the core language. A proposed core based upon these criteria follows.

```
program-unit     is   unit-heading
                      [external-facility]...
                      [local-specification]...
                      block
                      END

unit-heading     is   [PROGRAM program-name]
                 or   SUBROUTINE subroutine-name [(dummy-arg-list)]
                 or   FUNCTION function-name [(dummy-arg-list)]
                 or   library-heading

external-        is   USING external-facility-list
    facility

local-           is   FORM structure-name structure-definition END FORM
specification    or   IMPLICIT implicit-list
                 or   PARAMETER (constant-name=value[,constant-name=value]...)
                 or   type-name declaration [,declaration]...
                 or   SAVE [(local-name[=value][,local-name[=value]]...)]
                 or   INTRINSIC procedure-list  or  EXTERNAL procedure-list

block            is   [ [label] executable-stmt ]...

executable-      is   [IF (scalar-logical-expr)] basic-stmt
       stmt      or   DO [loop-control] block REPEAT
                 or   IF (scalar-logical-expr) THEN block
                      [ELSE IF (scalar-logical-expr) THEN block]...
                      [ELSE block]
                      END IF

basic-stmt       is   assignment-stmt
                 or   io-stmt
                 or   GO TO label
                 or   CALL [prefix-name.]subroutine-name[(actual-arg-list)]
                 or   CONTINUE or EXIT or RETURN or STOP [stop-code]

assignable-      is   [prefix-name.] oject-name[section] [.component]...
       object

section          is   subscript-range [,subscript-range]...

subscript-       is   subscript-expr
       range     or   [subscript-expr]:[subscript-expr]
```

The preceding syntax is presented in a form of BNF similar to that used in the Fortran 77 standard. Metasymbols are the (italicized) words "is" and "or" for defining productions, square brackets specify optionality, and three dots indicate repetition. Terminal symbols are in upper-case (or are special symbols) and lower-case constructs represent syntactic classes. Only the high-level syntax is presented here, with low-level syntax (e.g., name constructs) left undefined (in most cases simply refer to Fortran 77 for the details of the low-level syntax). The purpose here is to focus attention on the overall structure of the core language.

As indicated by the last few productions in the preceding syntax, this core design includes dimensionally contiguous arbitrary array sections as assignable objects. Expressions, constants, and functions may be array-valued. Programmer-defined data structures (FORM - END FORM) may also be objects of assignment and function values. Procedures may be recursive, and programmers have greater control, than in Fortran 77, over the precision properties of numeric objects.

A major new functionality of this core, over Fortran 77, is the extensive array facilities it provides. In addition to the above-mentioned features, these include dynamic local arrays. Fixed-length bit and character processing is provided by array operations on, respectively, LOGICAL arrays and arrays of single CHARACTERs. The array operations, and recursion, imply a stack-type dynamic storage capability; however this core design does not involve the more complicated requirements of heap storage management.

Another feature of this design is the identification (via the USING statement) of externally-defined facilities that must be available to the compiler. These include global data libraries that the program uses. Such libraries provide a global data mechanism without the storage association of COMMON; storage association is not required in this core design. The prefix-name in the above syntax is a qualification mechanism to disambiguate references to externally-defined facilities (such as global data objects).

Finally, the above syntax does not identify the set of intrinsic functions available in the core. This set could be quite rich, including a number of useful array intrinsics.

III. THE FORTRAN OBSOLETE FEATURES

Any features of Fortran 77 not in the core language would become obsolete features. Based on the above core design, the list of obsolete Fortran features is as follows.

1. Fortran 77 source form (column 6 for continuation; "C", "*" in column 1 for comments, etc.)
2. EQUIVALENCE stmt
3. COMMON stmt and BLOCK DATA program unit
4. Passing an array element or substring to a dummy array
5. arithmetic-IF stmt
6. computed-GOTO stmt
7. alternate RETURN
8. ASSIGN, assigned-GOTO stmts
9. statement functions
10. DOUBLE PRECISION
11. END= and ERR= clist specifiers
12. H,X and D edit descriptors
13. specific names for intrinsic functions
14. ENTRY stmt
15. DATA stmt
16. Fortran 77 DO stmt
17. FORMAT stmt
18. DIMENSION stmt
19. PAUSE stmt
20. COMPLEX data type (also an extension)
21. *len specifier for CHARACTER data type

These obsolete features are not removed from the language - they remain an integral part of the language, the important bridge to the previous version.

However, their explicit identification as "obsolete" can serve as an effective flag to users that the language very likely offers better ways to perform the same functions.   In fact, the existence of this explicit set will allow processors to inform users, at each compilation, of the existence of obsolete features in the code.   This could form the basis of a continuing programmer education opportunity that in the long term could well prove more effective in improving software quality than any other recent software development.

IV. CONCLUSION

In this model of language evolution changes occur only at discrete times ("revision" times). At a revision time certain obsolete features may be dropped from the language entirely, and certain core features may be newly classified as obsolete.   On the other hand, new features may come into the core - in particular, extensions that prove to adequately meet the core criteria. In a sense the extensions may be thought of as the "input space" for the language (i.e., new features are tried out here before moving into the core), with the obsolete features being the "output space".

One must be careful, however, in simply moving features about in this manner, lest the core lose much of its integrity and coherency.   In the final analysis the proposed new version of the core should, as a whole, conform maximally to the core criteria.

References

(1) Document X3J3/S6.79, ANSI technical committee X3J3, August 1981.

(2) Minutes, X3J3 meeting # 79, August 1981, Document 79(10) JLW-1.

DISCUSSION

Delves   I think that reactions from the audience in the past two days stem from the impression that X3J3 is giving insufficient attention in its deliberations, to one important design principle.   That principle is economy: economy both of new language features (what the user can write) and of exceptions (what the user cannot write).   The danger stems perhaps from the sub-committee structure used by X3J3, with different groups discussing and making proposals in different language areas, with no way of knowing, until the proposals reach committee level, what is going to be proposed elsewhere.   I can point to two examples of an apparent lack of economy in the current proposals:
1)   WHERE as a new control structure rather than extending the meaning of IF-THEN-ELSE (violating economy of features).
2)   The introduction of new data types (e.g. RECORDS) but with the restriction that arrays of these data types cannot be constructed (violating economy of exceptions).
I hope that X3J3 intends over the next year or so to review its proposals as a whole, with the advantages of "clean" design very much in mind.

Kahan   By the way of the shibboleth of compatibility "the sins of the fathers are visited onto the children, yea of the third and fourth generation".   I was once asked by a company for which I consult from time to time whether they should be compatible with something or other. We looked at codes representing enormous investments and found that they were never going to see a compiler again so why care about the compatibility of the compiler?   They care much more about the compatibility of the operating system.   Then there were codes which do see the compiler frequently because they are always being changed.   You said that you are willing to re-educate your programmers. Sometimes the greatest education of all is liberation from past misconceptions.   It may be that using a new and simpler language to code the same functionality would be advantageous, provided the languages compile to modules that communicate at execution time.

Sedgwick   In programming languages in the late 60's and early 70's there was a great deal of attention paid to 'extensible languages'.   Early examples of that might be Algol 68 and a late example would be Ada.   This attention was paid to the design of the extensions that would not necessarily conflict with one another.   We are able to define new operators and allow different users to select different subsets of operator definitions that they would like to work with.   It seems that the FORTRAN standards committee is moving in a different direction from this well established body.   I do not fully understand the differences, but it seems that you are adding a lot of extension modules, each having its own key-word statements, and it is going to be awkward to work with mixtures.   I can well understand that one extension might be interval arithmetic, another might be multi-precision arithmetic, but there is going to be somebody who is going to want to use them both.   And it is important that the whole mechanism for adding extensions allows you to do that.

Wagener   The architecture that I described is the germ of an idea.   The elements of each module have not yet been set.   There have been things discussed like   . macros, for example, that have not been adopted by X3J3.   Whether such a facility would go in the core or in some part of an extension module is not clear.   At one point we were talking about a single extension module and the specialized features that we did not want to put into the core would be in the extension module. Currently the thinking has been in terms of multiple extension modules.   We hope to develop the core plus the extension modules as a coherent, consistent language like Ada or Algol 68.   We may not be successful.   Any assistance and/or suggestions you want to give us to help us do that we would accept.   I do not know of anything under consideration which makes any extension module incompatible with the core.

Lake   I think that one thing that X3J3 has to think about very carefully is the form of its report.   Will it be provably consistent?   The second thing that I feel is that the whole user community has been let down by the fact that the idea of extensible languages has not been pushed forward in the way that is clearly required.   People do not just want to have a library routine, they want to be able to specify as a package - the library routine, the optimization strategy, the operators, etc.   They want to do this transformational work as well.

Dekker   It is important that the core is such that the modules can be described by mechanisms in it.   An example in Algol-68 is the matrix package TORRIX by van der Meulen.

Wagener   That is what X3J3 intends to do with respect to the application modules.   The application modules are to be built out of the core plus language extension module mechanisms.

Schlechtendahl   Our experience at Karlsruhe indicates that:
1)   at compile time you must be able to switch from one to another language;
2)   similarly at execution - you need to be able to enter completely separately environments dynamically.

Ford   The strategy of using a core plus modules structure for the next version of Fortran is inevitably a matter for anxiety and concern for software developers - indeed for any programmer who seeks to run his programs on many different machines. NAG, for example, runs its library on 33 different machine families. Our fear is that each manufacturer will simply implement the core plus only those modules relevant to the specific market places he wishes to exploit.   We shall return to the situation of substantial dialect problems (as in the late 60's and early middle 70's) between different Fortran systems, only far more vicious since many modules (for example those for precision, environmental enquiries, array manipulation,...) may simply not be available. We would be forced to use only the core, in our pursuit of transportability, and that core would be devoid of fundamental features required by numerical software (as these are perceived as not being essential for all scientific and technical computing). Why are you risking this situation?

Ford   Users have no control over what computer manufacturers include in their Fortran systems. Very few customers have sufficient clout to influence such matters.   The standard is an excellent social control mechanism. If there is a single integrated standard from our previous experience we know that the vast majority of computer manufacturers will implement the whole standard.   The core plus modules design of the standard certainly permits - perhaps even encourages - the fractured approach to implementation of the standard discussed above.   Please use the control mechanism of the standard to ensure one integrated language specification, all of which will be implemented by computer manufacturers.

Wagener   We appreciate this input. Language development is an evolutionary process. We cannot expect to get it exactly "right" all at once. What we can expect is to get it better and better as a function of time.

THE RELATIONSHIP BETWEEN NUMERICAL COMPUTATION
AND PROGRAMMING LANGUAGES, J.K. Reid (editor)
North-Holland Publishing Company
© *IFIP, 1982*

LANGUAGES AND HIGH-PERFORMANCE COMPUTATIONS[*]

D. Kuck, D. Padua, A. Sameh, and M. Wolfe

Department of Computer Science
University of Illinois at Urbana-Champaign
Urbana, Illinois 61801

In this paper we discuss aspects of the process of
solving a numerical problem on a parallel or vector
computer. The most important stage in this process
is the first--the design of a stable numerical algorithm that is suitable for the machine's architecture. We also consider the important role of automatic restructurers in solving the problem of
software transportability among various parallel and
vector computers.

1. INTRODUCTION

Several languages have been proposed or implemented for programming array and
vector processors. They are all based on one of the traditional sequential languages, usually Fortran. In the course of this paper we discuss briefly five
representative languages of the 1970s. Some are closely tied to the architecture
of the machine such as the CDC Fortran vector extensions for the CYBER 205, while
others are proposed independently of an existing machine such as IBM's Vectran.
We compare these languages, illustrate their use on a simple example, and indicate their deficiencies in handling a parallel algorithm.

Even though a programming language plays an important role in realizing a "good"
performance of a program on a parallel or vector computer, it is neither the only
nor the most important factor. Figure 1 shows the classical three levels of distinct activities that are needed in order to map a numerical problem onto a machine. (For the sake of illustration, assume that we are dealing with a parallel
computer of p identical processors.) It is obvious that at each of the three
levels depicted, many possibilities exist when one moves down one level. We will
discuss how to move along the path marked by the stars to reach the region of
"good" performance for the parallel machine under consideration. We define such a
region as follows. Let the time required to execute a program on the parallel machine be τ_p, while the time required to execute its sequential counterpart on only
one of the p processors be τ_1. Then the region of "good" performance is that for
which τ_1/τ_p is not too small a fraction of p.

Level 1

By a combination of physical, mathematical, and numerical analysis, a <u>crucial</u>
first-step is the development of an algorithm. Assuming that the numerical analyst is aware only of the broad features of the machine's architecture, such as

---

[*]This work was supported in part by the National Science Foundation under Grant
Nos. US NSF MCS79-18394 and US NSF MCS80-01561.

D. Kuck et al.

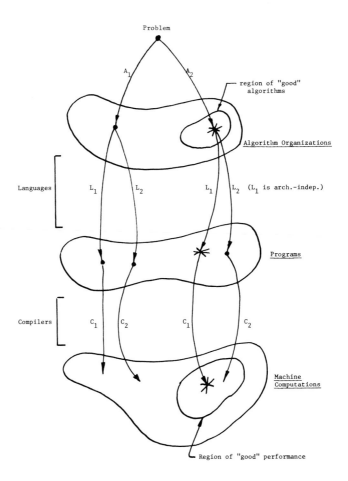

Figure 1

the type of the interconnection network , the numerical algorithm produced should:

    (a)   be numerically stable,

    (b)   achieve maximum utilization possible of the available p processors, and

    (c)   have the cost of its interprocessor communications be at most of the same order as the cost of its arithmetic.

For real physical problems, this may amount to organizing a number of such distinct algorithms into a coherent whole.

Today, many of the numerical analysts developing algorithms for array and vector computers concern themselves with the subsequent levels in Figure 1, e.g., language details of the machine together with the available compilers and other software. In the future, such considerations should not be necessary except in the broadest terms. In the past decade, a good deal of research has been done on parallel numerical algorithms. Much of this work, however, is not widely used for one of two reasons. Either it was done for a specific machine and its software and hence is of little value elsewhere, or it was done more abstractly and the lack of "good" languages and compilers has prevented its practical implementation on the various available high-performance machines.

Level 2

Assume, now, that a good algorithm organization has been selected. Next, a language must be chosen. Ideally, the language should be architecture-independent and have expressive powers that are well-suited for numerical algorithms in general. Thus, the structure of the program written should depend mainly on the user's perception of the algorithm and the syntax of the language. Thoughts about the architecture and its constraints should not have to enter the process except perhaps as broad guidelines.

Level 3

Once a program has been written in a given language, a well-designed compiler, ideally suited for the details of the machine's architecture, translates the program into machine language that will deliver high performance. We expect the compiler to make such choices as proper storage schemes for arrays and correct mappings of unlimited resource programs onto fixed-resource machines.

Since the process depicted in Figure 1 depends heavily on the machine, it is clear that the problem of software transportability among the various parallel and vector computers is much more complex than in the case of sequential machines. Consider, for example, the case where we have two programs (written in the same language L) implementing two different algorithms for solving the same problem on two machines of different architecture, see Figure 2. Compiling P2 for machine 1 will most probably result in poor performance. Such inevitable deterioration of performance may be minimized by the use of a software restructurer that can be tuned to several basic architectures. We envision that such a piece of software will change the structure of P2 to make it as suitable as possible to the details of the compiler C1. Even though we may come close to the point H1 of the highest performance possible, we are aware that (except in very simple programs) H1 cannot be reached using only the information embodied in P2.

We should state here that whenever one uses an automatic restructurer to change the computational graph of the original stable numerical algorithm, the numerical stability of the altered computational graph should be investigated by an automatic error analysis package such as that provided in [MiSp78], or [LaSa80].

In the remainder of the paper we discuss these three levels in the reverse order.

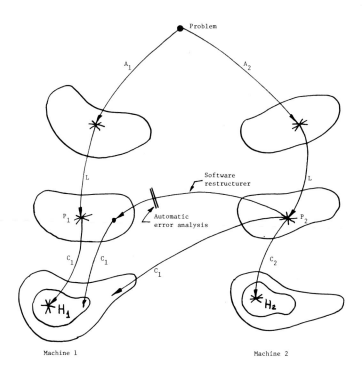

Figure 2

Language	Cyber 205 Fortran	Cray Fortran	BSP Fortran	Vectran	Actus
Base Language	Fortran	Fortran	Fortran	Fortran	Fortran
Subscript Simultaneity	1-dimension	1-dimension	multi-dimension	multi-dimension	1-dimension
Memory Access Allowed in Above	Stride-1	Any Stride	Any Stride and Regular Partition	Any Stride and Regular Partition	Any Stride
Recurrences	Reductions	Reductions	Many	Many	None
Vector IF Statements	Using META Assembler Calls	Using Merge Functions	Yes	Yes	Yes
Array Valued Functions	Yes	No	Yes	Yes	Yes
Automatic Vectorization	Limited	Yes	Yes	---	---

Table 1

Section 2 covers aspects of restructuring, Section 3 surveys some of the available languages, and Section 4 discusses an algorithm for tridiagonal systems which supports our earlier assertion that the design of the algorithm is the most important step in using a high-performance machine.

2. RESTRUCTURING

In order to illustrate how a software restructurer may make the transportation of programs among machines of different architectures less painful and in some cases quite successful, we consider the following simple programming problem.

Suppose we are given an m×n array A of rows a_j^T, j = 1, 2, ..., m, where m is even, and a multiprocessor of k = m/2 identical processors. The problem is first to compute the inner products $P(i) = a_{2i-1}^T a_{2i}$, i = 1, 2, ..., k, forming a vector P, followed by computing X = 2 * P + Y/P, where Y is a given vector of order k.

An implementation of the most obvious algorithm for this architecture is shown in Program 1.

```
            real array    A[m,n], P[k], X[k], Y[k]

            define  Row [i] = A[i,*]

            for all  i := 1  to  k  do
S1                          P[i] ← inner prod (Row [2i-1], Row [2i])
S2                          X[i] ← 2 * P[i] + Y[i]/P[i]

            end do
```

Program 1 -- Given Program

Here, the k = m/2 inner products are computed simultaneously and independently, one inner product per processor.

If we wish to run Program 1 on an array machine rather than the above multiprocessor, several changes need to be made. In Programs 2 and 3, we illustrate the transformations that a software restructurer may affect so as to produce Program 4.

```
            real array    A[m,n], P[k], X[k], Y[k]
            for   i ← 1,k    do
S0                   P[i] ← 0
                     for   j ← 1,n   do
S1                          P[i] ← P[i] + A[2i-1,j] * A[2i,j]
                     end do
S2                   X[i] ← 2 * P[i] + Y[i]/P[i]
            end do
```

Program 2 -- Expansion of Inner Product Function

```
            real array    A[m,n], P[k], X[k], Y[k]
            for   i ← 1,k    do
S0                    P[i] ← 0
            end do
```

```
          for    j ← 1,n    do
                 for    i ← 1,k    do
  S₁                    P[i] ← P[i] + A[2i-1,j] * A[2i,j]
                 end do
          end do
          for    i ← 1,k    do
  S₂             X[i] ← 2 * P[i] + Y[i]/P[i]
          end do
```

Program 3 -- Distribution and Interchanging of Loops

```
          real array    A[m,n], P[k], X[k], Y[k]
          index    Even = <2,4,...,m>
  S₀      P ← 0
          for    j ← 1,n    do
  S₁             P ← P + A[Even-1,j] * A[Even,j]
          end do
  S₂      X ← 2 * P + Y/P
```

Program 4 -- Vector Subscripts

Next, suppose that the program is to be run on a machine with a limited amount of local memory such as the CRAY-1. Since the CRAY-1 has eight 64-word vector registers, we will have to deal with strips of the array A each of width 64. Program 5 shows the result of the various transformations that a software restructurer may apply to Program 4 so as to produce a code suitable for the CRAY-1.

```
          real array    A[m,n], P[m], X[m], Y[m]
          index    Even[i] = <2 + (i-1) * 64,4 + (i-1) * 64,...,min(i*64,m)>
          for    i ← 1, ⌈m/64⌉    do
  S₀             P[Even[i]] ← 0
                 for    j ← 1,n    do
  S₁                    P[Even[i]] ← P[Even[i]] + A[Even[i]-1,j] * A[Even[i],j]
                 end do
  S₂      X[Even[i]] ← 2 * P[Even[i]] + Y[Even[i]]/P[Even[i]]
          end do
```

Program 5 -- Blocked Loop

A related problem is that of long vector start-up time for machines such as the CDC STAR-100 and to a lesser extent the CDC CYBER 205. For such machines, it is crucial to operate on vectors that are as long as possible. Thus, if in our example $m = 10$ and $n = 1000$, a program similar to Program 5 is definitely not suitable, as the inefficiencies incurred in computing the reduction operations in inner products will be negligible compared to the start-up overhead for very short vectors. Thus, in transporting Program 1 to a vector machine the transformations affected by the software restructurer will have to depend on the relative values of m and n. Note that any decision based on relative values of m and n may not be possible until run-time.

Another important factor in deciding whether the software restructurer should avoid the reduction operations in inner products is the capability of the memory of the machine to which the program is to be transported. For example, if the compiler normally stores rows across the memory units it may be difficult to access columns, and vice versa.

Automatic restructuring transformations have been implemented for various architectures. The Parafrase system can be configured to produce code and analyze the results for several types of architectures including pipeline processors [KKLW80], and multiprocessors [PaKL80].

In Section 1 (Level 1), we mentioned that the numerical analyst should be aware of interprocessor communication costs in designing an algorithm, for it is certain that a compiler cannot restructure properly (for a given architecture) an algorithm that is sufficiently ill-suited for such an architecture. On the other hand, it may be extremely difficult for a numerical analyst, or even a computer architect, to assess in detail the impact of a memory and an interconnection network on a particular algorithm's performance. However, if the numerical analyst provides a reasonably good algorithm-machine match, it is possible to have a restructuring compiler map a program onto a machine in such a way that the data movement and computation times are matched in order of magnitude. Such a restructuring technique has been presented and demonstrated to be effective on a number of algorithms [Kuhn80].

3. PROGRAMMING LANGUAGES

Several languages have been designed, implemented, or used in the 1970s. Most of these languages are based on Fortran. We will briefly discuss five representative languages (see Table 1) and illustrate their use on the example problem in Section 2.

3.1 CDC Fortran Vector Extensions

This language was originally designed in the 1960s for the CDC STAR and is now used for the CYBER 205 [CoDC78], [Kasc79]. The compiler of the CYBER 205 performs limited vectorization. Due to the stringencies of the CDC machines' memories, only one parallel dimension is allowed and each element in this direction must be accessed and processed (stride 1). Some reduction functions are supplied in the language, as well as array valued functions. Vector IF statements are also allowed using META (the CYBER 205 assembler) in-line calls.

Program 6 is the analogue of Program 1 written using CDC Fortran vector extensions. Notice that the storage scheme for two-dimensional arrays must be specified; here, the rows of A are stored across the memory units. A dot product procedure is available in the language and its arguments have vector subscripts. Thus, $A(I,1;N)$ accesses memory beginning at $A(I,1)$ and fetches N contiguous elements. Since A is stored row-wise, this selects the Ith row of A. The array A must be stored in this way for the function call to work.

If the columns of A are, by default, stored across the memory units, the analogue of Program 4 becomes more complicated, see Program 7. Since the memory can only access arrays with stride 1, we have to double the dimension of P and perform roughly twice as many multiplications as necessary. Note that only the even elements of P contain meaningful results.

```
      REAL A(M,N), P(K)
      ROWWISE  A
      DO  1  I = 1, K
         P(I) = Q8SDOT(A(2*I-1,1;N), A(2*I,1;N))
1     CONTINUE
```

Program 6 -- CDC Fortran

```
      REAL A(M,N), P(M)
      P(2;M-1) = 0
      DO  1  J = 1, N
         P(2;M-1) = P(2;M-1) + A(1,J;M-1) * A(2,J;M-1)
1     CONTINUE
```

Program 7 -- CDC Fortran

3.2 Cray Fortran

The compiler of this language, developed for the CRAY-1, performs more vectorization than that of the CYBER 205. The language has no explicit vector extensions,
and no recurrence operators. It provides the compiler, however, with directives
that aid in vectorizing complex loops. Vector IF statements in the form of
merges are possible. As is the case in the CYBER 205, this requires redundant
computations.

Using Cray Fortran, Program 1 can be presented in the form of Program 8, see
Section 4, page 5 in [Cray80]. This program is complicated by the fact that vector registers of size 64 must be dealt with, and because of the absence of machine
instructions for vector reductions. It is written to achieve maximum speed, and
in this form, three of the four inner loops will be vectorized by the Cray CFT
compiler, namely, loops 2, 4, and 5. Loop 2 is used to initialize a vector register of min(64,N) elements. This is followed by the computation of the number of
iterations for loops 3 and 5. The value of KOUNT is the number of 64 word blocks
remaining, and LAST is the number of words in the last partial block.

```
      REAL A(M,N), P(K), REG(64)
      DO  1  I = 1, K
         DO  2  J = 1, MIN(64,N)
2           REG(J) = A(2*I-1,J) * A(2*I,J)
         JS = J - 1
         KOUNT = (N-JS)/64
         LAST = (N-JS)-64 * KOUNT
         DO  3  JB = 1, KOUNT
            DO  4  J = 1, 64
4              REG(J) = REG(J) + A(2*I-1,JS+J) * A(2*I,JS+J)
3           JS = JS + 64
         DO  5  J = 1, LAST
5           REG(J) = REG(J) + A(2*I-1,JS+J) * A(2*I,JS+J)
         P(I) = REG(1)
         DO  6  J = 2, MIN(64,N)
6           P(I) = P(I) + REG(J)
1     CONTINUE
```

Program 8 -- Cray Fortran

(The CFT compiler has been modified recently to perform the above vector reductions automatically.)

Program 9, the analogue of Program 4, on the other hand, does not contain inner product reductions; hence, it is much easier to write in Cray Fortran where the inner loop is vectorized via CFT.

```
        REAL A(M,N),P(K)
        DO  1  I = 1, K
            P(I) = 0
1       CONTINUE
        DO  2  J = 1, N
            DO  2  I = 1, K
                P(I) = P(I) + A(2*I-1,J) * A(2*I,J)
2       CONTINUE
```

Program 9 -- Cray Fortran

3.3 BSP Fortran

Burroughs BSP Fortran [Burr75] was supported by a very ambitious vectorizer. The language allows n dimensions of parallelism (two supported in the hardware). Regular subarrays can be accessed with any stride in any dimension. Many hardware supported recurrence operators exist in the language. A WHERE clause is used to conditionally execute a single assignment statement or block of assignment statements. Array-valued functions are also available.

The BSP Fortran version of Program 1 is quite similar to the CDC Program 6. Since the conflict-free BSP memory allows us not to worry about row-wise versus column-wise storage, Program 10 (the analogue of Program 4) is much simpler than CDC Program 7.

```
        REAL A(M,N), P(K)
        P(*) = 0
        DO  1  J = 1, N
            P(*) = P(*) + A(1:M-1:2,J) * A(2:M:2,J)
1       CONTINUE
```

Program 10 -- BSP Fortran

Here, the commonly used * notation indicates all elements of an index position. The notation A(2:M:2,J) indicates the array consisting of the even elements 2 through M of column J.

3.4 Vectran

Vectran is a language proposed by IBM [PaWi75] independently of any specific architecture. Parallelism in n dimensions can be expressed and regular subarrays can be accessed. Many recurrence operators are provided in the language. Vector IF statements in the form of AT or WHEN clauses are available, as well as array-valued functions.

Again, the Vectran analogue of Program 1 is similar to Program 6. In Program 11, the analogue of Program 4, the IDENTIFY statement is used to partition previously declared arrays. Upper limits of each subscript position are given between slashes; lower limits of 1 and increments of 1 are implicit.

```
        REAL A(M,N), P(K)
        IDENTIFY/K,N/AODD(I,J) = A(2*I-1,J)
        IDENTIFY/K,N/AEVEN(I,J) = A(2*I,J)
        P(*) = 0
        DO  1  J = 1, N
            P(*) = P(*) + AEVEN(*,J) * AODD(*,J)
```

```
1     CONTINUE
```

Program 11 -- Vectran

3.5 Actus

Actus is a Pascal-based language that was proposed without a specific machine intended for its implementation [Perr79]. Actus allows only one-dimensional parallelism, and the same dimension must be parallel throughout the whole program (the designer had in mind a machine with a somewhat weak memory system). Any stride is possible along one dimension. No recurrence operators are provided. Since no inner product function is included, a serial program would have to be written for Program 1.

The analogue of Program 4 appears as Program 12. The declaration of A indicates that the first index may be accessed in parallel, while the second index is for serial access only. I is an index from 2 to M in steps of 2 and SHIFT is used to index A. The language also provides ROTATE to wrap an array around on itself.

```
        VAR
            A:ARRAY[1:M,1..N] OF REAL;
            P:ARRAY[1:K] OF REAL;
        INDICES
            I = 2:(2)M;
        BEGIN
            P(*) :=0;
            FOR J := 1 TO N DO
                P(*) := P(*) + A(I,J) * A(I SHIFT -1,J)
        END;
```

Program 12 -- Actus

A revised Actus language which removes some of these limitations is presented in [Perr81].

4. AN ALGORITHM

Above, we discussed an elementary example, namely, that of inner products. In order to illustrate our assertion that a crucial first step in producing an efficient code for a class of parallel machines is a well-designed parallel algorithm we consider the simple problem of solving a positive definite tridiagonal system $T\underset{\sim}{x} = \underset{\sim}{f}$, where $\underset{\sim}{T} = [b_{i-1}, a_i, b_i]$ is of order n, on a parallel machine with $1 < p \ll n$ linearly connected processors.

The favorite sequential algorithm is, of course, the LDL^T factorization

$$\underset{\sim}{T} = \underset{\sim}{L} \underset{\sim}{D} \underset{\sim}{L}^T, \tag{4.1}$$

where L is unit lower triangular, and D is diagonal, followed by solving the three systems $L\underset{\sim}{z} = \underset{\sim}{f}$, $D\underset{\sim}{y} = \underset{\sim}{z}$, $L^T\underset{\sim}{x} = \underset{\sim}{y}$. A program is shown below.

```
PROCEDURE  decomp (n: integer;
        a,b:   ARRAY [1..n] of real;
        VAR l,d:  ARRAY [1..n] of real);
VAR j: integer;
BEGIN
        d[1] := a[1];
        FOR j := 1 to n - 1 DO
            l[j] := b[j]/d[j];
```

```
        d[j+1] := a[j+1] - b[j] * ℓ[j]
          ENDDO;
    END;
PROCEDURE   solve (n: integer;
      d,ℓ,f: ARRAY [1..n] of real;
      VAR x:  ARRAY [1..n] of real);
VAR j: integer;
BEGIN
      x[1] := f[1];
      FOR j := 2 to n DO x[j] := f[j] - ℓ[j-1] * x[j-1] ENDDO;
      FOR j := 1 TO n DO x[j] := x[j]/d[j] ENDDO;
      FOR j := n - 1 DOWNTO 1 DO x[j] := x[j] - ℓ[j] * x[j+1] ENDDO;
    END;
```

Supplying this sequential algorithm to an automatic restructuring compiler, hoping for an output that is a high speed and efficient parallel algorithm, is unwise. First, if the restructuring compiler cannot reduce continued fractions to linear recurrences, the factorization part (4.1) can only be handled sequentially. Second, even if the restructuring compiler can handle continued fractions, the factorization process often results in a very ill-conditioned second-order linear recurrence in products of the elements of $\underset{\sim}{D}$, with the elements of the solution suffering from over· or underflow [SaKu78]. In order to circumvent these pitfalls, we propose the following stable and efficient parallel algorithm which achieves a speedup of $O(p)$ [Same81].

Let $\underset{\sim}{T}$ and $\underset{\sim}{f}$ be partitioned as $\underset{\sim}{T} = [\underset{\sim}{E}_{i-1}, \underset{\sim}{T}_i, \underset{\sim}{E}_i]$, $i = 1, 2, \ldots, p$, and $\underset{\sim}{f}^T = (\underset{\sim}{f}_1^T, \underset{\sim}{f}_2^T, \ldots, \underset{\sim}{f}_p^T)$, where $\underset{\sim}{T}_i$ and $\underset{\sim}{f}_i$, $i = 1, 2, \ldots, p - 1$ are each of order $q = \lceil n/p \rceil$, and $\underset{\sim}{E}_i = [b_{iq} \underset{\sim}{e}_q, 0]$ in which $\underset{\sim}{e}_i$ is the ith column of the identity $\underset{\sim}{I}_q$. Since $\underset{\sim}{T}$ is positive-definite, it follows that each $\underset{\sim}{T}_i$ is also positive-definite. Premultiplying both sides of $\underset{\sim}{T} \underset{\sim}{x} = \underset{\sim}{f}$ by the block-diagonal matrix $\underset{\sim}{C}^{-1} = \text{diag}(\underset{\sim}{T}_1^{-1}, \underset{\sim}{T}_2^{-1}, \ldots, \underset{\sim}{T}_p^{-1})$, we obtain the block tridiagonal system, see Figure 3,

$$\underset{\sim}{W} \underset{\sim}{x} = \underset{\sim}{g}, \qquad (4.2)$$

where $\underset{\sim}{W} = [\underset{\sim}{V}_i, \underset{\sim}{I}_q, \underset{\sim}{U}_i]$, $i = 1, 2, \ldots, p$, and $\underset{\sim}{g}^T = (\underset{\sim}{g}_1^T, \underset{\sim}{g}_2^T, \ldots, \underset{\sim}{g}_p^T)$. Here, $\underset{\sim}{U}_i = [\underset{\sim}{u}_i, 0]$, $\underset{\sim}{V}_i = [0, \underset{\sim}{v}_i]$, and $\underset{\sim}{g}_i$ are given by the linear systems $\underset{\sim}{T}_1[\underset{\sim}{u}_1, \underset{\sim}{g}_1] = [b_q \underset{\sim}{e}_q, \underset{\sim}{f}_1]$ $\underset{\sim}{T}_j[\underset{\sim}{v}_j, \underset{\sim}{u}_j, \underset{\sim}{g}_j] = [b_{(j-1)q} \underset{\sim}{e}_1, b_{jq} \underset{\sim}{e}_q, \underset{\sim}{f}_j]$, $j = 2, 3, \ldots, p - 1$ and $\underset{\sim}{T}_p[\underset{\sim}{v}_p, \underset{\sim}{g}_p] = [b_{(p-1)q} \underset{\sim}{e}_1, \underset{\sim}{f}_p]$, in which $\underset{\sim}{e}_i$ is the ith column of $\underset{\sim}{I}_r$, where $r = n - q(p-1)$. Now it is not difficult to see that equations iq, iq+1, for $i = 1, 2, \ldots, p - 1$, form an independent system of $2(p-1)$ equations,

$$\underset{\sim}{S} \underset{\sim}{z} = \underset{\sim}{h} \qquad (4.3)$$

see Figure 4. In fact, it can be shown $\underset{\sim}{S}$ is nonsingular with all of its principal minors being positive. Consequently, the system (4.3) can be solved via Gaussian elimination without pivoting. Once we obtain $\underset{\sim}{z}$, i.e., obtain x_{iq} and x_{iq+1}, $i = 1, 2, \ldots, p - 1$, the rest of the components of $\underset{\sim}{x}$ are obtained by direct substitution in (4.2). Thus,

$$\underset{\sim}{x}_1 = \underset{\sim}{g}_1 - x_{q+1} \underset{\sim}{u}_1,$$

$$\underset{\sim}{x}_k = \underset{\sim}{g}_k - x_{(k-1)q} \underset{\sim}{v}_k - x_{kq+1} \underset{\sim}{u}_k, \quad k \geq 2$$

and

$$\underset{\sim}{x}_p = \underset{\sim}{g}_p - x_{(p-1)q} \underset{\sim}{v}_p. \qquad (4.4)$$

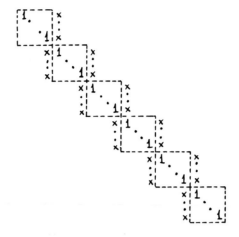

Figure 3

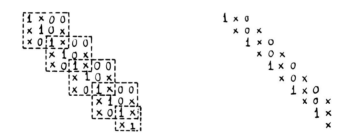

$\underset{\sim}{S}$ before and after Gaussian elimination

Figure 4

The total time consumed by this algorithm, including time for interprocessor communication is given by $17q + 22p - 42$. This time is almost minimized when $p = \sqrt{n}$. In this case it is reduced to $39\sqrt{n}$, resulting in a speedup of roughly $\sqrt{n}/5$.

A program describing this algorithm in an architecture-independent language is given below.

```
PROCEDURE   newdecompsolve (n: integer;
                            a,b,f:  ARRAY [1..n] of real;
                            VAR x: ARRAY [1..n] of real);
CONST   p = numberprocessors;
        q = n/p;              (* assume n = p*q  *)
VAR
      d,ℓ: ARRAY [1..p,1..q] of real;
      g,u,v,e:  ARRAY [1..p,1..q] of real;
      i,j: integer
      (* elements of 3×4 system to be solved *)
      r22, r24, r34:  ARRAY [1..p] of real;
      r12, r21, r31, r32: real;
DEFINE  (* mapping functions, as in BSP FORTRAN, VECTRAN *)
      a0: ARRAY [1..p,1..q] AS  a0[i,j] = a[q*(i-1) + j];
      b0: ARRAY [1..p,1..q] AS  b0[i,j] = b[q*(i-1) + j];
      f0: ARRAY [1..p,1..q] AS  f0[i,j] = f[q*(i-1) + j];
      x0: ARRAY [1..p,1..q] AS  x0[i,j] = x[q*(i-1) + j];
BEGIN
      FORALL i := 1 TO p DO        (* parallel *)
            decomp (q, a0[i], b0[i], d[i], ℓ[i]);
            IF  i > 1
            THEN
                  e[i,*] := 0;
                  e[i,1] := b0[i-1,q];
                  solve (q,d[i], ℓ[i], e[i], v[i]);
                  ENDIF;
            IF  i < p
            THEN
                  e[i,*] := 0;
                  e[i,q] := b[i,q];
                  solve (q, d[i], ℓ[i], e[i], u[i]);
                  ENDIF;
            solve (q, d[i], ℓ[i], f0[i], g[i]);
            ENDDO;
      (* solution of (p-1) 3×4 systems *)
      r34[1] := u[1,q];
      FOR  i := 2 TO p-1 DO        (* serial *)
            r22[i] := 1;
            r24[i] := u[i,1];
            r34[i] := u[i,q];
            r12 := r34[i-1];
            r21 := v[i,1];
            r31 := v[i,q];
            (* gaussian elimination *)
            r22[i] := r22[i] - r21 * r12;
            r32 := -r21 * r12;
            r34[i] := r34[i] - (r32/r22[i]) * r24[i];
            ENDDO;
      r22[p] := 1 - v[p,1] * r34[p-1];
      (* back substitution *)
      x0[p,1] := g[p,1]/r22[p];
      FOR  i := p-1 DOWNTO 2 DO          (* serial *)
            x0[i,q] := g[i,q] - r34[i] * x0[i+1,1];
            x0[i,1] := (g[i,1] - r24[i] * x0[i+1,1])/r22[i];
```

```
        ENDDO;
    x0[1,q] := g[1,q] - r34[1] * x0[2,1];
    xnext := x0[2,1];
    FOR j := 1 TO q - 1 DO
           x0[1,j] := g[1,j] - u[1,j] * xnext
           ENDDO;
    FORALL i := 2 TO p - 1 DO
           xprev := x0[i-1,q];
           xnext := x0[i+1,1];
           FOR j := 2 TO q - 1 DO
                  x0[i,j] := g[i,j] - u[i,j] * xnext - v[i,j] * xprev;
                  ENDDO;
           ENDDO;
    xprev := x0[p-1,q];
    FOR j := 2 TO q DO
           x0[p,j] := g[p,j] - v[p,j] * xprev
           ENDDO;
    END;
```

This concocted language includes features of several of the programming languages surveyed in Section 3.

1. We used a mapping function (define) similar to the IDENTIFY statement in Vectran.

2. We used a FORALL construct to indicate parallel loops, and these loops need not be (and are not, in our program) innermost loops. We could have written this program with vector statements, but the FORALL is a more natural expression of the algorithm. For example, this program would be difficult to write in VECTRAN, and even more difficult to understand.

5. CONCLUSION

We have discussed the three levels of the process of solving a numerical problem on high-performance machines, as well as the important role of automatic restructuring compilers in addressing the problem of software transportability among these machines. Such automatic restructurers may also be useful in gaining insight regarding how a suitable algorithm may be further "fine-tuned" for a given architecture.

We have also reviewed the languages available for existing high-performance machines. Almost all of these languages are based on Fortran. It seems that in spite of all rational arguments predicting its demise, Fortran will endure for many years in the scientific programming community. Whether or not it will prevail, depends upon many factors. Two important factors are:

(a) how much will (can) Fortran be changed and extended as time passes, and

(b) how successful and widely used will powerful automatic Fortran restructuring compilers become.

REFERENCES

[Burr75] Burroughs Scientific Processor (BSP), Vector Fortran Preliminary Specification, (Revision 0), 1975.

[CoDC78] Control Data Corporation (CDC), Star Fortran Language, (Version 2), Reference Manual, 1978.

[Cray80]   CRAY-1 Computer System, <u>Fortran (CFT) Reference Manual</u>, 1980.

[Kasc79]   M. J. Kascic, Jr., "Vector processing on the CYBER 200," in the Notes of the Univ. of Michigan, <u>High Speed Computation: Vector Processing</u>, Engineering Short Course, Summer 1979.

[KKLW80]   D. J. Kuck, R. H. Kuhn, B. Leasure, and M. Wolfe, "The structure of an advanced vectorizer for pipelined processors," <u>Proc. of COMPSAC 80, The 4th International Computer Software and Applications Conf.</u>, Chicago, IL, pp. 709-715, Oct. 1980.

[Kuhn80]   R. H. Kuhn, "Transforming algorithms for single-stage and VLSI architectures," <u>Proc. of the Workshop on Interconnection Networks</u>, West Lafayette, IN, pp. 11-19, April 1980.

[LaSa80]   J. L. Larson and A. H. Sameh, "Algorithms for roundoff error analysis-- A relative error," <u>Computing</u>, Vol. 24, pp. 275-297, 1980.

[MiSp78]   W. Miller and D. Spooner, "Software for roundoff analysis, II," <u>ACM Trans. on Math. Software</u>, Vol. 4, pp. 369-387, 1978.

[PaKL80]   D. A. Padua, D. J. Kuck, and D. H. Lawrie, "High-speed multiprocessors and compilation techniques," Special Issue on Parallel Processing, <u>IEEE Trans. on Computers</u>, Vol. C-29, No. 9, pp. 763-776, Sept. 1980.

[PaWi75]   IBM Palo Alto Scientific Center, <u>The Vectran Language: An Experimental Language for Vector/Matrix Array Processing</u>, G. Paul and M. W. Wilson (Eds.), Rpt. No. G320-3334, Aug. 1975.

[Perr79]   R. H. Perrott, "A language for array and vector processors," <u>ACM Trans. on Programming Langs. and Systs.</u>, Vol. 1, No. 2, pp. 177-195, Oct. 1979.

[Perr81]   R. H. Perrott, "Language design approaches for parallel processors," <u>Proc. of the Conf. on Analysing Problem Classes and Programming for Parallel Computing</u>, Nürnberg, Germany, W. Händler (Ed.), pp. 115-126, Springer-Verlag, 1981.

[SaKu78]   A. H. Sameh and D. J. Kuck, "On stable parallel linear system solvers," <u>J. of the ACM</u>, Vol. 25, No. 1, pp. 81-91, Jan. 1978.

[Same81]   A. H. Sameh, "Parallel algorithms in numerical linear algebra," <u>Proc. of the 'Design of Numerical Algorithms' for Parallel Processing</u>, CREST Summer School, Bergamo, Italy, June 1981.

<u>DISCUSSION</u>

<u>Boyle</u>   What about hardware details?   Are they useful to you?

<u>Sameh</u>   In my opinion, the user should not be burdened with worrying about hardware details.   The compilers should do this.   But we need a language with sufficient features to write the program.

<u>Hennell</u>   You suggested starting with a good sequential algorithm and evolving to a good parallel algorithm, but it seems to me that sometimes you have to start afresh looking for a parallel algorithm.

<u>Sameh</u>   I merely meant that it is useful to start by examining the available sequential algorithms.

Swartztrauber   It looks as if the tridiagonal algorithm gets less efficient as the number of processors increases.

Sameh   Exactly.   The optimal number is $\sqrt{n}$.

Swartztrauber   So you do not achieve the theoretical complexity of log n for that algorithm.

Sameh   No, it does not achieve log n.

Swartztrauber   Perhaps the point is that you are working in a very realistic world, with a machine having a certain interconnection scheme.   If you could define the machine to suit the problem then you could achieve the theoretical complexity of log n.

Boyle   Have you considered algorithms for completely independent parallel machines (MIMD machines) such as the Intel 432, where different instructions can be issued in different processors simultaneously?

Sameh   Not in any detail.   In any case my observation of algorithms in numerical linear algebra is that you really do not need such machines.

Boyle   Are they actually worse?

Sameh   The control is worse.

Reid   I am worried about having to restructure algorithms to suit each particular machine.   We are just too short of man-power for this.   We should be looking for common features, such as the ability to process vectors, and expressing algorithms in languages that can represent such features.

George Paul   One of the things we tried to do many years ago in Vectran was to look for common forms of parallelism and to develop a language which bridges them.   In fact Sameh remarked that Vectran is independent of any specific architecture.   This is a very important factor and X3J3 is trying to do this for Fortran 8x.

Wilson   I agree.   It bears repeating that the different vector extensions and parallel languages that have been tried out tend to have features for treating vectors and arrays as a whole.   These are not new things which X3J3 has invented; rather X3J3 is aiming to systematize a number of extensions from different sources.

Kahan   I think that in this discussion of the intent to make languages independent of architectures, we are doing something that has already been done (program portability) in the course of making our languages and our algorithms independent of the vagaries of arithmetic.   If we insist upon a level of abstraction in our languages that hides too much of the architectural detail, then we will be able to utter in our languages only that which is common to all the machines on which we want our programs to run.   We will then be working with the least common denominator, which is a hypothetical machine that is far worse than any of the individual machines for which the program is intended. The price of building a different machine should be paid by the builders and users of that machine.   It should not oblige the whole community to pay by limiting their thoughts to those that can conveniently be implemented on all machines no matter how different they may be.

Wilson   I would like to emphasize the importance of having functions (e.g. the dot product in Fortran 8x) as intrinsics with names known to the compiler. This way the compiler can generate in-line code tailored for the hardware.   A library of external functions is not a satisfactory alternative.

Hehner   But it is entirely possible to compile library functions as in-line code.   It might be useful to have some means of indicating to the compiler when this should be done.

Wilson   Anything is possible, but in the definition of Fortran "intrinsic" has this meaning.

Meissner   It has nothing to do with the definition of the Fortran language. It is just that this is what has been implemented historically.

Wilson   The essential thing is that there is a mechanism for getting in-line code.

Moler   I am in favour of having a function for dot products, since these are well understood and used in a wide range of applications.   However this is not so for some of the other Fortran 8x functions, such as EOSHIFT.   I am interested in matrix calculations so would only use a few of these functions. Even here, I would like to see some modifications. For instance I am interested in where the maximum element of a vector is located. For my applications I would like all the BLAS ⌈Basic Linear Algebra Subprograms, see Lawson et al, ACM Trans. Math. Software 5 (1979), 308-325⌉, but would not want to impose them on the whole Fortran community. People have vector machines and need vector operations, but there has not been enough experience with this particular set to justify standardizing it.

Meissner   It should be appreciated that the list of intrinsic functions for Fortran 8x is tentative.   The plan is to spend 1981-2 working on the core content, then begin integrating the facilities and smoothing out the syntax, and freeze the thing sometime in 1985.   Violent reactions are very beneficial, because they tell us where the worst problems are.   But don't just rail at us. Start writing details down. As editor of FOR-WORD I would, for instance, welcome letters on exactly how much of the array features should be in the core.   This is something X3J3 must decide soon.

SESSION 6
PROGRAM STRUCTURE

Chair
J.C. Adams

Discussant
E.L. Battiste

THE RELATIONSHIP BETWEEN NUMERICAL COMPUTATION
AND PROGRAMMING LANGUAGES, J.K. Reid (editor)
North-Holland Publishing Company
IFIP, 1982

NUMERICAL PACKAGES IN ADA

S J Hammarling and B A Wichmann

Division of Numerical Analysis and Computer Science
National Physical Laboratory
Teddington, Middlesex TW11 01W, UK

This paper considers the suitability of the programming language Ada, [DoD (1980)], for numerical computation and discusses the problems that arise when implementing numerical packages and libraries in Ada. In particular issues concerning separate compilation of packages and procedures, communication of numeric types between packages and the problems of supplying user routines to packages and procedures are raised.

INTRODUCTION

Although the Ada language was not designed primarily for numeric work, it meets most of the obvious requirements for numerical computation. Current indications are that the impact of Ada will extend further than the military real-time systems for which it was designed and could rival Fortran and Cobol in their respective fields. This paper attempts to assess Ada in the area of numerics albeit, in part, from the viewpoint of one of the language designers. Some of the issues raised were considered in Cox and Hammarling (1980) with respect to preliminary Ada as part of the test and evaluation phase for Ada.

The Fortran language has been enhanced gradually to meet user demands whereas the Ada design started from scratch, attempting to provide general machine-independent solutions wherever possible. It is therefore difficult to assess Ada because it is largely untried and contains several new features for which there is no parallel in Fortran 77. These features include:

1. User defined numeric types.
    The user specifies the precision required. The language specifies the minimal properties of such types. Attributes of these types are accessible in a machine independent manner.

2. Secure use of pre-compiled libraries.
    The types of the parameters passed to library procedures are checked.

3. Packages.
    Groups of procedures and associated data can be conveniently placed in a package, this being the natural unit of program development and specification.

4. Generics.
    Generic packages or procedures may have types, functions and procedures as parameters. Thus generic program units provide a mechanism whereby the text of a package or procedure can be made general. This general form can then yield specific packages and procedures by instantiating the generic. For example, a generic package that has a type as parameter is instantiated by giving the package an actual type.

5. Overloading.
    Procedures or functions which serve a common purpose can be given the same name. The actual procedure or function called will depend upon the types

    of the parameters involved.

6. Default parameters and keywords.
    Input parameters may be given default values and may then be omitted from
    a call and parameters may be referenced by keyword.

7. Error handling.
    The exceptions mechanism of Ada provides a controlled method of recovery
    from run time errors.

All of these features have appeared in other languages; however with Ada one has
the first opportunity to use them together for numeric work. The success of the
language in this respect cannot be judged until several large numerical packages
have been developed and used in Ada.

1. USER DEFINED NUMERIC TYPES

An evaluation of the numeric types and operations provided in Ada is hindered by
the lack of any generally accepted language design criteria. There appear to be
four views as follows:

(a)   The problem is one of hardware design rather than software design and we
       should aim to get hardware standards agreed and widely used.

(b)   In view of (a), programming languages should say nothing about the
       properties of floating point computation and merely provide a minimal
       syntactic framework. One can call this the classical Fortran view since
       the Fortran standard says nothing about accuracy.

(c)   Programming languages should at least allow access to the basic parameters
       which give the accuracy of the various types provided. Use of these
       parameters can then guarantee portability.

(d)   Programming languages should provide real types with clearly defined
       properties, adequate for numerical analysis (without resort to the machine
       manual).

The specification of Ada demanded that the user should be able to state the
precision required [DoD (1978)]. In Ada this facility is provided by requiring
that each real data type specifies the accuracy (either relative in decimal
digits for floating point, or absolute grain' of resolution for fixed point).
Given this fact, it is impossible to assign any meaning to the specification of
real data types unless its consequences for the accuracy of operations is
defined in the language. This rules out options (a) and (b) above.

The remaining question is therefore how to specify the accuracy of both fixed
and floating point operations without burdening the manual with excessive
detail. (Ada is not designed primarily for numeric computation and hence this
aspect should not be considered out of proportion to the rest of the language.)
The method chosen was that of Brown (1977) for the following reasons:

1.   One can deduce the rules of classical error analysis [Wilkinson (1963)]
       from it and hence it is adequate for floating point.

2.   The majority of machines in serious use conform to Brown's model.

3.   The method can be extended without difficulty to handle fixed point types,
       several floating point types and conversions between numeric types (as
       well as the conversion of literals in the program text).

4.   The method is reasonably simple.

The Brown model has an important property which is essential for its use in Ada. If a programmer writes

<u>type</u> F <u>is</u> <u>digits</u> 5;

then the implementation is free to provide a hardware type to implement F which has <u>at</u> <u>least</u> 5 decimal digits of precision. The Brown model is permissive in this sense as the axioms give error bounds which are not necessarily minimal for an actual implementation. In fact, the number of floating point types provided by hardware is often only 1 or 2, and so the choice of F above may be limited to hardware types of say 7 or 14 digits precision. This implies that at least for the choice of the hardware type, the closeness of the Brown model to the hardware is not critical.

A few changes have been made to the Brown model in its application to Ada. These are as follows:

(i)   The radix is fixed at 2 which implies that a decimal (floating point) machine cannot support the Ada model. This does not seem to be serious. Also, radix 8 and 16 machines cannot be matched very accurately by the Ada model. This is only of concern to numerical analysts who wish to take extreme care by, for instance, ensuring that constants are in the correct range to avoid the "wobbling" precision problem. A consequence of fixing the radix is that one knows that 0.5 is exact but 0.1 is not.

(ii)   The exponent size is four times (the binary) mantissa length. Fixing a minimum for this ratio ensures a reasonable range for scientific computation so that programmers do not need to take excessive precautions to avoid overflow and underflow. In a few cases, such as quadruple length on the 360 and double length on the Honeywell, it is necessary to reduce the <u>Ada</u> value of the mantissa length because of the very small exponent range. In most other cases, the ratio of four is exceeded, such as with the IEEE floating point standard, Stevenson (1981).

(iii)   The division operation is supported (in the sense of Brown) which means that machines such as the Cray 1 for which division is performed by reciprocation and multiplication, cannot implement the Ada model properly. The reason for this change is one of simplification and the fact that Ada compilers have yet to be written so that the "optimization" of replacing division by reciprocation followed by multiplication need not be permitted. (Compilers can perform the optimization, but only if the model is preserved, which is not true in general.)

(iv)   Conversion of literal expressions to real values and constants are also supported operations. This is quite a severe test on radix conversion routines, but there seems no reason why these operations should not provide the same precision as internal operations.

(v)   Fixed point and numeric conversions are supported, which is a logical extension. Note that since fixed point is likely to use a radix of 2, the model numbers of both fixed point and floating point overlap to provide very tight error bounds on conversions. This could be important as computations within a loop may well involve conversions.

Some controversy has centred on the provision of fixed point facilities in Ada. Although not needed in many applications, there is an important class for which fixed point is vital. Signal processing at very high speeds or on machines without hardware floating point require numeric facilities not addressed by either integers or software floating point. The requirement makes the language larger for everybody, but the strong type mechanism ensures that the facilities are not used implicitly (or by accident).

The Ada model clearly has defects. It is less precise than that of a particular machine (such as one conforming to the IEEE standard [Stevenson (1981)]), it does not specify the exact conditions pertaining to overflow and underflow, nor does it prohibit the "surprising" effects of an overlength accumulator. On the other hand, the fact that the Ada language specifies minimal properties for real arithmetic makes portability feasible, defines clearly what optimization is permitted, and gives a framework for further development. An important property of the Ada model is that optimization of real computations is possible, even when the compiling machine and the target machines have different floating point hardware. (In fact compilers are likely to evaluate static expressions by a rational arithmetic package.)

2. ADA FACILITIES FOR NUMERICAL COMPUTATION

Since one of the authors (BAW) was a member of the Ada design team, an objective evaluation of Ada for numerical work is clearly not possible. Moreover, until several large library systems have been written in Ada, a satisfactory evaluation of the language cannot be completed. Fortunately Rice has been collecting a list of desirable features of a language for numerical work which has been widely discussed at IFIP working group 2.5 meetings [Rice (1980)]. Although not offically adopted by the working group, this list provides a basic set of criteria for an evaluation exercise, especially since each feature is assigned a weight for both ease of implementation and desirability. In using this list to assess Ada, one realises that it was written with current practices in mind and it has a strong Fortran flavour.

2.1 Evaluation against the Rice report.
This section lists each of the features of the Rice report together with a response based upon the Revised Ada manual [DoD (1980)]. In most cases, the response is straightforward, but in a few situations, the requirement is met by a language feature that needs some explanation.

The two figures in each heading give the ease of implementation and importance on the scale 0 .. 100, as given in Rice (1980), the larger value implying a greater need. The sub-section headings used correspond to those in the Rice report. For example, our 2.3.4 correponds to Rice's topic 3.4.

2.1.1 Access to machine parameters (95/50)
A set of machine parameters is available for all the basic data types in Ada. For floating point, these give the parameters of the version of the Brown model used to give the semantics of floating point types in the language definition. Packages can easily be written to give other sets of parameters such as those recommended by Ford (1978).

2.1.2 Separate compilation (80/90)
The separate compilation facilities of Ada are very powerful and require that type checking is performed across separately compiled units before program execution. See below for a more detailed discussion.

2.1.3 Dynamic workspace allocation (50/60)
Workspace allocation in Ada combines the best features of Algol 60 and Pascal. Hence on entry to a subprogram, an array can be declared whose size depends upon a dynamically calculated value. Access types (pointers) are used to allocate and deallocate space for objects whose requirements are not tied to the block structure of the program.

2.1.4 Hiding names (70/30)
The only name that should not be hidden from the programmer is that of the package STANDARD which defines the basic data types of the language. All other names are made available by accessing a library package (which must be done explicitly by "with NAME_OF_PACKAGE;"). If two procedures share common data,

this can be hidden from users of the two procedures by placing the data in the body (and not the specification) of the package.

2.2.1 Complete internal specification (70/75)
The Ada language requires that type checking is performed by the compiler. With an array, this implies that the number of dimensions, type of the components and the type of each index matches correctly with the subprogram specification.

2.2.2 Array building and subarray extraction (55/35)
A slice' consisting of a sequence of elements from a one-dimensional array can be manipulated as a variable:

```
A( 1 .. 4 ) := ( 1.0, -3.4, B(1), B(2) );
```

However, the facility is not generalised to more than one dimension because it places an additional overhead on accessing array elements.

2.2.3 Array operations (65/55)
The predefined operations available in Ada do not include the mathematical array operations. However, Ada permits the user to extend the definition of "+" etc, to accept arrays as operands and give arrays as results. The set of operator symbols is fixed as is the priority of each operator. The user defined operators are written as functions.

2.2.4 Range indices (60/65)
For an array A, the value A'FIRST(1) gives the first subscript value for the first index position. Hence the range of values for the first index position is A'FIRST(1) .. A'LAST(1). The prime (') is used to access attributes of objects.

2.2.5 Array constants (70/25)
Array constants are available in Ada, including the ability to set all the remaining elements with a fixed value, e.g.

```
B : TABLE := TABLE'( 2| 4| 10 => 1, others => 0 );
```

2.3.1 Matrix Algebra (60/75)
Since Ada is a general-purpose language rather than a language designed for numerical computation, no matrix facilities are built into the language. It would be straightforward to implement a package to perform such calculations, including the use of infix notation for matrix operations (see 2.2.3, above).

2.3.2 Matrix-Vector Algebra (60/75)
Similar remarks apply as for 2.3.1. However, the Greek sigma sign is not available in Ada source text (although the full ASCII set is used). If vectors and matrices are different types in such a system, then explicit code must be written to regard a vector as a 1 by N matrix.

2.3.3 Submatrix selection (55/40)
This is only available in a simplified form in Ada, see 2.2.2.

2.3.4 Inverse matrices and the solution of linear equations (50/60)
Again, such operations can be provided by a package.

2.3.5 Matrix-Vector constants (75/30)
Available, see 2.2.5.

2.3.6 Specialised matrix structures (40/40)
The built-in data structures in Ada include arrays and records. Other structures can be defined by the user in terms of the base data types. If it is required to define a type MATRIX whose storage details are hidden from the user, then the Ada private type mechanism should be used. Functions can be provided to access

the matrix values which would depend upon the storage method used (say, upper triangle stored for a symmetric matrix).

2.4.1 Tabular output (Arrays 90/60; functions 85/55)
All input-output in Ada is performed by a package. No special language features are introduced for input-output. In consequence, if the current package is regarded as being inadequate, then it can be supplemented by further packages or even replaced.

2.4.2 Printer plotting (80/70)
Use an additional package, see above.

2.4.3 Graphical system input/output (50/75)
Use an additional package, see above.

2.4.4 Data storage access (40/35)
There are significant problems in defining facilities in this area. Current systems are often machine dependent.

2.5.1 The finite operators (90/80)
Operators for vectors or matrices, such as MAX, MIN and summation, can readily be supplied as an Ada package.

2.5.2 Differentiation (Symbolic 50/65; numerical 60/50)
Symbolic and numerical differentiation would need to be provided by extensive packages.

2.5.3 Integration (Symbolic 30/40; numerical 70/50)
Similar remarks apply as for differentiation.

2.5.4 Infinite series summation (65/40)
To sum such a series, one would need a function to give the values of the terms. Since Ada does not provide for function parameters, one would need to use the equivalent facility of generics.

2.6.1 Natural definitions (90/90)
The problem mentioned in the Rice report of regarding a function M(X, A, B) as a function of just X exists in Ada also. The use of an intermediate function seems reasonable and allows the use of generics.

2.6.2 Functions as program variables (30/65)
Functions variables do not exist in Ada. However, the example given by Rice is quite easy to code in Ada using generics and also makes the distinction between functions and functors (ie produces a function). The example is, given F, to define a function H such that:

H(X) = F(X) + A*(X**2 + 3*SIN(X))

In Ada this becomes:

-- declaration of REAL, A : REAL, and SIN(X : REAL) return REAL.

```
generic
    with function F( X : REAL ) return REAL;
function H( X : REAL ) return REAL is
    begin
        return F( X ) + A*( X**2 + 3.0*SIN( X ) );
    end H;

function NEW_H1 is new H( ABS );
function NEW_H2 is new H( COS );
```

The point about this example is that in Ada there is a clear distinction between H which is a functor and NEW_H1 which is a function produced from ABS and H. Unlike the mathematical formulation, the fact that F is a parameter of H is explicit, whereas SIN is not a parameter.

2.6.3 Arrays of functions (80/55)
Ada uses the same notation for function call and array access. Hence a text written for access to an array can be changed to call a function. A function cannot, however, appear on the left hand side of an assignment statement. A more natural and powerful method of hiding the implementation details of a data type is to use private types.

2.6.4 Mathematical typing (55/40)
The mathematical classification of functions into analytic, polynomial, differentiable etc, must be programmed by the user.

2.7.1 Global declaration (80/50)
As with Algol 60, variables in Ada can be declared globally. This is not a recommended practice because the entire program using these variables would have to be nested within the main program. It is much better to declare variables in a package so that procedures and other packages requiring this data can explicitly access it by "with PACKAGE_NAME;". This method provides a facility similar to named common in Fortran except that only the package name is needed in each routine as details of the variables (and their types) are retrieved by the compiler from the separate compilation of the package.

2.7.2 Partitioned global declarations (70/30)
See 2.7.1 above.

2.7.3 Subprogram argument lists (55/35)
A variable of any data type can be passed as a parameter in Ada. The type of the parameter must match the specification of the subprogram. The parameter modes are in, out and in out. An implementation is free to implement the parameter modes by reference or by copy. For an array parameter, the attribute notation can be used to access details of the array so that no additional parameters need be passed.

2.7.4 Internal procedures (65/50)
As with Algol 60, internal procedures are permitted, so that the internal structure of the procedure is not visible from outside. Also, a package can have internal procedures so that a set of procedures (and data etc) can share implementation details which are hidden from users of the package.

2.7.5 Variable argument lists, defaults (40/50)
The specification of a subprogram in Ada allows for default values for parameters of mode in. A call of such a subprogram can then omit these parameters (but not any remaining parameters). This facility makes it possible to provide a more user-friendly interface.

2.8.1 Elementary functions (95/95)
The elementary functions are not defined as part of the language. An explicit package is needed for this which the user must request by starting his program with "with MATH_LIB;".

2.8.2 Higher mathematical functions (90/75)
See 2.8.1 above.

2.9.1 Identification of error type (75/75)
The Ada language defines the program errors which must be detected before execution (such as type mis-matches). Run-time errors are called exceptions in Ada. They must also be detected, but the language defines a mechanism whereby

the user can trap such an event himself and take remedial action if necessary.

2.9.2 Identification of error location (65/80)
The content of error reporting by an Ada compiler is not specified in the language reference manual. One hopes that purchasers of compilers will insist on good diagnostics. If a program fails due to a run-time error for which there was no exception handler, then a system should produce a good post mortem (but this will depend upon the environment of the particular compiling system).

2.9.3 Message transmission (70/75)
Run-time errors can be handled by a package which can then print a message devoid of superfluous internal details.

2.9.4 Error recovery control (75/60)
This is part of the Ada exception mechanism.

2.10.1 Real numbers (100/100)
Real data types in Ada can be either fixed point or floating point. The user specifies the number of decimal digits of precision required in the case of floating point. An implementation is free to reject such a specification if the target hardware is unable to provide the requested precision. The semantics of both fixed and floating point operations in Ada is defined in terms of the Ada variant of the Brown model.

2.10.2 Integer numbers (100/15)
Ada provides integer data types in an analogous manner to the real data types.

2.10.3 Higher precision real arithmetic (95/90)
The language does not impose any limits on the precision of fixed or floating point data types. An implementation is likely to impose limits determined by the hardware. One would hope that a few implementations will provide a very high precision sofware system for experimental work.

2.10.4 Complex variables (70/55)
Complex variables are not defined as part of the Ada language but can easily be specified in Ada. For instance, using the conventional cartesian form:

```
type COMPLEX is
    record
        RE : REAL;
        IM : REAL;
    end record;
```

However, type COMPLEX could be defined in terms of polar coordinates. In fact, one can define type COMPLEX as a private type so that its internal representation is hidden from the user. Functions would then be necessary to access the X coordinate or polar angle etc.

One difficulty that Ada faces in this area is that standardization of a complex data type is advantageous in order to allow different mathematical packages to communicate effectively. Such standardization is required in other areas, input-output being the only case in the current language where a package is defined as part of the language.

2.10.5 Character strings (60/50)
Basic facilities for string handling are defined in the language. Facilities such as concatenation of varying length strings (which requires extensive support) must be provided by additional packages.

2.10.6 Combinations
Combinations of data types can be built up in Ada using arrays and records.

2.11.1 Machine parameters (95/50)
Extensive facilities are available here, see 2.1.1.

2.11.2 System environment (90/35)
Some aspects of the environment are defined in the language such as the maximum value for an integer (SYSTEM.MAX_INT). Other aspects such as performance are not defined as part of the language.

2.11.3 Special machine facilities (50-80/20-40)
The Ada language does not make explicit provision for special machine facilities because it would hinder program portability. The semantics of real data types allows (but does not force) compilers to use extra length registers if they are available. Machine code statements can be written in Ada, and hence exotic peripherals or calls on special instructions can be done this way. Users are advised to localise machine dependencies in special packages to aid portability.

2.12.1 Two levels of precision (90/85)
The user specifies the precision required in decimal digits for floating point. An implementation is free to give more precision than that requested, corresponding to the hardware available. Hence an arbitrary number of levels of precision can be supported by the Ada language.

2.12.2 Extended processor precision (60/30)
As noted above, an implementation is free to use such a processor. A programmer could request that use is made of such features by means of a pragma (a compiler interpreted comment).

2.12.3 Dynamic precision control (80/70)
Precision control in Ada is static, corresponding to the ability of current hardware.

2.12.4 Special arithmetic (40/20)
Facilities such as interval arithmetic can be programmed in Ada by defining ones own type, say:

```
type INTERVAL is
    record
        MIN : REAL;
        MEAN : REAL;
        MAX : REAL;
    end record;
```

where REAL is a previously defined type. Operations such as "+" and "*" can then be provided, written as functions.

2.13.1 Declarations
All declarations in Ada are explicit, except that a loop parameter is declared by its use in the loop statement. Hence the request for implicit declaration as in Fortran is not met by Ada.

2.13.2 Reserved words
There are 62 reserved words in Ada, few of which are obvious candidates as identifiers.

2.13.3 Characters sets
The Ada character set is ISO (ASCII). Upper and lower case are regarded the same in identifiers because of potential confusion, especially with equipment which may not be able to process both cases. Greek characters are not available in programs outside character strings (and comments).

2.13.4 Names (80/65)
Identifiers in Ada are of arbitrary length (but effectively limited by the line length). They start with a letter and are followed with letters and digits. Medial underscores are permitted and are significant. Blanks are not significant except in separating reserved words and identifiers.

2.14.1 Problem solving statements (60/70)
There is no general problem solving ability in the language. It is up to the programmer to specify the solution in the conventional manner.

2.14.2 Problems (40/75)
The solution of a numerical problem depends critically upon the skill of the programmer in Ada, in a similar manner to conventional languages such as Fortran. As an example, the SIN and COS functions have been coded in Ada using fixed point using the algorithm of Cody and Waite (1980) [see Wichmann (1981)].

2.15.1 Dynamic storage allocation (40/25)
Storage allocation is defined in the language and is not subject to direct user control. Hence one cannot "assign 500 words for array ABAR". However, arrays are allocated storage on entry to the block/subprogram in which they are declared. The storage is automatically released on exit from the block/subprogram.

2.15.2 Block allocation/deallocation (80/70)
As for 2.15.1.

2.16.1 Arrays, vectors and matrices (Numeric arrays 95/95;
                    as separate types 85/70; special storage format 60/45)
See 2.3.1 and 2.3.2.

2.16.2 Basic data structures (75/65)
Lists, queues and stacks can be defined using records and access types for pointers. The internal details of such types can be hidden by using private types.

2.16.3 Trees and graphs (25/15)
Such types can be defined in the same way as lists etc, see 2.16.2.

2.2 Summary of Evaluation against Rice report
Of the 62 requirements listed in the report, 35 are met in a simple straightforward manner by Ada. An additional 15 can be met by special packages written in Ada itself (for instance, for matrix algebra or graphics). A further 8 requirements are only partly met by Ada (for instance, generalised subarrays are not possible but slices are allowed). The remaining 4 requirements are divided into three concerned with function parameters and one which is not addressed by Ada, namely dynamic control over precision. Implicit declaration of variables is not permitted in Ada.

The issue of function parameters is an awkward one. In many Algol like languages, function variables are either insecure or require extensive run-time checking. Algol 60 has function and procedure parameters, but these require run-time checking in general. Neither of these options were regarded as appropriate for Ada. Moreover, since Ada was not primarily for numerical computation (for which the need is apparent), the set of requirements for Ada excluded function parameters. Hence Ada does not admit the style of programming which makes extensive use of function/procedure parameters.

The Knuth Algol 60 input-output procedures make use of procedure parameters for error control. In Ada, one would use exceptions for this purpose, which is both more natural and more efficient. Many of the other major uses of function parameters can be handled by generics in Ada. (See section 3.6.)

3. SEPARATE COMPILATION AND PACKAGES

3.1 Implementing numerical packages in Ada.
When designing a suite of routines to solve a class of problems the natural facility to use in Ada is the package. Examples of existing Fortran "packages" and libraries for solving numerical problems are:

```
EISPACK - Eigenvalue problems.
          Smith et.al. (1976).
LINPACK - Linear equation and least squares problems.
          Dongarra et.al. (1979).
NOSL    - NPL optimization subroutine library.
          Gill et.al. (1979).
NAG     - Numerical algorithms library.
          Ford et.al. (1979).
BLAS    - Basic linear algebra subroutines.
          Lawson et.al. (1979).
```

In Ada these could all be designed as one or more packages and in this section we illustrate some of the issues involved in designing such packages and we hope that this will promote discussion of the issues so that a coherent approach towards the design of numerical packages in Ada can be reached.

Suites such as the BLAS, LINPACK and EISPACK, which do not contain calls to user supplied routines, are likely to be more straightforward to design than those suites and libraries that require user supplied routines. Nevertheless there are several crucial issues to be resolved.

3.2 Precision of the arithmetic and floating point types.
When floating point arithmetic is available the implementation must include the predefined type FLOAT and may include types such as SHORT_FLOAT and LONG_FLOAT. Thus we can implement packages using types similar to the Fortran REAL and DOUBLE PRECISION whose precision will vary from machine to machine.

Suppose that we consider a package for the "single precision" BLAS routines. We first give the specification part of the package, this being the part that the user must be able to see. Note that in Ada -- is the start of a comment.

```
package BASIC_LINEAR_ALGEBRA is
    type VECTOR is array ( INTEGER range <> ) of FLOAT;
    --
    -- <> Is an unconstrained range.
    -- Now give the functions and procedures.
    --
    function DOT( N : INTEGER; X, Y : VECTOR ) return FLOAT;
    --
    -- Description of DOT.
    -- DOT <== X(1)*Y(1) + X(2)*Y(2) + ... + X(N)*Y(N).
    --
    procedure AX_PLUS_Y( N : in INTEGER; A : in FLOAT;
                         X : in VECTOR; Y : in out VECTOR );
    --
    -- Description of AX_PLUS_Y.
    -- Y <== A*X + Y.
    :
    etc
    :
end BASIC_LINEAR_ALGEBRA;
```

Next we give the body of the package, this part may be hidden from the user.

```
with MATH_LIB;
package body BASIC_LINEAR_ALGEBRA is
   function DOT( N : INTEGER; X, Y : VECTOR ) return FLOAT is
   begin
      :
      code for DOT
      :
   end DOT;
   :
   etc
   :
end BASIC_LINEAR_ALGEBRA;
```

We can have similar, but distinct packages, for LONG_FLOAT and SHORT_FLOAT or these can be included in the same package and the functions and procedures can be overloaded. For example

```
   function DOT( N : INTEGER; X, Y : LONG_VECTOR ) return LONG_FLOAT;
```

Note that the Fortran BLAS have increment parameters associated with each vector so that equally spaced, but not necessarily contiguous, elements may be referenced but, for simplicity of presentation, we have omitted these increment parameters. The parameter N need not be included in the above example for DOT and AX_PLUS_Y because it could be obtained inside the routines by using the attribute 'LENGTH.

Ada also allows the definition of real types such as

```
type REAL is digits 10;
```

for which the compiler must select a floating point type having at least the specified precision, or report that such precision is not available. An alternative to using the predefined types is to make the package generic.

```
generic
   type REAL is digits <>; -- Matches any floating point type.
package BASIC_LINEAR_ALGEBRA is
   type VECTOR is array ( INTEGER range <> ) of REAL;
   :
   as before with REAL replacing FLOAT.
   :
end BASIC_LINEAR_ALGEBRA;
```

The package can then be "instantiated" for any required floating point type.

```
type MY_REAL is digits 8;
package MY_BASIC_LINEAR_ALGEBRA is new
         BASIC_LINEAR_ALGEBRA( MY_REAL );
```

BASIC_LINEAR_ALGEBRA is a template for which MY_BASIC_LINEAR_ALGEBRA is an actual instance.

Whether or not generics are used will partly depend upon the cost of instantiation. The following discussion is also relevent to this problem.

3.3 Mathematical types.
The decision on the use of real types affects the use of those entities that aepend upon the real type, such as vectors and matrices. It is obviously important for a user to have access to the types of the entities that are parameters of any routine that the user is using. The difficulty arises because if we have the two packages

```
package YOUR_PACK is
   type REAL is digits 8;
end YOUR_PACK;

package MY_PACK is
   type REAL is digits 8;
end MY_PACK;
```

then the REAL of YOUR_PACK is not the same type as the REAL of MY_PACK.

As a further example consider LINPACK. Many of the LINPACK routines call the BLAS, although there is no need for a user of LINPACK to be aware of this fact. Thus we should like the specification part to be:

```
package LINEAR_EQUATIONS is
   type VECTOR is array ( INTEGER range <> ) of FLOAT;
   type MATRIX is array ( INTEGER range <> ) of VECTOR;
   --
   -- Note: not array ( INTEGER range <>, INTEGER range <> ) of FLOAT;
   --       if we want a slice of a matrix to be a vector.
   :
   Description and specification of the LINEAR_EQUATIONS routines.
   :
end LINEAR_EQUATIONS;
```

We should like the package body, which the user does not need to see to successfully use LINEAR_EQUATIONS, to be:

```
with BASIC_LINEAR_ALGEBRA;
package body LINEAR_EQUATIONS is
   use BASIC_LINEAR_ALGEBRA;
   --
   -- This enables the names of the routines in
   -- BASIC_LINEAR_ALGEBRA to be used directly.
   :
   D := DOT( N, X, Y );
   :
end LINEAR_EQUATIONS;
```

This does not work because BASIC_LINEAR_ALGEBRA and LINEAR_EQUATIONS have vectors of different types. For this to work the specification of LINEAR_EQUATIONS would need to be

```
with BASIC_LINEAR_ALGEBRA;
package LINEAR_EQUATIONS is
   use BASIC_LINEAR_ALGEBRA; -- Now have access to type VECTOR;
   type MATRIX is array ( INTEGER range <> ) of VECTOR;
   :
   etc
   :
end LINEAR_EQUATIONS;
```

This means that the user also has to look at BASIC_LINEAR_ALGEBRA in order to look up the type VECTOR. Explicit type conversion could be used in the body of LINEAR_EQUATIONS, but this does not seem to be a very satisfactory solution and someone wishing to use both LINEAR_EQUATIONS and BASIC_LINEAR_ALGEBRA is then faced with two different types of VECTOR. The situation is just the same with the generic case.

A possible solution is to have a package containing the mathematical types to which all the numerical packages would have access.

```
package MATH_TYPES is
   type REAL is digits n;
   --
   -- n could be a specific value - perhaps machine dependent,
   -- or the package could be generic in the type REAL.
   --
   type VECTOR is array ( INTEGER range <> ) of REAL;
   type INTEGER_VECTOR is array ( INTEGER range <> ) of INTEGER;
   type MATRIX is array ( INTEGER range '> ) of VECTOR;
   :
   etc
   :
   type COMPLEX is
      record
         RE, IM : REAL;
      end record;
   function "-" ( X : COMPLEX ) return COMPLEX;
   :
   etc
   :
   type COMPLEX_VECTOR is array ( INTEGER range <> ) of COMPLEX;
   :
   etc
   :
end MATH_TYPES;

package body MATH_TYPES is
   :
   definition of functions etc.
   :
end MATH_TYPES;
```

If this approach is adopted then careful consideration needs to be given to the contents of MATH_TYPES. Of course MATH_TYPES could also be generic with vectors, matrices and so on, being defined in terms of a floating point type REAL. The other packages would then also need to be generic so that the user can supply his own REAL. This approach would enable the precision used by packages to depend upon the accuracy of the user s data, which is particularly appropriate in linear algebra applications where we usually have a good idea as to how many guard digits are necessary to return solutions that are as accurate as the user s data warrants. [See for example Wilkinson (1965)]. For instance, consider an example where MATH_TYPES is generic:

```
with MATH_TYPES;
generic
      ACC : INTEGER := FLOAT'DIGITS - 2; -- See type REAL below
   package LINEAR_EQUATIONS is
      type REAL is digits ACC + 2;
      --
      -- Use 2 guard digits. FLOAT will be the default precision
      -- because the default value of ACC is 2 digits less than the
      -- accuracy of FLOAT.
      --
      package MY_MATH_TYPES is new MATH_TYPES( REAL );
      use MY_MATH_TYPES; -- Can now use VECTOR, MATRIX etc.
      :
   end LINEAR_EQUATIONS;
```

If different precisions are required within a package then MATH_TYPES can be instantiated more than once.

3.4 The use of overloading in packages and procedures.
The Fortran BLAS routines have a number of different versions of each routine.
For example, the operation

$$y \; <== \; ax + y$$

has three realizations:

        SAXPY - Single precision.
        DAXPY - Double precision.
        CAXPY - Complex single precision.

In Ada these can all be overloaded with the same name, say AX_PLUS_Y, the
particular procedure chosen depending on the types of the parameters.
Furthermore it is possible to create the different versions from one coded
version by means of generics. For example:

```
with MATH_TYPES;
package body BASIC_LINEAR_ALGEBRA is
    use MATH_TYPES;
    generic
        type T is private;
        type T_VECTOR is array ( INTEGER range <> ) of T;
        with function "+" ( X, Y : T ) return T is <>;
                          --
                          -- <> Takes the default operator for a type.
                          --
        with function "*" ( X, Y : T ) return T is <>;
    procedure LOCAL_AX_PLUS_Y( N : in INTEGER; A : in T;
                               X : in T_VECTOR; Y : in out T_VECTOR );
        :
        Code for AX_PLUS_Y
        :
    end LOCAL_AX_PLUS_Y;

    procedure AX_PLUS_Y is
        new LOCAL_AX_PLUS_Y( REAL, VECTOR );
    procedure AX_PLUS_Y is
        new LOCAL_AX_PLUS_Y( COMPLEX, COMPLEX_VECTOR );
        --
        -- Since the operators are omitted the default "+" and "*" operators
        -- will be used for the types REAL and COMPLEX respectively.
        --
        :
end BASIC_LINEAR_ALGEBRA;
```

Such overloading can be used whenever the algorithm supports this action. There
should be no problems if the package, BASIC_LINEAR_ALGEBRA in the above example,
is not generic, but if the package is generic then the internal instantiations
cannot take place until the package itself is instantiated. Good compilers will
be needed to produce efficient instantiations in such situations.

3.5 User supplied routines.
In many areas of numerical computation the user has to supply information in
addition to the conventional input parameters. Examples are user supplied
functions, routines defining data such as observations in real time and routines
to form Az where A is a sparse matrix and z is a vector.

Examples of mathematical areas that require such information are zeros of
functions, optimization, data and function approximation, differential and
integral equations and quadrature.

In all these cases the user is expected to supply a routine which defines the information required by the package.

In Algol 60 these routines can usually be procedure parameters. In Fortran the equivalent facility of an EXTERNAL function or subroutine is only really convenient when the function or subroutine always has a set number of parameters from which the information can be calculated. For example, if we wish to find a zero of a function f(x) then it is convenient to use an EXTERNAL function when f(x) depends solely on x as with

$$f(x) = x - e^{-x},$$

but if the function depends upon data that itself needs to be calculated elsewhere such as

$$f(x) = ax - be^{-x},$$

where a and b are constants that depend upon some earlier calculation or experiment, then the situation is not so straightforward. The usual solutions to this problem are to use either

  i) COMMON, or
 ii) an additional workspace parameter, or
iii) reverse communication.

With reverse communication a routine designed to find a zero of a function would be written so that it performs one iteration per call, returning after each call to obtain the next function value from the user. Some mechanism for saving information between calls is necessary.

Reverse communication provides the most flexible solution, but it puts an additional burden upon the user. The use of COMMON in libraries, which is somewhat akin to the use of global variables, is rather controversial with libraries such as NAG allowing its use only to transmit information between the internal library routines [NAG (1973)].

In Ada there are no facilities for procedure parameters, instead functions and procedures can be generic parameters. Suppose that we have an optimization package containing a procedure designed to minimize a function of a single variable.

```
package body OPTIMIZATION is
   :
   with FUN_1_VAR;
   procedure UNIFUN( ...) is
      --
      -- UNIFUN minimizes a function of one variable.
      :
         FX := FUN_1_VAR.F( X );
      :
   end UNIFUN;
   :
end OPTIMIZATION;
```

Here the specification part of the package must be defined by the time OPTIMIZATION is compiled. For example

```
package FUN_1_VAR is
   function F( X : REAL ) return REAL;
```

The main disadvantages here are that the names of the user's package and function are fixed and that the user cannot minimize two separate functions without recompiling the body of the package FUN_1_VAR. This could be a considerable disadvantage for other library designers who wish to use OPTIMIZATION.

To avoid the disadvantages mentioned in the previous paragraph we must make UNIFUN generic.

```
generic
    with function FUN_1_VAR( X : REAL ) return REAL;
procedure UNIFUN( ...) is
    :
end UNIFUN;
```

We can now instantiate UNIFUN with any particular name and furthermore we can instantiate UNIFUN more than once in order to minimize different functions.

```
with OPTIMIZATION;
package USER_PACK is
    use OPTIMIZATION; -- Can now use names such as UNIFUN.
    :
    function FUN( X : REAL ) return REAL is
        --
        -- FUN has access to any previous information in USER_PACK.
        :
    end FUN;
    :
    procedure MY_UNIFUN is new UNIFUN( FUN );
    :
    MY_UNIFUN( ...); -- Call to UNIFUN to minimize FUN.
    :
end USER_PACK;
```

3.6 Separate compilation and instantiation

The Ada reference manual does not specify how generics are to be compiled in terms of speed and strategy and compilers are likely to vary between two extremes:

    a) For each instance the body is recompiled.
    b) Compilation is performed only once on the generic body.

The strategy adopted is likely to depend upon the generic parameters, but it is essential for numerical packages that compilers use b) when the only generic parameters are sub-programs, like Algol 60 procedure parameters. However a) is likely to be used with a type parameter, although this may not be the case with floating point types. In this case the generic may be compiled in all the machine s available floating point types and the required version picked up at instantiation.

3.7 Contents of packages

With libraries such as NAG and the NPL optimization suite careful consideration will need to be given to the contents of each package in order to minimize the number of generic parameters required. For example, the user requiring to minimize a function of a single variable will clearly not wish to instantiate a package that also requires a function of several variables as a generic parameter.

The number of generic parameters and the number of internal generic functions and procedures is also likely to affect the efficiency of compilation and the amount of part-compilation that will occur. We shall not discuss these issues here, but they are issues that will need further consideration.

4. IMPLICATIONS FOR FUTURE FORTRAN STANDARDS

We believe that some of the issues raised here in the context of Ada need to be considered by the Fortran standards committee X3J3. In particular, the adoption of some form of variable precision of the type being considered will involve similar problems on the inheritance of real types from one routine to another.

The traditional "separate compilation" philosophy of Fortran may have to be discarded and instead called subroutines may need to be (partly) compiled in advance and be on view to the calling subroutine.

Over the next few years, as Ada compilers become available, experience with programming in Ada will give us the opportunity to discover solutions to the problems. The experience may suggest features that are undesirable or that could have been better designed. Fortran has not always learnt from the experience of other languages, but we nevertheless hope that X3J3 will take this opportunity to use the Ada experience in order to influence the design of Fortran.

REFERENCES

BROWN, W. S. (1977) A realistic model of floating-point computation. pp. 343 - 360 in "Mathematical Software III" Ed. RICE, J. R. Academic Press, New York.

CODY, W. J. and WAITE, W. (1980) Software manual for the elementary functions. Prentice-Hall, New Jersey.

COX, M. G. and HAMMARLING, S. J. (1980) Evaluation of the language Ada for use in numerical computations. NPL Report DNACS 30/80, National Physical Laboratory, Teddington, Middlesex TW11 0LW, UK.

DoD. (1978) Steelman Department of Defense requirements for high order computer programming languages. US Department of Defense. (Un-numbered document, October 1978.)

DoD. (1980) Ada Reference manual for the Ada programming language. US Department of Defense. (Un-numbered document, June 1980.)

DONGARRA, J. J., MOLER, C. B., BUNCH, J. R. and STEWART, G. W. (1979) Linpack Users' Guide, SIAM, Philadelphia.

FORD, B. (1978) Parameterization of the environment for transportable numerical software. ACM Trans. Math. Software, 4, 100 - 103.

FORD, B., BENTLEY, J., DU CROZ, J. J. and HAGUE, S. J. (1979) The NAG Library "machine". Software - Practice and Experience, 9, 56 - 72.

GILL, P. E., MURRAY, W., PICKEN, S. M. and WRIGHT, M. W. (1979) The design and structure of a Fortran program library for optimization. ACM Trans. Math. Software, 5, 259 - 283.

HONEYWELL, INC and CII HONEYWELL BULL. (1979) Rationale for the design of the Green programming language. Honeywell, Inc., Systems and Research Center, 2600 Ridgeway, Parkway, Minneapolis, MN 55413, USA and CII Honeywell Bull, 68 Route de Versailles, 78430 Louveciennes, France.
This report is concerned with preliminary Ada, which was originally called Green.

LAWSON, C. L., HANSON, R. J., KINCAID, D. R. and KROGH, F. T. (1979) Basic linear algebra subprograms for Fortran usage. ACM Trans. Math. Software, 5, 308 - 323.

NAG (1973) NAG Reference Document No. R1/0, Numerical Algorithms Group, 7 Banbury Road, Oxford OX2 6NN, UK

RICE, J. R. (1980) Programming language facilities for numerical computation. Technical Summary Report no. 2033, University of Wisconsin, Mathematics Research Center, Madison, Wisconsin 53706.

SMITH, B. T., BOYLE, J. M., DONGARRA, J. J., GARBOW, B. S., IKEBE, Y. KLEMA, V. C., MOLER, C. B. (1976) Lecture Notes in Computer Science, Volume 6, Edition 2, Matrix Eigensystem Routines - Eispack Guide, Springer-Verlag, Berlin.

STEVENSON, D. (1981) A proposed standard for floating-point arithmetic. Draft 8.0 of IEEE Task P754. Computer, 14, 51 - 62.

WICHMANN, B. A. (1981) Tutorial material on the real data-types in Ada. US Army Contract Number DAJA37-80-M-0342, National Physical Laboratory, Teddington, Middlesex TW11 OLW, UK.

WILKINSON, J. H. (1963) Rounding Errors in Algebraic Processes, Notes on Applied Science No. 32, HMSO, London.

WILKINSON, J. H. (1965) The Algebraic Eigenvalue Problem. Oxford University Press, London.

DISCUSSION

DELVES. What costs are involved in a new instantiation of a generic module?
HAMMARLING. This is an important question and affects the kind of generic parameters that we would wish to use in practice. We have made some comments on this in the paper although they do not give a complete answer.
ROUBINE. The question must be refined in that such costs depend on - the quality of the implementation - the kinds of parameters - the tradeoff instantiation cost vs. efficiency of the instantiated unit. For a "reasonable" implementation we can expect that - subprogram parameters do not induce a replication of the unit - numeric type parameters induce only as many copies as there are different underlying hardware representations (this also includes enumeration types) - private type parameters are likely to induce copies, except for word size - array type parameters do not introduce copies, unless required by the index or element types, if they are themselves parameters.

SMITH. If a machine has precision 12 and one specifies two types - digit 6 and digits 15 - does one get more precision using 15 than using 6?
HAMMARLING. That is not guaranteed because digits 6 could use a precision higher than 12.

NELDER. If one uses a generic from one package in another package does one have to recompile everything?
ROUBINE. Not necessarily - that depends upon how the program has been written.

WAGENER. Can Ada programs use existing facilities (e.g. Linpack) without conversion to the equivalent Ada source form? Alternatively, is it feasible for existing programs written in other languages to use facilities developed in Ada? HAMMARLING. In theory Ada routines can use existing routines in other languages by using the INTERFACE pragma; however in practice there are likely to be severe difficulties. Draft 3 of the Ada-Europe portability guide recommends that the pragma INTERFACE should not be used. Of course, particular machines may provide facilities which readily allow interfacing, but a particular difficulty with using a Fortran package such as Linpack is the array storage problem. Similar difficulties are sure to exist when trying to call an Ada routine from another language. It is an area which deserves attention.
ROUBINE. As noted by Hammarling, Ada provides a pragma INTERFACE to indicate

that a given subprogram is implemented in another language, whose calling conventions must be obeyed. The converse depends on the existence of the same facility in the other language. A particular system may be nice enough to provide a uniform calling convention, but such a situation seldom occurs. To come back to Ada, the limitation is that subprograms do not take subprograms as parameters. However, a "friendly" implementation could conceivably extend the INTERFACE pragma to also apply to generic subprograms, and therefore achieve the desired functionality.

SCHLECHTENDAHL. It seems to me that some of the differences between packages in Ada and in the conventional Fortran libraries might be clearer when the question of binding time is investigated in more detail. By the term binding, I mean in general the replacement of a name by the corresponding object or of a reference to some generic function or procedure by its instance. In Ada binding is apparently concentrated at compile time, while with Fortran we usually think of binding of subprograms at "link editing" time. I might add that there are some systems which even allow the delay of subprogram binding until execution time.
HAMMARLING. Within an Ada compilation unit, binding is at compile time, since this is required for efficiency and for type checking. Note that Ada always checks that the type of every non-local is correct at compile time. Compilation units can be linked together by an Ada system in any compatible manner within a single library. For instance, a library may contain several subprograms which can be a main program. Also an Ada system can, but need not, offer a facility for providing more than one implementation of a package. Run time linking would not seem to provide any additional capability in Ada since compiling within a library is always required and there is no such thing as a free standing package.

THE RELATIONSHIP BETWEEN NUMERICAL COMPUTATION
AND PROGRAMMING LANGUAGES, J.K. Reid (editor)
North-Holland Publishing Company
© *IFIP, 1982*

Tasking Features in Programming Languages

Olivier Roubine
Centre de Recherche Cii Honeywell Bull
Louveciennes France
June 1981

Abstract: This paper traces the evolution of programming language design in the area of parallel processing. It describes how ideas that originated from the fields of operating system design and machine architecture for distributed computing have evolved to yield language features that are now available.

1. Introduction

The notion of parallel processing, i.e. the concurrent and, in general, independent execution of several programs on a computer system dates from the first "multiprogramming" operating systems. The idea of interleaving the execution of several programs as a way to optimize the utilization of the (then) expensive computer resources by permitting the CPU to perform some useful computation for one program while another one was blocked waiting for the completion of some operation (generally input-output). The term "task" appeared then to designate one program that could be executed independently (of other tasks).

The next step was the realization that a substantial reduction in the complexity of large programs could be achieved by breaking them into distinct activities that could have an independent execution. This independence was however only partial, as two separate needs emerged: communication between two independent activities, and synchronization, in order to permit one activity to be suspended until something happens in another. Parallel processes were of primary interest to operating system builders, and it is chiefly for them (and often by them) that came the first advances in the area.

Although the notion of parallel processes was fairly elaborate, the primitives to deal with them remained at a fairly low level, but even so, it was soon discovered that the manipulation of these entities was closely connected to the programming activity. For a substantial period of time, parallel processing had an ambiguous status, as an area of concern for systems people, while language designers (not necessarily a disjoint class) kept an eye on it.

In the late sixties, and mostly in the seventies, the idea of using parallel processes as a programming tool in certain domains (real-time, computer communication) made its way, thus contributing in shifting the emphasis from the system area to the programming language area.

In the next section, we trace this slow evolution through a coverage of the various features that have been proposed to deal with parallel processing. As an example of current ideas, section 3 contains a brief description of the tasking features in the language Ada. We then examine some specific machine architectures and their impact on the design of tasking features, especially in the area of numeric computation. We finally conclude by some speculations on the possible evolution of thoughts in this area.

2. Historical Overview

While the notion of independent tasks appeared with the first multiprogramming systems (e.g. OS360/MFT or MVT, circa 1964) [IBM1], the idea of a program unit whose execution could be interleaved with other similar units was introduced in the programming world by Conway as early as 1958 and was first extensively described in the literature in [Con 63]. Such program units are known as coroutines.

Consider two procedures A and B. The execution of B is started when A executes a call to B. Control is transferred to B which starts its execution as a normal subprogram. However, by executing the statement

            RESUME A;

B´s execution is suspended, and A´s execution proceeds past the place of the first call. A can in turn pass control back to B by a similar resume operation. This exchange can go on until B terminates by a return statement.

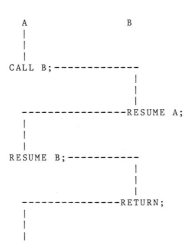

            fig. 1 Coroutines

Coroutines appeared in at least one important programming language : SIMULA 67, [Dah 70], where they were presented as features for "quasi-parallel" execution.

A more conventional (in the operating system sense) notion of parallel processing was introduced by the tasking features of PL/1 [IBM 2]. There, execution of a procedure could be started as a parallel activity by using the special option TASK.

```
        CALL B TASK;
```

In such a case, execution of B was started, but execution of A could proceed in parallel. Synchronization was achieved by means of <u>events</u> . Events are special variables, with a completion status. The primitive

```
        WAIT (EV);
```

causes the suspension of the task that executes it until the completion status of EV becomes true. This can be achieved explicitly by an assignment

```
        COMPLETION (EV) = "1" b;
```

or implicitly by the termination of an operation associated with the event.

These efforts to introduce parallelism in programming languages were not actively pursued by language designers. However, the time was not wasted as it permitted a variety of ideas to florish among operating system designers.

We will first mention a proposal by Dennis and Van Horn ([Den 66], who introduced in particular the notion of locks : a lock is a boolean variable that is manipulated by two primitives, <u>lock</u> and <u>unlock</u>. If a process attempts to lock a variable that is already locked, the process will not be permitted to proceed past the point until the variable is unlocked by another process.

By far the most important work in that domain was due to E.W Dijkstra who introduced the notion of semaphore in the early sixties ([Dij 68]). Essentially, a semaphore is a counter that can be indivisibly incremented or decremented by appropriate operations called (for historical reasons) V and P. In addition, to each semaphore is associated a queue of waiting processes. A P operation decrements the semaphore; if its value was positive it has no other effect. However, if the value of the semaphore is non positive, the operation is not permitted to complete, and the process is placed on the semaphore queue. A V operation increments the semaphore. In addition, if the semaphore queue is non-empty (this corresponds to the fact that the value of the semaphore was negative), one of the processes placed in that queue is released.

A semaphore can be used to control access to a common resource. Its chief advantage is that a process need not know what is being done by other processes. The programming of each process can thus be done independently.

Locks and semaphores quickly led to the notion of critical region : in a variety of systems, processes are mostly independent of each other, except in certain restricted areas where they need access to some information that is potentially shared between processes, and where access must be controlled so that only one process can access the information at a time. The concept is called <u>mutual exclusion</u>, and the term <u>critical region</u> applies to a region of program where mutual exclusion to certain objects must be obtained.

The next step along this path was concerned with the problem of mutual exclusion. It culminated in the work of Brinch Hansen ([Bri

72]) and Hoare ([Hoa 74]) and the resulting concept of <u>monitor</u>. A monitor is a collection of local data (which constitutes a "resource") and associated procedures and functions to access the data. These procedures can be called by various processes, except that access to the data is guaranteed to occur in mutual exclusion : only one of the procedures of the monitor can be active at a given time, and execution of a call is held up if the monitor is already "busy". This basic model is augmented by the notion of condition variable which is akin to the events mentioned above : a process can <u>wait</u> on a condition variable, and waiting processes can be released by a <u>signal</u> relative to that variable.

The operations that were described so far aimed principally at achieving synchronization and mutual exclusion between processes. The needs for such features stemmed partially from the desire to access common data, for the purpose of exchanging information from one process to the other. This communication function was thus handled indirectly. Other attempts have been made to provide primitives for interprocess communication.

One of the first comprehensive systems based on communication was developed by Brinch Hansen [Bri 70] for the RC 4000 computer. One of the basic notions of the system is that of a message buffer which is a receptacle for some information to be passed from one process to another. Primitives are provided to send a message to a particular process, to await (and receive) a message from a particular process, or to send or receive a reply. Each process possesses a message queue, which allows several messages to be pending.

Variations of such systems were proposed by Balzer [Bal 70], Walden [Wal 72], and Kahn [Kah 74].

A more recent scheme was proposed by Hoare [Hoa 78], and is known as CSP (Communicating Sequential Processes). The basic primitives are:

    X?Y   input Y from process X
    X!Y   Output Y to   process X

These primitives operate as follows: when an input command is executed by a process, the process is suspended if no corresponding output to it has been executed by another process; conversely, execution of an output command suspends the process that executes it if the destination process has not itself executed an input command with the other process as source. When both processes have reached matching input and output commands, the value becomes available to the destination process, and both processes can then resume their independent execution.

One can observe two main differences with a message system: there is a tight coupling between the source and destination processes, and communication is typed. In fact, the tight coupling also provides a means to express problems where synchronization is more important than communication.

Along similar lines, Brinch Hansen proposed the concept of Distributed Processes [Bri 78], with an important difference with CSP residing in the asymmetry of the communication: the sender designates the destination, but the receiver does not have to identify the source. Processes can contain special procedures, that are visible (i.e. can be called) from other processes, in addition to their own statements. Special guards can be provided to inhibit execution of certain statements, or delay it while a given condition is not true. When a visible procedure is called from another process, its statements can be executed by the called process. Execution is interleaved with that of the process own statements and that of other procedures that have been called, as follows: execution of a sequence of statements proceeds either to completion, or until it is deferred by a false condition, at which point, another sequence will be executed.

We conclude this summary by mentioning the work of Habermann and Campbell who introduced the notion of path expressions [Cam 74]. This work is substantially different from other proposals in the following sense: in most schemes, the computational aspects of a process are described by a sequential algorithm with synchronization or communication expressed by the use of special primitives inside the algorithm. Here on the contrary, processes are constructed from

(a) a set of "procedures" which define sequences of actions

(b) a separate description of how these procedures can be executed, by means of path expressions.

Path expressions are built from sequencing (e.g. "p;q", meaning execute p then q), selection ("p,q" meaning execute either p or q) and repetition ("<u>path</u> p <u>end</u>" meaning repeatedly execute p).

3. The Tasking facilities in Ada.

We will now describe in greater detail the tasking features of the programming language Ada [DoD 81], that has been developed for the Department of Defense. The treatment of parallelism in Ada can be considered to be representative of the state of the art in many respects : the tasking features have been designed to fit harmoniously with other features of the language, they are not directly borrowed from elsewhere, but rather build on existing work to provide a higher level treatment of the problems of programming parallel tasks. Furthermore, Ada is by no means a research language; it is neither a language built solely around parallel processing, nor a language in which tasking has been added almost as an afterthought.

The following features of Ada bear a certain resemblance with CSP and Distributed Processes. Tasks can be declared as program units with <u>entries</u> as their only visible components. Entries are similar to subprograms, except that their execution is synchronized, as in Distributed Processes.

Example:

```
     task READER_WRITER is
        entry READ  (V: out INTEGER);
        entry WRITE (V: in INTEGER);
     end;
     task body READER_WRITER is
        VALUE : INTEGER;
     begin
```

```
      accept WRITE(V : in INTEGER) do
         VALUE := V;
      end;
      loop
         select
            accept READ(V : out INTEGER) do
               V := VALUE;
            end;
         or accept WRITE(V : in INTEGER) do
               VALUE := V;
            end;
         end select;
      end loop;
   end READER_WRITER;
```

For synchronization, communication and mutual exclusion, tasks use the single concept of <u>rendezvous</u> : a task can <u>call</u> an entry of another task, e.g.

```
   READER_WRITER.WRITE(n);
```

When the call is performed, two things can happen: either the called task is ready to accept the call or not. If not, the calling task is suspended. A task where an entry is declared is ready to accept a call to that entry if it has encountered an <u>accept statement</u> for the entry; when such a statement is reached, either the entry has been called or not. If not, the task is suspended. In the case where both the calling task has reached the call point and the called task has reached an accept statement for the entry, a rendezvous is said to occur between the two tasks: the instructions appearing in the accept statement e.g.

```
   accept WRITE(V :in INTEGER) do
      VALUE := V;
   end;
```

are executed while the calling task is still held suspended. After the accept statement has been executed, both tasks can pursue their execution in parallel.

Primitives are provided for more complex interactions. First, it is possible to wait for one of several possible calls, the first one to occur being accepted:

```
   select
      accept READ (V : out INTEGER) do
         V := VALUE;
      end;
   or
      accept WRITE(V : in INTEGER) do;
         VALUE := V;
      end;
   end select;
```

Here, if any of the entries mentioned has been called already, one of the possible rendezvous is selected arbitrarily. If no such call has occurred, the task is suspended until the first acceptable call is issued, in which case the corresponding rendezvous is executed.

Boolean guards can be added to render the selection more precise. For example, if we want to permit only one read between two consecutive writes,

```
task body READER_WRITER is
    VALUE    : INTEGER;
    OBSOLETE : BOOLEAN := TRUE;
begin
    loop
        select
            when not OBSOLETE =>
                accept READ(V : out INTEGER) do
                    V := VALUE;
                end;
                OBSOLETE := TRUE;
        or
                accept WRITE (V : in INTEGER) do
                    VALUE := V;
                end;
                OBSOLETE := FALSE;
        end select;
    end loop;
end;
```

There is a major difference with the when clauses of Distributed Processes: here, the condition is evaluated once when the select is reached, and its value determines if calls to the following entry should be considered. The fact that the value of the condition might later change is irrelevant as long as the entire select statement is not reexecuted.

The select statement can also have a time-out, to indicate that the task should wait for a call for only so long. The following example illustrates the case of a value that remains meaningful only if used within 3 seconds.

```
loop
    select
        when not OBSOLETE =>
            accept READ(V : out INTEGER) do
                V := VALUE;
            end;
    or
            accept WRITE(V : in INTEGER) do
                VALUE := V;
            end;
            OBSOLETE := FALSE;
    or
            delay 3;
            OBSOLETE := TRUE;
    end select;
end loop;
```

A conditional clause can allow a rendezvous only if the call has already been registered: we can use this feature to give a priority to writes over reads:

```
loop
    select
        accept WRITE(V : in INTEGER) do
```

```
            VALUE := V:
        end;
    else
        select
            accept READ(V : out INTEGER) do
                V := VALUE;
            end;
        or
            accept WRITE(V : in INTEGER) do
                VALUE := V;
            end;
        end select;
    end select;
end loop;
```

Similar statements exist to perform conditional and timed entry calls.

Aside from the synchronization and communication aspects described above, one should mention an important point: tasks can be generalized by the introduction of task types which permit several instances of the same task to be created, possibly dynamically, and also to be inserted in data structures.

For example, one may declare

```
task type READER_WRITER is
    entry READ  (V̄: out INTEGER);
    entry WRITE (V: in INTEGER);
end;
```

(the body remaining as before), and instances can then be created by simple object declaration, e.g.

```
X,Y : READER_WRITER;
```

X and Y become the names of two tasks, and can be used as in

```
X.READ(I);
Y.WRITE(3);
```

With the possibility of treating tasks as objects, we depart from the traditional view of a process as a pure unit of execution, and start an evolution towards a view that is more data-oriented.

4.   Specific Architectures

Most of the proposals presented thus far were concerned primarily with parallelism as a programming concept: the existence of tasking features in a language allows the expression of elegant solutions to complex problems. Architectural considerations, such as the existence of multiple processors, is generally not a guiding factor, although some of the more recent proposals (CSP, Distributed Processes, Ada) do exhibit some concern for distributed architectures.

In this section, we consider two specific classes of problems where parallelism is of importance: those are distributed systems, and large machines for scientific computation.

4.1 Distributed Systems

Distributed architectures are increasingly popular, especially to build specialized computing equipment. We can consider the requirements to program a system consisting of a network of processors that can communicate with each other, but do not necessarily share memory.

Two approaches are possible: either one programs in terms of the machine, i.e. each processor has a specific program, and communication can take place between each node; or one programs in terms of the application, knowing that certain tasks can be run only on certain processors, and allow communication to take place between tasks (rather than between nodes), letting the implementation discriminate between local and distant tasks.

Of course, the latter approach is more flexible, and is generally preferable, at least as soon as some hardware extensibility is envisaged.

The requirements from a programming point of view are not essentially different from those we have seen already, communication and synchronization being the two important aspects. However, one must take into account the cost of certain interactions.

If we assume that no shared memory is available, and that information is exchanged between nodes by transmitting packets over communication links, then the tighter the coupling between two distant tasks, the more expensive. Accessing a non local variable may itself require a complex interchange of messages. Also, the duration of a message transmission being substantially longer than a processor cycle, it may be undesirable to keep a processor idle between a request and a reply. A form of typed message system would then correspond quite closely to such a situation : the two main operations would be

        send (message, destination)

which forwards the message to the node hosting the indicated task, without delaying the sender, and

        receive (message, source)

which awaits for a message of the given type from the designated task until such a message is available.

Such primitives would have to be augmented with operations to enquire about the status of a task, or of a communication, to place delays on a waiting time, or to wait for several possible senders or several message types.

4.2 Array and Vector Processors

Special architectures have appeared to handle the requirements of specific large numeric problems such as can be found in weather forecasting or fluid dynamics. The numerical methods used in such problems call for the computation of certain expressions at numerous points on a large "grid". It is often the case that the same computation is performed over different input data, and special machine architectures have been devised to operate efficiently in such conditions: array processors are built round a repetition of processing elements, each one performing the same operations, but on its own operands (example the Illiac IV [Bar 68]). In vector processors, different processing units can perform the various steps of a computation, by pipelining these units, and feeding a sequence of operands at a high rate, one can obtain the results at a much faster rate than on conventional machines (example Cray 1 [Rus 78]).

The problems posed to the language designer for such cases of parallelism are very different from the ones considered earlier, since here, the emphasis is on the parallelism of data, rather than the parallelism of control. Certain language features have been proposed (e.g. in [Per 79]), which allow the user to specify an extent of parallelism for collections of data, together with constructs that indicate over which indices a certain computation should be performed. As in other modern high level languages, the features should permit the expression to the properties of the problem, rather than the properties of the machine.

5. Conclusion

We shall close this overview of the treatment of parallel tasks in programming languages by some considerations on what may lie ahead of us.

The traditional algorithmic languages have made a clear distinction between Instructions and Data, and the treatment of parallelism has dealt mostly with instructions, in that the emphasis has been on expressing parallel algorithms that operate on some potentially shared data. The noticeable exception is the introduction of parallelism in scientific computers, where the emphasis is on data.

On the other hand, in the programming area, we have witnessed a trend towards closer connections between the design of algorithms and the design of data structures, culminating in the concept of an abstract data type.

Now, coming back to the notion of a task, such an entity has been traditionally viewed as purely active, in the sense that a task describes the execution of some instructions. Specialized concepts, and in particular that of a monitor, have been introduced to describe some passive entities with special properties.

One can justifiably wonder if the same considerations that have led to the introduction of abstract data types, namely the close connection between data and operations that can be performed on it, will not ultimately prevail in the parallel area as well. The distinction between an active and a passive entity would then become less clear, but one would be able to manipulate "active data structures". Let us take as an example a radar management system.

Each track corresponds to some evolving information, corresponding in particular to some position and speed estimates. Thus information is regularly updated either on the basis of refreshed external information, or on the basis of some extrapolation, so as to display the actual and projected course. A traditional organization would consist of various tasks to perform the various functions (input translation, computation, display) for all tracks. A much simpler organization consists in representing each track as a separate task, keeping its internal data up-to-date, and communicating with the rest of the system when appropriate: thus, the rest of the system does not have to worry about the evolution of the information, and the track management task does not have to worry about other tracks in the system, nor about how the information is used. Such an organization adheres to the principle of minimizing interfaces. Of course, the use of such techniques would require not only appropriate features in programming languages, but also a substantial modification of our approach to program development, and possibly of the level of confidence that deals with parallel processes when it comes to the question of performance.

In the area of tasking, as in more traditional areas in programming languages, the main efforts are oriented towards providing higher level primitives that allow the user to express the solutions in terms of the concepts of the problem, rather than those of the machines.

References

[Bal 70] Balzer, R.M., "Ports, A Method for Dymamic Interprogram Communication and Job Control", The Rand Corporation, Report R605ARPA, 1970.

[BAR 68] Barnes, G.H., et al. "The ILLIAC IV Computer", IEEE Transactions on Computers, vol C-17, August 1968.

[Bri 70] Brinch Hansen, P., "The Nucleus of a Multiprogramming System", Comm.ACM vol 13,4, April 1970.

[Bri 72] Brinch Hansen, P., "Structured Multiprogramming", Comm.ACM, vol 15,7, July 1972.

[Bri 78] Brinch Hansen, P., "Distributed Processes: A Concurrent Programming Concept", Comm.ACM vol 21,11, November 1978.

[Cam 74] Campbell, R.H., and Habermann, A.N. "The Specification of Process Synchronization by Path Expressions", Lecture Notes in Computer Science, vol 16, Springer Verlag, 1974.

[Con 63] Conway, M.E., "Design of a Separable Transition Diagram Compiler", Comm.ACM vol 6,7, July 1963.

[DoD 81] US Department of Defense. "Reference Manual for the Ada Programming Language", (Government Printing Office, Number L008-000-00354-8), November 1981.

[Dij 68] Dijkstra, E.W., "Cooperating Sequential Processes", in "Programming Languages", (F. Genuys, ed.) Academic Press, 1968.

[Dah 70] Dahl, O.J. Myhrhaug, B., and Nygaard, K., "SIMULA 67 Common Base Language", Publication S22 Norwegian Computing Center, Oslo, 1970.

[Den 66] Dennis, J.B., and VanHorn, E.C., "Programming Semantics for Multiprogrammed Computations", Comm.ACM vol 9,3, March 1966.

[Hoa 74] Hoare, C.A.R: "Monitors: An Operating System Structuring Concept", Comm.ACM vol 17,10, October 1974.

[Hoa 78] Hoare, C.A.R., "Communicating Sequential Processes", Comm.ACM vol 21,8, August 1978.

[IBM 1] "IBM System/360 Operating System: MVT Guide", Form GC28-6720.

[IBM 2] "IBM System/360 Operating System: PL/I Reference Manual" Form C28-8201-1.

[Kah 74] Kahn, C., "The Semantics of a Simple Language for Parallel Programming", Proc. IFIP Congress 74, North Holland, 1974.

[Per 79] Perrot, R.H., "A Language for Array and Vector Processors", ACM transactions on Programming Languages and Systems, vol 1,2, October 1979.

[Rus 78] Russel, R.M., "The Cray-1 Computer System", Comm.ACM vol 21,1 January 1978.

[Wal 72] Walden, D.C., "A System for Interprocess Communication in a Resource Sharing Computer Network", Comm.ACM, vol 15,4, April 1972.

*   *   *

DISCUSSION

Feldman Can one start several processes together?

Roubine The only problem is with passing initial parameters. Once these have been passed through an accept, computations can proceed in parallel.

Manfred Paul Is there a means for V- and P- operations in Ada to identify particular processes queued up with regard to a certain semaphore? And, how does one program fairness strategies in Ada?

Roubine In general, it is not possible to explicitly state "accept a call only from this task". However, the important concept of entry families can be used in most cases for such a purpose, and is in particular indicated to program fairness.

We must assume that the number of possible callers is known when the accepting task is activated. Instead of a single entry, the accepting task declares a family (essentially an array) of entries:

    entry E(1 .. N)(... -- formal parameters)

The idea is that each member of the entry family is dedicated to one caller. Fairness can then be programmed with appropriate guards in a select statement

```
loop
   select
      when ... =>
      accept E(1)(...) do
      ...
      end;
   or when ...=>
      accept E(N)(...) do
      ...
      end;
   end select;
end loop;
```

George Paul Is there a way to associate task affinity to a specific processing element in Ada?

Roubine There is an address specification mechanism, applicable to any object:

    for T **use at** ...;

The interpretation of the address given after **at** is implementation dependent, and can thus express a particular processor identity in a network.

Rice You describe a view of parallel programming that arises from multi-processing in operating systems and is to be extended to real time control applications. Since operating systems is real time control of a particular type, this extension is natural. However, this has little relevance to parallel numerical computation as currently practiced on CDC, Cray, ICL and similar hardware.

More distressing, this view (and the related facilities) may be inadequate for numerical computation as it will be practiced on multi-processor (as opposed to multi-processing) machines in the future. The current fashion in language design (e.g. ADA) is to make it impossible (or, at least, very difficult) to do things the designers find undesirable or unfamiliar. Thus, we might find ourselves using Fortran for numerical computation on multi-processor machines not because it is suitable, but because its laissez-faire philosophy allows us to do what needs to be done.

I give a concrete non-numerical example which is a simple model of how some major numerical computations will be done on multi-processor machines. The objective is to find the word "nugget" written in gold from a body of text. We assume that once the word "nugget" is found, it is a substantial additional task to determine if it is written in gold. The algorithm follows:

Program: Chief-golddigger

1. Determine the number N of processors needed to carry out the task.

2. Request N processors from the system and recieve K processors called Golddigger_I for I = 1 to K.

3. Divide text into K parts and distribute to all Golddigger_I along with Nugget_search program and instruct them to start.

4. When one Golddigger_I finds "nugget"

    A. Stop all Golddigger_I
    B. Retrieve the word "nugget" from Golddigger_I
    C. Request one processor from the system; call it Assayer
    D. If no processor is received, save Golddigger_K data and status; rename Golddigger_K Assayer
    E. Distribute nugget_assay program to processor Assayer and instruct to start

5. When Assayer is finished:

    If "nugget is written in gold

        Then discharge all processors and retire
        Else discharge processor Assayer
            start Golddigger_I for I = 1 to K-1
            If Assayer was Golddigger_K, reset and rename Golddigger_K
            start Golddigger_K

6. Go to 4

I make three observations about this computation. First, programs must be able to capture full control of processors. Even if step 2 is not dynamic within a system, portability requires this capability as this program may want N=327 processors but be running on a machine with only 256 so that only K=255 are available. Second, the program stops at Step 4.A occur at completely arbitrary and unpredictable points within the program Nugget_search. Third, at Step 4.B (as usual in real time control) it makes no difference which processor found "nugget" or if "nugget" was found simultaneously at 4. Further, no record need be kept as to where it was found. Of course, processor Golddigger_I should make a note that he found the "nugget" so if it turns out to be written in gold he can approach the Chief_golddigger at retirement time.

A fourth observation is that in some numerical computations it would make no difference at Step 4.B if the letters of "nugget" were taken from different processors. That is, if three real numbers are in simultaneous contention to be stored in X, one might well choose the i-th digit at random from the three contendors rather than decide which of them to store.

My specific questions are:

1. Will Ada support, hinder, prevent, or be neutral to expressing the parallel control of the Chief_golddigger program?

2. If Ada does not support this type of computation, are there other languages that do?

Roubine This interesting problem can be coded fairly easily in Ada, except for one particular aspect: tasks cannot be assigned dynamically to a processor. In the present situation, we have to assume that each processor has an executive which knows about the possible tasks that can be performed. On the other hand this has the advantage that the data of a Golddigger does not have to be saved explicitly.

The other salient point is the use of the FAILURE exception to regain control of a processor.

We assume a central dispatcher that can distribute processors:

```
type ACTION is (DIG|ASSAY|...);
type PROCESSOR_EXEC;
type PROCESSOR is access PROCESSOR_EXEC;
task type PROCESSOR_EXEC is
   entry START_DIGGING (DIGGER_ID : out...);
   entry START_ASSAYING (ASSAYER_ID : out...);
   entry INIT (WHO : PROCESSOR);
end;

task DISPATCHER is
   entry AVAILABLE (PR: in PROCESSOR);
   entry GET (PR: out PROCESSOR);
end;

task body DISPATCHER is
   ...
begin
   loop
      select
         accept AVAILABLE (PR: in PROCESSOR) do
         -- store the value of PR
         end;
      or when EXISTS_FREE_PROCESSOR()=>
         accept GET (PR: out PROCESSOR) do
            PR:= ...;
         end;
      end select;
   end loop;
end DISPATCHER;

task body PROCESSOR_EXEC is
   ME: PROCESSOR;
   ...
begin
   accept INIT (WHO : in PROCESSOR) do
      ME := WHO;
   end;
   loop
      DISPATCHER.AVAILABLE (ME);
      select
         accept START_DIGGING (.....) do
```

```
             -- forks a new digger
          end;
     or accept START_ASSAYING (.....) do
             -- forks a new assayer
          end;
       end select;
   end loop;
end PROCESSOR_EXEC;
```

A digger task receives the string to work on, together with the index attributed to it (it is supposed to know who it works for).

```
task type WORKER is
   entry SET_UP (STR: TEXT; INDX:...);
   -- to pass the text to work on, and the index
   -- used to identify the digger
   entry ASSAY (NUGGET : WORD; GOLDEN: out BOOLEAN);
   -- order to assay the given nugget; GOLDEN
   -- will be set to true if test is positive
   entry RESUME;
   -- called to indicate that digging should go on
   entry STOP;
   -- to terminate work;
end WORKER;

task body WORKER is
begin
   accept SET_UP (STR: TEXT; INDX:...) do
      WORK := STR;
      ME := INDX;
   end;
   NEXT_WORD := FIRST_WORD;
   while STRING_NOT_EXHAUSTED() loop
   <<BODY>> begin
      if IS_NUGGET (NEXT_WORD) then
         CHIEF.FOUND (NEXT_WORD);
      end if;
   exception
      when WORKER'FAILURE=>
         loop
            select
               accept ASSAY (NUGGET : WORD;
                            GOLDEN : out BOOLEAN) do
                  T := new ASSAYER_TASK;
                  T.TRY (NUGGET,GOLDEN);
               end;
               -- will wait for next order
            or accept RESUME;
               goto BODY; -- will retry current word
            or accept STOP;
               raise; -- will terminate
            end select;
         end loop;
   end;
   NEXT_WORD := get_next_word;
   end loop;
   CHIEF.FINISHED (ME);
end WORKER;
```

```
task body CHIEF is
begin
    -- obtain K<N processors
    N_PROC := N;
    for I in 1 .. N loop
        select
            DISPATCHER.GET (PROC(I));
        else
            N_PROC := I;
            exit;
        end select;
    end loop;
    -- activate diggers on the K processors
    for I in I .. N_PROC loop
        PROC(I).START_DIGGING (DIGGER(I));
    end loop;
    for I in I .. N_PROC loop
        DIGGER(I).SET_UP(-- string ref,...);
        STILL_WORKING(I) := TRUE;
    end loop;
    -- We are through with initialization.

    loop
        -- now wait for someone to find a nugget
        select
            accept FOUND (WORD_REF:...) do
                -- record the word ref (does not have to
                -- know who found it)
                for I in I .. N_PROC loop
                    raise DIGGER(I)'FAILURE;  -- STOP ALL DIGGERS!
                end loop;
                -- note that we are still inside rendezvous
            end;
            -- OK. Now we try to get a new processor.
            select
                DISPATCHER.GET (A_PROC);
                A_PROC.START_ASSAY (ASSAYER);
                ASSAYER.TRY (NUGGET, ANSWER);
            else
                DIGGER (N_PROC).ASSAY (NUGGET, IS_GOLD);
            end select;
        -- we wait for the answer by remaining
        -- in rendezvous. After the rendezvous, we
        -- can look at the answer
            if IS_GOLD then
                for I in 1.. N_PROC loop
                    if STILL_WORKING(I) then
                        DIGGER(I).STOP;
                    end if;
                    -- note that we need not worry about last digger
                end loop;
                exit;    -- Ada word for retire
            else
                for I in I .. N_PROC loop
                    if STILL_WORKING(I) then
                        DIGGER(I).RESUME;
                    end if;
                end loop;
            end if;
        or
```

```
          accept FINISHED (DIGGER_ID) do
             STILL_WORKING (DIGGER_ID) := FALSE;
          end;
          ACTIVE_DIGGERS := ACTIVE_DIGGERS-1;
          exit when ACTIVE_DIGGERS = 0;
       end select;
    end loop
 end CHIEF;
```

In conclusion, I think Ada does support this kind of computation rather well; in fact it fares substantially better than most languages I can think of.

W. Kahan Should every computer science student be exposed to numerical ideas?

Roubine Yes – it is desirable (as long as I don't have to go back to school).

THE RELATIONSHIP BETWEEN NUMERICAL COMPUTATION
AND PROGRAMMING LANGUAGES, J.K. Reid (editor)
North-Holland Publishing Company
© *IFIP, 1982*

Language Support for Floating Point

Stuart Feldman

Bell Laboratories
Murray Hill, New Jersey 07974

ABSTRACT

Although computers were first designed to do numeric computations, most programming languages offer poor support for floating point. We examine current support for floating point, and consider what facilities ought to be provided in future languages, in light of the proposed IEEE Standard for Binary Floating Point Arithmetic [1]. We also consider the changes that are needed to available languages to make minimal use of the IEEE arithmetic.

1. Introduction

Electronic computers were invented to do numeric computations. Soon, floating point arithmetic was simulated in software, then supported in hardware to speed up programs. The first programming languages were designed to ease the evaluation of formulas. and algorithms. Yet most languages today seem to pay lip service at best to floating point.

This state of affairs is unfortunate but not surprising. Past hardware implementations of floating point have been varied and (to be polite) idiosyncratic. Super-computers have sacrificed precise handling of exceptions and sometimes correctness of results to achieve throughput. Inexpensive machines have sacrificed range or accuracy to reduce hardware costs. Important commercial systems include:

System	Radix	Overflow Value	Total Bits	Special values	Representation
IBM 360 etc	16	2^{252}	32,64,128		sign and magnitude
Honeywell 66	2	2^{127}	36,72		two's complement
CDC Cyber	2	2^{1070}	60	indefinite, infinity	one's complement
DEC VAX	2	2^{127}	32,64,128	reserved	sign and magnitude (implicit normalization)

Any programming language that catered to the details of one of those implementations would lose any claim at universality. Therefore, language designers provided syntax for floating point constants, declarations for floating point variables, and symbols for the most popular floating point operations (without specifying the range or precision of those numbers or the result of the operations), but nothing further.

Overall, languages have provided weak syntactic support for floating point, and almost no detailed semantics. We will consider what must be done to upgrade old languages and design new ones in view of the problems presented by floating point arithmetic. The changes to handle special cases and types would in most cases be straightforward. If the underlying arithmetic matched the needs of the Standard, then specifying the precise semantics of arithmetic in a language would force few changes to compilers. But intelligent handling of exceptions may require difficult changes to languages, libraries, compilers, and operating systems.

This paper is prompted by the publication of the latest (Draft 8.0) "Proposed Standard for Binary Floating-Point Arithmetic"[1] by the Floating-Point Working Group of the Microprocessor Standard Committee of the IEEE Computer Society. This draft (hereafter called 'the IEEE Standard') describes a well-considered system for doing careful computation. Although it has not yet been formally accepted as a standard, this proposal has already influenced the design of floating point units for micro- and mini-computers, and can be viewed as one of the de facto industry standards. (The much-copied IBM hexadecimal formats are also an industry standard, alas). We will also consider some potentially valuable ideas not included in the Standard.

In this paper, we describe the aspects of floating point of most immediate concern to programming language designers. We will then see how the more popular languages meet these needs. We then consider the features a language should contain to support quality floating point. Finally, we will consider the changes that would have to be made to available languages to provide minimal practical support of IEEE-style arithmetic.

2. The IEEE Standard

Although the IEEE Standard was written by a microprocessor standards committee, it is not a hardware definition, but a description of a system for supporting floating point arithmetic. Since most users will program in a higher-level general language rather than in a machine-specific language, it makes sense to consider the language with its required library and compiler as part of the support of the IEEE Standard. We ignore the most important aspects of the IEEE Standard such as the introduction of gradual underflow and the requirement that arithmetic operations yield best possible results and only discuss those with implications for the language and system designer. We examine the low-level aspects of the language that deal with the floating point details, not the algorithmic structure, of scientific computation. This is not an essay on numeric computation, so we also ignore desirable language features such as arrays and procedure parameters that are not directly associated with arithmetic.

'*The* IEEE Standard' is actually many standards, since it allows the implementor leeway in certain areas. Thus, provision of traps is optional, and the exponent range in extended formats may be richer than required. Conceivably, there can be many different standard-conforming implementations, though it is expected that only a few major groupings will be selected.

2.1. Formats

The IEEE Standard defines two 'basic' formats and two 'extended' formats:

Format	Maximum Exponent	Bits in Significand	Total Bits
Single	127	24	32
Double	1023	53	64
Single-Extended	≥ 1023	≥ 32	≥ 44
Double-Extended	≥ 16383	≥ 64	≥ 80

Floating point numbers are represented as a pair of values, an exponent and a significand; we avoid the more traditional word 'mantissa' since we make no assumptions about the location of the binary point. The value of the number is $significand \times radix^{exponent}$. The IEEE Standard recommends support of the extended format corresponding to the widest basic format supported, but any set of formats that includes Single is technically consistent with the IEEE Standard. Note that the formats of the basic types are defined completely, but the extended formats need only satisfy certain constraints. As with most non-IEEE floating point systems, unnormalized numbers can be represented in the extended formats, so there may be many different representations of a value. There are no unnormalized numbers in the basic formats. (Numbers with very small magnitudes are saved in a 'denormalized' form in the basic formats, but the representation is unique.)

The basic formats are intended for stored values: Single is sufficient for almost all real world data (only spectroscopic numbers habitually have more than seven decimal digits of precision), and Double for holding intermediate values needed for numeric reasons. The extended formats were intended to diminish the incidence of overflow, underflow, and roundoff while evaluating expressions; it is safe to do many additions and a moderate number of multiplications of basic format quantities using an extended intermediate, since underflow or overflow can result only if the final result is unrepresentable, and it is unlikely that much significance will be lost. These features of the extended formats are also needed for more complicated intermediate quantities, as we will discuss below.

Each format also contains non-numeric values. There are two infinities with well-defined algebraic properties. There are also two zeroes, with $\dfrac{1}{-\infty} = -0$. In ordinary arithmetic operations, $+0$ and -0 are indistinguishable.

The IEEE Standard also has a class of values called NaN (not-a-number). Each NaN contains a quantity in the field normally used for the significand, so there can be more than eight million Single NaNs, and even more for other formats. NaN's may appear as the results of invalid operations or from non-arithmetic sources such as input and initialization. If a NaN is used as an operand, a trap may be precipitated or the value of a NaN operand may be propagated into the result. The interpretation of a NaN is implementation defined; suggested uses for NaNs include naming the operation that first produced a non-number, marking uninitialized memory, indicating unavailable statistical data, and pointing to alternate representations or further descriptions of values.

2.2. Operations

The IEEE Standard specifies that the following operations are supported to full precision:

> sum
> difference
> product
> quotient
> remainder
> square root
> binary floating format → a different binary floating format
> binary floating ↔ decimal
> binary floating format ↔ integer
> comparison

Conversion between extended binary and decimal formats need not be to full precision. The comparison operations are tricky since trichotomy does not hold: two numbers can compare 'less than', 'greater than', 'equal to', or 'unordered', so it is possible that neither $a < b$ nor $a \geq b$ be true. (Any comparison against a NaN is unordered, as is any comparison against infinity in the 'projective mode' mentioned below.)

Although not required by the Standard, a language ought to support a richer set of operations for scientific computing. Negation, absolute value, and raising quantities to integer powers ought to be mandatory. The language or library ought to support the elementary functions (logarithmic, exponential, trigonometric, hyperbolic) and their inverses.

Operations related to the floating point rather than real arithmetic nature of computation are also needed, as outlined in the IEEE Standard, or by IFIP WG2.5 [2], or by Brown and Feldman [3], including

> environment inquiries
>   (actual precisions and ranges of types)
> categorization of value
>   (ordinary number, NaN, infinity, -0, $\cdots$)
> extraction of exponent and significand
> synthesis of number from exponent and significand
> scaling (multiplication by a power of the radix)
> copy sign
> nearest neighbor
> inherent absolute uncertainty
> inherent relative uncertainty

2.3. Modes

The IEEE Standard has user-settable modes of operation. There is a default setting of each mode. These modes are viewed as part of the global state of the program.

The rounding mode determines the result when a value cannot be represented exactly in the specified format. There are four directions of rounding: round to nearest, round toward zero, round up, and round down. The default is round-to-nearest, since that mode gives the best approximations on the average. The last two forms are expected to be used for interval arithmetic.

There are two modes for handling infinity: affine and projective. The modes only differ in the handling of exceptions and the rules of comparison involving infinity. In affine mode, $-\infty < any\ finite\ quantity < +\infty$; in projective mode (the default), comparisons against either infinity are 'unordered'.

The IEEE Standard recommends that a normalizing mode be provided. This will prevent the creation of unnormalized values that might otherwise be invalid or adversely affect the result of certain operations. Otherwise, if normalizing mode is not in force, invalid operations and underflows may be caused by the appearance of unnormalized values.

The other five modes specify whether a default action or a trap should be executed when one of the exceptions discussed in the next section occurs.

2.4. Exceptions

Floating point computations are fraught with exceptional conditions: discontinuous functions, functions invoked with arguments out of their domain, functions whose range is larger than the floating point system can accommodate, or results that suffer loss of significance. Intelligent handling of these exceptional conditions is essential to the construction of robust software. A program must be able to tell that something untoward has occurred without making excessive tests. Robust software must be able to continue after an error, sometimes substituting a meaningful value for the meaningless one that might have been produced.

The IEEE Standard defines what is done by default when any of the following conditions occurs:

invalid operation (domain error, e.g. $\sqrt{-1}$, $\frac{0}{0}$)

division of a finite quantity by zero
overflow (magnitude of result too big)
underflow (magnitude of result too small)
inexact result (roundoff occurred)

Execution continues after a specified value (usually a NaN or an infinity) is substituted as the result of the operation. This mode of operation may be especially useful for pipelined processors that are unable to abort an operation.

A flag and a mode are associated with each condition. The flag indicates whether that condition has occurred since the last time the flag was reset by a trap handler or the user program. The quantity held by the flag after an exception is implementation-defined; it may be as simple as a Boolean value, or it might record the cause of the exception.

A more interesting implementation option permits trapping when an exception occurs; the mode associated with the condition names the trap handler to be activated when an exception is detected. If trapping mode is selected for the condition, a user-specified procedure is invoked with information about the operation, intended precision of the result, and either the operands or a form of the result that sprang the trap; that procedure can substitute a value for the operation and continue, reset the flag, or take any other action it likes.

3. Other Floating Point Ideas

The IEEE Standard contains no totally new features: multiple levels of precision have been available for twenty years in Fortran, CDC machines have had indefinites and infinities almost as long, DEC machines have reserved operands, and so forth. What distinguishes the IEEE Standard is its careful choice of what is important. It represents an excellent set of compromises between ideal real arithmetic and the costs of implementation. But it does not incorporate some possibly good ideas in floating point, such as:

There are systems with unbounded precision, or with a finer-grained set of alternative formats.

Machines that compute in decimal rather than binary may be important in the future. There are representations that are efficient in storage and that promise good performance.

Some floating point systems have symbols for tiny values $\pm \epsilon$ similar to infinitesimals: they are not zero but have magnitude smaller than any representable floating point number.

Machines have been designed in which the type of each operand is associated with the datum or its address rather than with the operation code. These tagged architectures are not directly related to floating point arithmetic, but they offer the possibilities of dynamic mixed mode arithmetic and changes in the forms of declarations.

Possibilities such as these have been incorporated in real machines or are under active investigation by research groups or standards committees. These ideas need to be considered, though we will concentrate on the already daunting task of encompassing IEEE Standard arithmetic.

4. Current Language Support

A language suited to careful floating point computing ought to provide access to the features of the floating point system listed above, including several formats (traditionally called 'precisions'), non-numeric constants and values, many needed operations, and flexible handling of exceptions. The full language definition ought to specify the domain of values and the interpretation (not just the name) of the operations, at least in terms of parameters accessible in the language. Most languages actually used for computation fall far short of this ideal.

4.1. Fortran

Fortran[4] is the language used for most heavy numeric computing. It contains two formats, but the Fortran Standard only requires that the precision of a double precision datum be greater than that of a real (single precision) datum. Many operations are named, including an extensive set of elementary functions. Conversions are present between different floating formats and between floating formats and integers. Conversions to and from decimal are available through the input/output statements in Fortran 77 and in many extensions to Fortran 66. Unfortunately, the rules for formatted output mean that double precision constants may be written in a form that cannot be reentered in a subsequent compilation.

Standard Fortran has no way to discuss exceptions. (The original Fortran had statements to test for overflow, but this facility disappeared years ago when hardware traps were introduced). Most large Fortran systems have library procedures for handling underflow, overflow, and library exceptions, but these differ greatly from machine to machine.

4.2. Pascal

Pascal[5] has only one floating point type. It has symbols for the operations but few of the needed functions. There is no mention of exceptions.

4.3. Basic

Standard Basic has one arithmetic type. Many of the necessary operations are present. There are inflexible rules for exceptions (underflow to zero, abort on overflow).

4.4. Algol 68

Algol 68[6] specifies names for an unlimited number of floating point formats (**real**, **long real**, **long long real**, and so on), but most implementations provide just a single type. The language has a reasonable set of operation symbols, and their meaning can be extended using mechanisms in the language. Some environment parameters are defined in the standard prologue. There is no linguistic facility for exception handling.

4.5. PL/I

PL/I[7] has both floating binary and floating decimal declarations. The programmer may specify the number of digits in each datum, but the compiler may use any implementation type that is sufficiently precise, even one of a different radix. There is no way to specify a needed exponent range. The language comes with a reasonable set of operations and a good mathematical library.

The PL/I ON-condition is a powerful mechanism for responding to errors and exceptions, but generally there is no access to the operation or operands that caused the trouble, so it is hard to make intelligent use of the facility.

4.6. ADA

The new language ADA[8] approaches floating point in a serious spirit, but gets many of the details wrong. The description is based on Brown's[9] model of floating point computation. The user can specify the precision of a datum, in decimal digits; the compiler rounds this up to the nearest convenient binary precision. A minimum range is implied by the precision. (It is possible to specify the exponent range using a **range** specification, but no lower bound on the exponent can be stated if the interval contains zero). ADA environment enquiries yield parameters describing the precision and range of both model and machine numbers. But the underlying model is just a one-parameter family, so the enquiries do little good.

ADA has an elaborate exception handling mechanism. It seems to be designed to cope with serious difficulties, since an exception terminates execution of the procedure in which it occurred, and control proceeds to the handler associated with that procedure. The handler gets no information about the circumstances of the exception (operator, operands, or statement), only the name of the condition. It is easy to abort the routine in which problems arose, but it is almost impossible to substitute a value for the interrupted operation and continue computing.

5. Language Needs

It is clear that none of the standard algorithmic languages provides adequate support for floating point computation, though all the above have been used to do important calculations. Let us instead ask in more detail what features we would like a language designed for this purpose to have.

5.1. Types

The thorniest questions about types relate to extended. In some implementations, extended formats will be more expensive to save and to restore than basic formats, and the amount of register or hardware stack available for them will be limited. The plausible choices are

> ignore extended entirely
> use for compiler temporaries, but hide from programmer
> offer them as just another type

In the last case, users should be aware that it may be costly to declare more than a few extended format variables.

If the language definition does not determine the full details of precision and exponent range, then this information must be made accessible to the program both for determining data types and parameters to control the computation. Otherwise it becomes difficult to write portable programs, ones that apply to more than one instance of the language. (The extended formats are not fully specified in the IEEE Standard, so environmental enquiries may be needed to discover exactly how much precision and range these types provide in a particular implementation.)

5.2. Constants

The possible floating point values must be representable in the program source, data input, and in program output. Ordinary decimal scientific notation is suitable for representing most constants of interest. The obvious notation (1.2×10^{7}) suggests evaluating an expression requiring an exponentiation and a multiplication rather than just denoting the value of a single number, so this form is often abbreviated, to $1.2_{10}7$ in Algol 60 and $1.2E7$ in Fortran.

But the context or notation must imply not only the value but the format in which it is to be represented. In some languages it is possible to infer a satisfactory type from context such as the types of neighboring operands or the declared type of a corresponding formal parameter. Otherwise, the constant form itself must carry the type. In Fortran, a constant is double precision if and only if it contains an exponent field marked with a letter D. This notation is compact, but it does not generalize neatly to many types (though this is the direction currently recommended for Fortran 8X[10]). A usually better approach is to apply a cast (type coercion) to the constant, which would otherwise be evaluated in the widest format available: *double* ($1.2e7$). The language in question must permit casts in contexts that require a constant. The IEEE Standard does not say what happens when converting decimal values with more digits than necessary; a language ought to be graceful in this regard.

In special cases, decimal notation may be inappropriate, and it may be preferable to present the number in binary (or octal or hexadecimal) form. Examples: decimal to binary conversion is not required to be perfect in extended format, but constants for some mathematical functions may need to be exact. If an algorithm is suspected of cycling, it may be useful to examine the exact bit patterns it produces. Binary notation may also be a way to satisfy the very rare need to express an unnormalized extended number.

The problems become more grievous when special constants are needed. If the type of a constant cannot always be determined from context, it must be possible to denote the format of non-numeric values. Thus, it may be necessary to carry type information with the infinity symbols (∞_{Double} or *double* (∞)). If the keyboard and output device do not include the ∞ symbol, some other notation for infinity will be needed. A plausible notation might be '!!', with the ugly extension *double* (!!) (or perhaps !*double*!). A language that supports the notion of tiny values will also need a notation for the special ϵ symbols.

In some cases, it is sufficient to indicate that some NaN is to be used as a constant or input value, or to write out an indication that some NaN was stored, in which case a simple form like 'NaN' or ?? might suffice. In other cases, the number and variety of NaNs present more complicated problems. An obvious typographical notation might be NaN_{134}, but in more restricted environments, some notation like '?134?' might be sensible. If so, there need to be rules for truncating a NaN when it must be printed in a narrow output field. Clearly, the notation adopted will depend on the intended use of the indicator field; a decimal value of an address may be a useless representation of the operation that caused an invalid result.

5.3. Variables

The basic decision is what formats a user may declare. See the discussion in section 4.1 above.

If data were self-describing, it would be unnecessary to declare the type of the arguments since this information could be discovered dynamically. If the type information is available, the language ought to permit declaration of local variables with types computed from the types of other variables. For example, a local variable with twice the precision of an argument might be needed to evaluate products without roundoff. Such an environment might simplify the declaration of many generic procedures, since most of the instructions would be independent of the data types being manipulated, though difficult problems may need to be solved to determine the precisions and ranges necessary for general calculations.

5.4. Expressions

The usual questions about operators in programming languages involve precedences, associativity, and overloading. The traditional rules for precedence are satisfactory. The rules for what rewritings ('optimizations') are permitted must be made clearer than they are in most current languages. A careful programmer needs to know the order in which certain operations will be performed. It is unacceptable to violate parentheses. (The profit to be gained by reordering floating point operations is probably small anyway. Optimization of pipeline references and addressing sequences is usually far more important.) Overloading or punning (use of an operator to mean different things in different contexts) is common, since the operator that adds two Single numbers is different from the one that adds two Double numbers, but the same symbol + is usually used in both cases. In some languages (e.g., Algol 68) the programmer can specify the multiple meanings of the operator; in others (e.g., Fortran) the rules are fixed. The major question relevant to floating point is: what precision should be used for evaluating the expression? The plausible answers are

The precision of the operands, all of which must be the same type. If mixed mode expressions are needed, the programmer must supply explicit coercions to the desired type.

The widest available precision (e.g., Double-extended). This rule is simple, loses the least precision, and will be efficient on many machines, though very inefficient on others. (Some of the IEEE-style microprocessors do all their arithmetic internally in Double-extended, for example. Moving such long operands in and out of memory might however be expensive.) Such a rule may not be portable if different implementations use different widest precisions for evaluation.

The higher of the precisions of the operands of each operator ('local type rule'). This rule is the one used in many languages; the type of each operand is discovered in the course of a depth-first walk of the expression tree.

The highest precision appearing in a top-level expression (one appearing on the right side of an assignment or as an argument to a procedure or control construct) ('global type rule'). This rule seems in many ways the most natural, since it is then clear that the products in *single · single + double* and *double + single · single* are both to be evaluated to double precision. The global type rule seems to come closest to the informal rules of mathematical notation. The only way this rule conflicts seriously with the expectations of Fortran programmers is with integer division, since $1.1+1/2 = 1.6$; perhaps a special integer-division operator (as in Algol 60 and Pascal) ought to be introduced.

6. Exception Handling

Exceptions are the aspect of floating point processing that is least well handled by conventional languages. A trap implies a drastic change in the flow of control, and is not easy to specify. Yet the careful recognition and perhaps repair of exceptional conditions is absolutely necessary for robust software. If overflow and underflow must be avoided because the results are unpredictable and likely catastrophic, it is necessary to restrict the range of validity of certain programs greatly, and to insert frequent range checks. If occasional exceptions can be detected and repaired intelligently, the programming can be greatly simplified.

The IEEE Standard provides a several ways to handle exceptions: In non-trapping (default) mode, one can compute, then check that the answers are valid (not a NaN, for example), or one may examine the flags to see if any unfortunate conditions were detected. If traps are provided (they are optional in the IEEE Standard), handler may substitute a mathematically useful value or apply a different algorithm in a special case.

Flags and modes are described as part of the global state in the IEEE Standard; this view may be reasonable for flat languages without nested scopes such as Fortran but may not be natural in a block-structured language. When the language designer considers the environment in which a procedure executes, he should consider exception handling as well as storage and scope issues.

Floating point traps are really quite benign, since they are always due to actions of user programs and involve no security problems or interesting system interactions. Even so, floating point traps seem to cause system-level problems. To make best use of the facilities of the IEEE Standard, saving and restoring trapping modes and flags ought to be cheap, and invoking a trap handler ought not to be expensive. These conditions often do not hold because of hardware or system misdesigns or over-generalizations: On many systems most or all these flags are under privileged system control, and a full supervisor call may be needed to perform as benign operation as disabling a floating point trap. The floating point flags are sometimes part of the privileged processor state and hence inaccessible to a user program. The full system fault handler may be entered, with expensive context switching, just to discover that a multiplication has overflowed and that a process needs to make a minor repair. Although a data-generated fault can be handled by the same system mechanism that handles memory parity errors, it ought not to be.

6.1. Operations on Flags and Modes

In an implementation of the IEEE Standard, one must be able to test, clear, save, and restore the flags. The operations of saving and restoring flags are essential since they cannot easily be simulated if the flags encode complicated values such as the location of the instruction or data that caused the problem. In the simpler case where the flags are Boolean quantities, the flags occupy a total of five bits.

Four bits are needed to encode the rounding, infinity, and normalizing modes. It is useful to be able to save and set them; testing and clearing can be handled using those operations. (Two more bits may be needed to encode the precision control mode, which is defined to enable a machine possessing only wide registers to correctly simulate arithmetic in the shorter formats.)

The trapping modes are more complicated, since a mode is either disabled or specifies a trapping routine. It will be necessary to save, restore, enable, and disable these modes. Fortran and some other important languages have no way to store a procedure-valued quantity (pointer to a trap handler), so a significant extension might be needed. It might be useful to be able to test the value of a mode, to discover whether it is enabled, and perhaps to compare trap handlers, but many languages lack the facility to compare procedures or procedure pointers. Because of these complexities and ones to be described subsequently, it may be wise to restrict the range of the trap handlers to some small menu rather than to permit them to be arbitrary procedures.

6.2. Inheritance of Flags and Modes

The problem of mapping single hardware states like the flags and modes of the IEEE Standard, and the comparison condition codes of many processors, onto a sensible language model is difficult. Most procedures will be unaffected by the details of the modes, since they will either be designed to work satisfactorily regardless of mode, or they will have been written in ignorance of the subject. But some routines will be written to take advantage of the rules. A run should not terminate because a routine hunting for minima selects a value near a pole or outside the domain of the function being minimized; the minimizer just wants to look elsewhere. A routine that is attempting to compute intervals in which results must lie may need to compute certain expressions twice, first rounding up, then rounding down. Such a routine may need to invoke certain procedures and require them to round different ways on each call.

There are conflicting alternatives. During testing, a program should probably abort as soon as an exceptional condition is detected. A program written naively for short-term or experimental use may need the same security. Other programs will be written with the IEEE Standard's default modes in mind, and may not function in other modes. Still others may be written to obey the spirit but not the letter of the IEEE Standard: a cosine routine will obey the directed roundings better by computing to extra accuracy then rounding the final value according to the specified rule, than by rounding the result of each of the operations of its usual algorithm in a specified direction.

When a procedure invokes another procedure, what mode and flag values should the called procedure start with? What mode and flag values should the calling procedure have upon return? These questions are analogous to the choice of method for passing parameters to procedures in algorithmic languages. The obvious models are call by value and call by value-result. One view is that a procedure is a simple operation that should not have side effects; that would suggest passing by value. But then it would be impossible to write a procedure for initializing traps, since they would revert upon return from the initializer. Similarly, the calling procedure could not interrogate a flag to discover whether for example an overflow had occurred. Examples such as these lead me to think that the proper model is passing by value-result, which is equivalent to having a global value.

A knottier question arises when a procedure-valued entity (perhaps an argument) is passed or saved by one procedure and used within another. In which environment should the saved entity be executed? For example, imagine a program that passes a function to a general-purpose zero-finder. This question is analogous to that of variable binding in programming languages: does the passed procedure operate as if it had been called directly at the time it was saved or passed (static scope) or as if it had been called directly by name by the called procedure (dynamic scope)? Complicated issues of interpreting the environment arise (if a non-local variable is referenced, is it the deepest such variable on the dynamic stack, or the nearest such variable in the lexicographic scope?). For modes and flags, it seems more likely that shallow binding is wanted; this again is consistent with the picture of the modes and flags being global quantities. In the example of the zero-finder, the minimizer usually wants the function evaluator to return a value (even a NaN) rather than abort the run, even if the caller of the minimizer is operating in a debugging (and hence trapping) environment. Most programs will not care about this issue, but the few that do will make many calls with controlled environments. Perhaps a more general procedure invocation mechanism is needed, which might be helpful in other contexts:

call *s* (*arguments*) **in environment** (*specification*)

Appendix. Changes To Fortran to Meet the IEEE Standard

The above discussions consider an ideal: what should a language permit to give complete access to intelligent floating point, particularly as embodied in the IEEE Standard? A question of immediate interest is the minimal set of changes needed to make current languages usable with the IEEE Standard. To summarize, language changes absolutely required are small, though libraries will require augmentation.

The most pressing issues are constants and type declarations: there should be some way of expressing the infinities and NaNs. Ugly but sufficient mechanisms could be added to many languages: In Fortran, one might define a few new 'dot' constants named .INF. and .NAN., in analogy to .TRUE. and .FALSE. If the special values need not appear in DATA or PARAMETER statements, then one need only define (possibly intrinsic) functions that returned these values. The IBM Fortran REAL*8 notation could easily be extended to cover the extra types. For constants, either the type coercion functions could be used, or a simple but ugly notation (1.0DE6 for Double-Extended million). Syntactic problems only arise because there are contexts (such as static initialization) in which true function calls are not allowed, but in which a syntactically similar type conversion ought to be permitted.

Flags and modes can be supported in a stopgap fashion by defining intrinsic functions to manipulate those values. No difficult scope issues arise in current Fortran because there are no nestings of environments. The only syntactically thorny problems arise from the need for a type to hold implementation-defined values of flags and from the need to save procedure pointers when manipulating trap handlers in languages that do not have procedure-valued variables. A simple but useful implementation would restrict flags to Boolean values and provide a small range of trap handlers. A more complicated one would store such values in integer variables.

The IEEE Standard defines operations for binary-to-decimal conversion, remainder and square root, so any language needs to provide these operations through operators or through library functions. It is required that the operations be performed to IEEE Standard precision, so the typical square root routine and input/output number converters would have to be rewritten to meet those stringent requirements.

References

1. A Proposed Standard for Binary Floating-Point Arithmetic, *Computer Magazine* (March 1981), pp. 51-62.

2. Ford, B., Parameterisation of the Environment for Transportable Numerical Software, ACM Trans. on Math. Software **4**, 2 (June 1978) 100-103.

3. Brown, W. S. and Feldman, S. I., Environment parameters and Basic Functions for Floating-Point Computation, ACM Trans. on Math. Software **6**, 4 (December 1980) 510-523.

4. *American National Standard programming language Fortran*, (ANSI X3.9-1978), (New York: American National Standards Institute, Inc., 1978).

5. Jensen, K. and Wirth, N., *Pascal User Manual and Report*, second edition, (Springer-Verlag, New York Heidelberg and Berlin, 1975).

6. van Wijngaarden, A. et al, eds, Revised Report on the Algorithmic Language Algol 68, ACM SIGPLAN Notices **12**, 5 (May 1970) 1-71.

7. *American National Standard programming language PL/I*, (ANSI X3.53-1976), (New York: American National Stnadards Institute, Inc., 1976).

8. ADA Reference Manual, in Ledgard, H., *Ada, An Introduction,* (Springer-Verlag, New York Heidelberg and Berlin, 1980).

9. Brown, W. S., A Simple but Realistic Model of Floating Point Computation ACM TOMS, to appear (December 1981).

10. ANSI X3J3 Fortran Committee, Proposals Approved for Fortran 8x, internal document S6/79 (August 27, 1981).

Discussion

George Paul

During the development of the IEEE standard, was it envisioned that the NaN's might be used to specify uninitialized data?

Kahan

Yes. There are two types of NaN's, those which are "silent", which arise but go unnoticed by the program unless queried, and those which raise a flag to the program. The second type can be used for uninitialized data values.

Feldman

The NaN's would be an excellent way to represent uninitialized data. A careful program ought to check for NaN-valued data before computing.

Gentleman

One may need the value of a NaN. Missing data usage would be common in statistics. A NaN (data not provided) bit pointer to meaningful fitted values would be useful.

Nelder

One might need several kinds of missing values (several types of NaN's).

Meissner

Language support should provide separate facilities for detecting the sign of zero, instead of the "traditional" arithmetic-oriented methods such as the Fortran sign function or signed-zero output.

Feldman

Certainly. The categorization function I propose would enable a program to distinguish plus and minus zero by an explicit test. One should not resort to tricks to tell the difference.

Fritsch

Many CDC installations make another use of -0. When a blank field is read with a numeric format (I, F, ...), a -0 is generated. Some applications distinguish between this and $+0$, providing default values when the input field is blank.

Feldman

It is true that CDC users use -0 to denote such special values as missing data, but this is a very nonportable and dangerous facility to depend on. The IEEE Standard's NaN is a far better way to do this bookkeeping.

Meissner

The language implementor (depending on the cost of trapping mode) can maintain a software copy of each hardware trap, flag, mode, NaN, etc. This software copy is obviously easy to save, identify, and "raise".

Arrays of exceptions are scary to contemplate. Lake's earlier paper implies that arrays of Boolean NaN's might be used.

Dekker

I think it is important to determine global states such as affine or projective modes and the various rounding modes at compile time and not at run time. Otherwise, it might be difficult to obtain efficient execution.

As to the flags for the various exceptional conditions, I think that separate sets of flags are needed for parts of programs which are to be executed in parallel or collaterally (that is, in unspecified order).

Brent

There is no need for floating-point exceptions to be as expensive as are other exceptions (page fault, I/O, clock, addressing, ...). Typically, machines have system and user modes, and on an exception in user mode the context is saved and a switch made to the system mode. For floating-point exceptions this is unnecessarily expensive. All that needs to happen on a floating-point exception is a return jump to a known address in the user's address space. The operating system has no need to know that an exception has occurred. If hardware was designed this way, floating-point exceptions could be handled economically (that is, at the cost of only a few multiplications).

Feldman

I agree; I was trying to make that point.

OVERALL SESSION DISCUSSION

<u>Roberts</u> At present computer languages are mostly developed independently, and similar facilities are implemented in arbitrarily different ways. In particular, this makes it unnecessarily difficult to call packages written in one language from those written in another.

A modular approach might now be appropriate, which would be largely orthogonal and additional to what has been done so far. Classify the purposes that a language has to fulfill into areas (for example, those concerned with data structures, arithmetic, logic, inter-procedure communication, mathematical functions, input/output, system interaction) and try to insure that different languages use a common structure and nomenclature within each area. Elsewhere I have called this approach "comparative linguistics". Bodies analogous to X3J3 might be established to propose standards for these inter-language areas.

If such uniformity could be achieved, different languages would naturally use the same methods for communication between subprograms or procedures and the combination of packages would be easier.

<u>Roubine</u> The idea of standardizing the underlying implementation of specific language features runs contrarily to the general trend of providing abstract language definitions (for example, Ada does not enforce a particular parameter passing mechanism). If one wishes to achieve such a conviviality of languages, what seems to be needed is more a set of standardized interfaces, than a legislation on basic language features: the idea is not to say that a language must support such capability in such standard way for its own sake, but rather that the implementation must enforce the standard interface only when the use of another language is called for.

In another manner - you want a language-building kit. Existing languages are too big, people say. The kit menu would be even bigger.

<u>Feldman</u> The real problem is one of implementation. Can one pass a Fortran subroutine to an Algol 68 program?

<u>Kahan</u> We should be leaning on experiences with existing languages.

<u>Ris</u> Kahan suggested during discussion that perhaps rather than one large, "general-purpose" language we would be better off with a number of simpler application oriented languages. The practical realization of this idea requires efficient linkages among modules compiled from different source languages.

The present state of the world is somewhat different. This meeting has been contemplating large languages such as Fortran 8X, Algol-68, Ada, PL/I and APL. Each of these takes a view that the others do not exist, and in fact need not exist. Such view becomes difficult only when we wish to combine large bodies of code previously written in different languages. That combination is difficult principally because the large languages have large execution-time environments written by large teams who have enough intra-communication difficulties as it is.

The issue is an implementation issue more than a language issue. However, language standardization efforts must recognize that they effectively contribute to further polarization by encouraging manufacturers to provide (single) <u>language</u> support rather than <u>system</u> support.

<u>Nelder</u> Hammarling commented on the dangers of mutual inconsistencies between packages written in Ada. This problem is not, of course, a particular defect of Ada, but arises whenever a language allowing the definition of structures and

operators by the user is employed.   Gower and I, in a paper to ISI conference
of 1969, pointed out the dangers to the statistical community that would arise
if standards were not defined for structures and operators of interest to
statisticians.

The language babel would become worse, and much of programming in these new
languages would consist of unpacking other structures and repacking them into
one's own form.   (In Fortran, one works permanently in the unpacked state).

Things are no better, and no worse, in Ada than in, say, Algol 68, except
perhaps that identically defined types in Ada are not identical.   The use of
extensible languages imposes a need for standardization at a new level (that is,
structure and operator definition) among user communities if libraries are to be
really useful.

SESSION 7
OPEN SESSION

Chair
B. Einarsson

Discussant
F.N. Fritsch

*THE RELATIONSHIP BETWEEN NUMERICAL COMPUTATION
AND PROGRAMMING LANGUAGES, J.K. Reid (editor)
North-Holland Publishing Company
© IFIP, 1982*

A COMBINED LEXICAL, SYNTACTIC, AND SEMANTIC APPROACH
FOR IMPROVING NOTATION

Mark B. Wells

Computer Science Department
New Mexico State University
Las Cruces, New Mexico 88003
U.S.A.

The purpose of this paper is to show how ideas from lexical,
syntactic and semantic analyses can combine to produce a
programming language having significantly improved notation
over classical languages. An important aspect of this
approach is that because of the independence of the three
analyses and the generality of each, the resulting translation scheme will be no more complex than a classical compiler.
In fact the syntax of a recent language (Wells (1980)) using
this approach is in some sense half as complex as Pascal.

INTRODUCTION

A premise of this work is that notation is important. Of course notions are more
important than notations! But in today's complex world of information it is essential that we be able to capture concepts with concise notations at one level of
abstraction so that we may proceed to higher levels without attention to the
lower-level details. This is especially true in numerical computation. For
example, the simple notation

$$\int_a^b f$$

captures an important concept that we can use within higher-level expressions
without concern over how the integral is itself to be calculated.

A look at programming language development over the past twenty-five years shows
little improvement in notational capabilities. The original Fortran had prefix
and infix operator symbols available to express operations on a very restricted
class of operands and a prefix notation for expressing subroutine calls. Today's
Pascal, for instance, has a few more operator symbols and the class of operands
has been expanded somewhat, but little other improvement can be noted.

One reason for this slow rate of improvement is, I am sure, a historical lack of
hardware facilities to support advanced notation. But the major reason is, I
believe, a prevailing attitude among many scientific computer users that (1) they
don't really need more powerful notations and (2) the existence of such notations
surely must complicate the language they must learn and use. In response to these
I argue:

   (1) The calculations being performed on today's much faster machines are far
more complex than conceived (say) twenty years ago. We need all the help we can
get to cope with this complexity -- abstracting via powerful yet natural notation
can be an important tool in this respect.

   (2) Our increased understanding of programming languages and their compilers

today make it possible to improve notation with minimal increase in language complexity. In fact there are certain aspects of language which even become simpler in the light of the general and orthogonal approach advocated here.

I discuss the three areas of language separately to emphasize their essential independence, a property which is responsible for any simplicity that the approach has. Of course it is the combined effect of the features from the three areas that yields the powerful notation.

LEXICAL CONSIDERATIONS

It is convenient to view the translation process as consisting of three separate serial translations: lexical analysis (scanning), syntax analysis (parsing), and semantic analysis (analyzing):

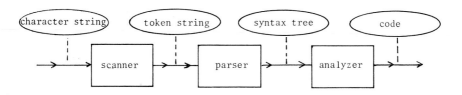

As shown, the first of these (the scanner) translates a string of characters to a string of tokens.

Lexical properties of the language that are relevant to the construction of advanced notations are:

   (1) use of a large ("printable") character set, and

   (2) use of control characters such as space, backspace, upspace and downspace that allow a two-dimensional relationship to be expressed among printed characters.

It used to be that special input devices were needed to attain both of these properties (Wells (1961)), but today both programmable character sets and facilities to move the cursor vertically as well as horizontally one character position are available on a number of raster-scan terminals.

A character set only twice as large as standard ASCII still can use an 8-bit character encoding yet greatly increases the notation potential. The extra 96 characters allow the Greek alphabet and most of the common mathematical symbols such as $\int$, ∂, $\sqrt{}$, ∇, $\|$ to be made available. As will be seen later, the extra characters do not add any essential complexity to the language, they merely increase the vocabulary of identifiers and operator symbols.

This author has studied two-dimensional constructions for several years (Wells (1961, 1972, 1975)). After playing with many more elaborate schemes, I have now developed a simple linearization scheme, easily incorporated into the scanner, that allows use of the three most common two-dimensional constructs: sub-superscripting, displayed division, and binomial-coefficient-like notation. The scheme is recursive, so quite elaborate notations can be constructed:

$$e^{-\frac{i \cdot x^2}{h}}$$

[Most examples in this paper are written with half-line vertical spacing, although current raster-scan technology permits only full-line spacing. There are printers available, however, that do permit half-line spacing.] Certain natural, hence easily remembered, restrictions are placed on the forms to preclude notational ambiguities that would complicate interpretation by either the scanner or human reader.

Examples of the linearizations to token strings accomplished by the scanner are

$$A_i^2 \qquad \text{becomes} \qquad A(i)\hat{}(2)$$

$$\binom{2n}{n} \qquad \text{becomes} \qquad ((n)\circ(2n))$$

$$\frac{A+B}{C} \qquad \text{becomes} \qquad ((C)\backslash(A+B))$$

It is important to remember that these linearizations are independent of later interpretations by the parser or analyzer. The extra parentheses are inserted in the last two forms to insure natural interpretation by the parser.

SYNTACTIC CONSIDERATIONS

Input to the parser is a token string; output is a syntax tree, most likely a reordered token string representing a prefix Polish representation of the program. Parsing depends solely on the rules of grammar, given either in BNF notation or by more readable syntax diagrams accompanied with certain rules of precedence. It is completely independent of any data types that will later be associated with the constructs by the analyzer. Even the usual function evaluation notation, f(X), cannot be distinguished as such by the parser -- it is merely the juxtaposition operator applied to two operands, the identifier f and the parenthesized identifier X.

Besides the existence of the juxtaposition operator, which plays a prominent role in building notations (recall that subscripting gets translated to juxtaposition), an important syntactic property is the generality of use of the operator symbols. Most of the symbols can be used as prefixes, infixes or suffixes and represent distinct operators in each case. These operators are classified into three precedence classes -- relative, additive and multiplicative -- so that natural notation can be utilized without requiring parentheses. The syntax diagrams for the three classes differ only in the operator symbols that may be used as the prefixes, infixes or suffixes. The syntax diagram for sum, which uses the additive operators is

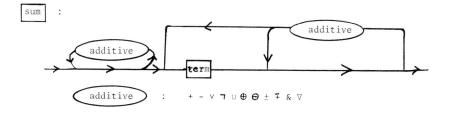

It is convenient to have a fourth higher-precedence class of <u>evaluative</u> operators. Because this class includes juxtaposition as an infix operator, the true symbolic operators of the class -- $\int$, $\int$, !, #, etc. -- can, to avoid ambiguities, only be used as prefixes and suffixes.

In an unparenthesized expression containing operators of the same class prefix operators associate from right to left and infixes and suffixes from left to right. For example, the expression ++X++Y is interpreted as $((+(+X))+)+Y$. Examples of translations to syntax trees accomplished by the parser are

$\int_{a,b} f$ →
```
                    juxtaposition
                    /          \
                prefix ∫        f
                  /
               (a,b)
```

+X++Y   →
```
                                            infix +
                                           /       \
                              suffix +              Y
                             /
                    prefix +
                   /
                  X
```

It is important to remember that these translations are independent of meanings attached to the trees by the analyzer.

SEMANTIC CONSIDERATIONS

Input to the analyzer is a syntax tree; output is a code. Whereas the scanner works with characters and the parser with tokens, the analyzer works with types. It is the types of the operands that determine the meaning to be attached to a given operator of the syntax tree. To get everything started, there are certain literal forms (constants) in the language that automatically have types associated with them. For example, type <u>real</u> is associated with a constant 6.2, type <u>string</u> with a constant "cat", and type <u>function</u> with the literal <<X² + 17>>. In addition, there are forms that associate a type of the user's design, say type <u>complex</u> or type <u>matrix</u>, with certain quantities. All these types then propagate upwards through operations and expressions and determine the type of variables being assigned the values of the expressions. The analyzer, in fact, simulates the execution of the program expressed as a syntax tree, only it is using types (classes of values) instead of the values that will be used when the generated code is finally executed.

All operations appear in abstract data-type modules. An operation in a module is either primitive or given by a source-program function. If primitive, the propagation of types and the code associated with the operation are built into the analyzer. If a function, then the types propagated are the output types of the function and the generated code is merely a call to the function. In either case, the module referenced depends on the types of the operands, and within the module the operation referenced depends on the syntax-tree operator. Thus, each operator occurrence, including each juxtaposition, may have a distinct meaning based on the types of the operands. There can be a real addition, complex addition and matrix addition, for instance, each indicated with an infix + symbol. Several important operations, such as function evaluation (which is indicated by the juxtaposition of a function and a list of parameters), are primitive and serve as a basis for more advanced notations.

An important semantic consideration is the fact that functions are data objects of the language and may be passed as parameters or assigned as values or variables. Function literals are evaluated in the environment of their definition which may in general be different from the environment of their call. In a block structured language, such as I advocate, this retention of environments is especially important since operations are functions in some abstract data-type module. In particular, this allows notation such as

$$\| M \|_p$$

where the evaluation of the "norm" operation is delayed until both M and p are known. (This delayed evaluation is also illustrated by the example below.)

NOTATIONS POSSIBLE WITH THE COMBINED APPROACH

Consider the notation

$$\cdots \int_{r,s} f \cdots$$

[The notation $\int_r^s f$ is in fact also possible but its analysis is more complicated.] The scanner translates this to

$$\int (r,s) f$$

and then the parser to

```
              juxtaposition
              /          \
     prefix ∫             f
           /
         (r,s)
```

or in reality to a linearized preorder form of this tree. If r and s are real numbers then the prefix $\int$ operator gets referred to the data-type module REAL. The associated operation (a function) accepts r and s as actual parameters and returns a function as output. The juxtaposition then calls for the evaluation of this latter integration function with f (itself a function--the integrand) as actual parameter. The integration function evaluates in an environment which includes the previously input values of r and s. It thus has all the information necessary to calculate the integral, which is presumably a real number whose type could be used to determine the meaning of any expression containing it.

Other notations from numerical mathematics which could be constructed and given natural meanings using this approach are

$$\partial_{x,y} f \qquad e^{-\alpha^2 \frac{\pi \cdot n}{\ell}^2} \qquad \cos\frac{n \cdot \pi \cdot X}{\ell}$$

$$\Sigma_a^b f \qquad \lim_{a \to 0} f(x) \qquad \nabla^2 \varphi$$

$$\{^n_r\} \qquad \Delta X \qquad e^{i \cdot x} \qquad \int_0^\infty f$$

REFERENCES

1   Wells, M. B., The Modcap Programming Language, LASL Report LA-UR-80-898, Los Alamos, N.M., August 1980.

2   Wells, M. B., "Madcap: A scientific compiler for a displayed formula textbook language", CACM 4:1, pp. 31-36, 1961.

3   Wells, M. B., "A review of two-dimensional languages", SIGPLAN Notices, 7:10, pp. 1-10, 1972.

4   Wells, M. B., An Algorithm for the Recognition of Typed Two-Dimensional Mathematical Expressions, LASL Report LA-6138-MS, November 1975.

DISCUSSION

G. Paul (IBM): Would an integer data type for 0 and 1 in your example $\int_0^1 f$ imply a summation (as opposed to an integration carried out in integer mode)?

Wells: No, reals and integers have the same type. Complex or some other distinct type could be treated differently at the option of the user.

G. Paul: You do have summation?

Wells: Yes, nearly all the special (non-alphabetic) characters are available as operators.

M. Delves (Univ. of Liverpool): How can you tell that (0,1) is a pair of reals? It looks like a structure to me.

C. Lawson (JPL): How do you know (0,1) is a pair of reals and not a complex constant?

Wells: Structures use angle brackets, and there is a different notation for complex constants. The type propagates from constants, which are real in this case.

M. Paul (Tech. Univ. of Munich): Consider your graph for additive terms. The parsing is not unique here.

Wells: This is only a part of the full syntax. It is an unambiguous LR(1) grammar.

M. Paul: Not in this example!

Wells: Within a precedence class, the order of evaluation is left to right.

Lawson: Can you give us insight as to the status of this language? Are people using it?

Wells: There was a preliminary version working at LASL a number of years ago, which had some users. We are currently implementing a version of the modern language at NMSU. The editor, scanner and parser are essentially complete. The more difficult analyzer remains to be done.

THE RELATIONSHIP BETWEEN NUMERICAL COMPUTATION
AND PROGRAMMING LANGUAGES, J.K. Reid (editor)
North-Holland Publishing Company
© *IFIP, 1982*

PROGRAM TRANSFORMATION AND LANGUAGE DESIGN

James M. Boyle[1]

Applied Mathematics Division
Argonne National Laboratory
Argonne, IL 60439

Program transformations are very useful in the development of mathematical software. They reduce the number of routines that must be written by hand, while at the same time making it possible to produce routines tailored to user requirements. During transformation, one program evolves into another in a sequence of small steps. They must be performed easily and correctly, and they must yield an efficient program when completed. Programs written in Fortran can be transformed, but ones written in an improved language could be transformed more easily. Among the needed language improvements are simplicity, extensibility, and access to hardware-level features.

1. Introduction

Program transformation is a technique for creating many versions, or *realizations*, of a single master, or *prototype*, program. This statement can be looked at in another way: Suppose one has defined an abstract notation suitable for expressing algorithms in a particular area of numerical analysis or computer science. Program transformations enable one to specify easily how to convert that notation into programs in some existing programming language. Moreover, programs written in an abstract language often can be converted into executable programs in several different ways, each tailored to some specialized environment. Program transformation techniques are especially applicable to the production of numerical software, in part because suitable abstractions are already at hand from mathematics and in part because tailored realizations of a numerical algorithm are often needed in application programs.

2. Some Transformations of a Program in the Fortran 66 Language

In this section I present two examples of program transformations useful in developing mathematical software. They illustrate how program manipulations that are tedious (and hence error-prone) to perform by hand can be automated using transformations. They also illustrate a general property of transformations: *a program transformation takes a prototype program that is abstract with respect to the manipulation it performs and produces a program that is concrete with respect to it.* Such a transformation can be thought of as a *realization function* whose domain and range are programs.

2.1. Prototype Program

The realization functions discussed here are applicable to a large class of prototype programs. Their effect can be illustrated, however, by applying them to a

---

[1]This work was supported by the Applied Mathematical Sciences Research Program (KC-04-02) of the Office of Energy Research of the U.S. Department of Energy under Contract W-31-109-Eng-38.

subroutine adapted from the Linpack collection of routines for solving systems of
linear equations [7]. The abstract prototype form of this program,
$CPOFA_{rb} = \textbf{rb}(CPOFA)$, is shown in Figure 1. This prototype was obtained from the Linpack routine CPOFA by applying a transformational realization function **rb** that
replaces calls to the Basic Linear Algebra Subroutines (CDOTC in this example) by the
equivalent in-line Fortran statements [5]. The program uses the Cholesky factorization to construct from a positive-definite complex Hermitian matrix **A** an upper triangular matrix **R** such that $\textbf{R}^*\textbf{R} = \textbf{A}$ and **R** has (real) positive diagonal entries.

At first glance, the prototype form of CPOFA may not appear to be very abstract.
Nevertheless, it is abstract with respect to the realizations discussed in the following
sections in several important ways. First, it uses the complex data type, which is
abstract not only for constructing the realization of this algorithm that factors real
matrices, but also for constructing ones that factor complex matrices whose elements
are represented by explicit pairs of reals.

Moreover, the syntactic form of the program is more abstract than that of the
usual Fortran program. The statements that are in the range of a loop are nested syntactically in the DO-END and FOR-END constructs. (FOR introduces a loop that may be
executed zero times.) The upper bound of the FOR loop is explicitly J-1, facilitating
manipulations involving the loop bounds. And finally, the error exit for the case of a
non-positive-definite matrix is represented by an EXIT from the CHOLESKY: BLOCK-END

```
SUBROUTINE CPOFA(A,LDA,N,INFO);
PREAMBLE
    DECLARE
        INTEGER LDA,N,INFO;
        COMPLEX A(LDA,1);
        COMPLEX T;
        REAL S;
        INTEGER J,K;
        INTEGER I08;
        COMPLEX CDOT07;
    END;
END PREAMBLE
CHOLESKY: BLOCK;
    DO J = 1, N;
        INFO = J;
        S = 0.0E0;
        FOR K = 1, J-1;
            CDOT07 = (0.0E0,0.0E0);
            FOR I08 = 1, K-1;
                CDOT07 = CDOT07 + CONJG(A(I08,K))*A(I08,J);
            END;
            T = A(K,J) - CDOT07;
            T = T/A(K,K);
            A(K,J) = T;
            S = S + REAL(T*CONJG(T));
        END;
        S = REAL(A(J,J)) - S;
<-----CHOLESKY: IS EXITED IF (S .LE. 0.0E0
                          .OR. AIMAG(A(J,J)) .NE. 0.0E0);
        A(J,J) = CMPLX(SQRT(S),0.0E0);
    END;
    INFO = 0;
CHOLESKY: END;
RETURN;
END;
```

Figure 1. $CPOFA_{rb}$—Structured Linpack CPOFA with BLA Calls Replaced.

construct. All of these syntactic abstractions facilitate manipulating the statements of the program using transformations.[2]

2.2. Packed Hermitian Matrix Transformation

The realization function discussed in this section produces a new program from CPOFA$_{rb}$—one that uses only half the storage used by CPOFA. It reduces the storage required by converting the representation of a complex Hermitian (or real symmetric) matrix from a square, two-dimensional array of n^2 words to a triangle packed in a one-dimensional array of $\frac{1}{2}n(n+1)$ words. This realization function comes in four varieties: **utcp**, **utrp**, **ltcp**, and **ltrp**, representing upper-triangle-by-column packing, upper-triangle-by-row packing, lower-triangle-by-column packing, and lower-triangle-by-row packing, respectively. Subroutine CPPFA of Linpack corresponds to **utcp**(CPOFA)[3].

For this discussion, the subroutine CPPFA$_{rb,utrp}$ = **utrp**(**rb**(CPOFA)), illustrated in Figure 2, is more interesting than the Linpack CPPFA, because the subscript expressions in it are non-linear. (Although less efficient in some environments than CPPFA, CPPFA$_{rb,utrp}$ might be used in an application program that, for other reasons, had to use the upper-triangle-by-row packing.) The non-linear subscript expressions arise because a reference to A(I,J), I≤J in the square array maps to a reference to APR((2*N+1-I)*I/2-N+J) in the triangular array. Note that the program in Figure 2 does not perform repeated multiplications and divisions in order to compute subscript values; rather it obtains them by repeated additions that correspond to constructing the values using a difference table.

How does the realization function **utrp** produce such a program? If one were to examine **utrp** under a magnifying glass, one would see that it is actually the composition of two functions. The first of these consists of a single TAMPR transformation rule that is a generalization of the following:

```
<var> {
    A(<expr>₁,<expr>₂)
==>
    APR((2*N+1-(<expr>₁))*(<expr>₁)/2-N+(<expr>₂))
}
```

It converts references to elements of the square array to references to the corresponding elements of the packed array.

The second function consists of a large number of transformation rules that simplify (optimize) the resulting program. During transformation, the optimized program evolves from the prototype in a large number of small, correctness-preserving steps (approximately 1500 in this example). The strategy of the optimization is to try to move the subscript computations earlier and earlier in the program. When a subscript computation reaches the top of a loop, it can be initialized ahead of the loop and updated just before the end of the loop. The transformation rules construct an updating assignment by computing the symbolic difference of two successive subscript values and simplifying the resulting expression. Subscript expressions that are polynomials of any degree can be "linearized" by applying this simplification until closure occurs, *i.e.*, until no further simplification takes place. Thus, this example illustrates an important point: *program transformations can be used to implement complicated program optimizations*.

---

[2]In practice, this syntactically abstract, or *structured*, form of the program is itself usually obtained from an executable form of the program by applying a set of structuring transformations. This approach facilitates testing, since the prototype is executable and can be tested before other realizations are constructed.

[3]Subroutine CPPFA antedates the development of the **utcp** transformation and was adapted from CPOFA by hand.

```
SUBROUTINE CPPFA(APR,N,INFO);
PREAMBLE
    DECLARE
        INTEGER N,INFO;
        COMPLEX APR(1);
        COMPLEX T;
        REAL S;
        INTEGER J,K;
        COMPLEX CDOT07;
        INTEGER I20,I15,I09,I14,I13,I11,I10,I08;
    END;
ENDPREAMBLE
I20 = N + 1;
I15 = 1;
CHOLESKY: BLOCK;
    DO J = 1, N;
        I09 = 0;
        I14 = 1;
        I13 = J;
        INFO = J;
        S = 0.0E0;
        FOR K = 1, J-1;
            I11 = J;
            I10 = K;
            CDOT07 = (0.0E0,0.0E0);
            FOR I08 = 1, I09;
                CDOT07 = CDOT07 + CONJG(APR(I10))*APR(I11);
                I10 = I10 - I08 + N;
                I11 = I11 - I08 + N;
            END;
            T = APR(I13) - CDOT07;
            T = T/APR(I14);
            APR(I13) = T;
            S = S + REAL(T*CONJG(T));
            I13 = I13 - K + N;
            I14 = I14 - K + I20;
            I09 = I09 + 1;
        END;
        S = REAL(APR(I15)) - S;
<-----CHOLESKY: IS EXITED IF (S .LE. 0.0E0
                        .OR. AIMAG(APR(I15)) .NE. 0.0E0);
        APR(I15) = CMPLX(SQRT(S),0.0E0);
        I15 = I15 - J + I20;
    END;
    INFO = 0;
CHOLESKY: END;
RETURN;
END;
```

Figure 2. CPPFA$_{rb,utrp}$—CPOFA$_{rb}$ Using a Packed Hermitian Representation.

The transformations **utcp**, **ltcp**, and **ltrp**, of course, also have the same form. Moreover, in each case the second transformation (the simplification) is the same; only the first, which introduces the appropriate subscript mapping, differs. This sharing of a complicated component among several realization functions illustrates an important aspect of program transformation: *new realization functions can often be constructed by combining existing transformations in new ways*. Doing so saves both time and money, and it increases reliability.

2.3. Complex to Programmed-Complex Transformations

Program transformation is closely related not only to optimization but also to the programming concept of *abstract data type*. A simple abstract data type familiar to writers of mathematical software is the complex data type. It can be defined, of course, in terms of pairs of real numbers, just as it is in mathematics.

A realization function that implements this abstract data type is **pc**, the complex to programmed-complex realization function. One use for this function stems from the lack of a double precision complex data type in standard Fortran. By composing **pc** with a real to double precision realization function **dp**, a complex prototype program can be converted to a realization that performs double precision complex arithmetic. Another use is to obtain control over complex arithmetic in single precision. For example, a program may contain instances where the real or the imaginary part of a complex variable is known to be zero (or one). Similarly, it may contain instances where only the real part of a complex expression is required. In these cases, simplification may permit the programmed-complex realization to perform fewer floating point operations than does the code for the complex data type produced by a compiler.

S. J. Hague [9] (see also [8]) has written a preliminary set of transformations defining the realization function **pc** for TAMPR. They represent the real and imaginary parts of a complex variable Z by a pair of real variables (ZR,ZI). The program WPOFA = **pc**(**rb**(CPOFA)) is shown in Figure 3.

A subtle point that must be considered in the design of the transformations for **pc** concerns the temporary variable ZTMP10 that appears in Figure 3. A complex assignment must behave as if the real and imaginary parts of the variable being assigned to were altered simultaneously. Thus, when the real part of that variable is required for the evaluation of the expression to be assigned to the imaginary part, the original (not the altered) value of the real part must be used. The transformation that converts a complex assignment to a pair of real assignments must be designed to introduce a temporary variable automatically in this circumstance.

This preliminary implementation of **pc** is designed to avoid substantially reorganizing the program. Except for the necessary introduction of ZTMP10, it simply splits each assignment in the original program into a pair of assignments. Efficiency could be increased (perhaps at the expense of clarity) by making it introduce more assignments to temporary variables, so that the computation of subexpressions that appear more than once would not be repeated. For example, the transformation for the complex division operator could be altered to compute the denominator only once. (In any event, it would be better to use a form of complex division that avoids over- and under-flow where possible.)

An example of the possibility of saving arithmetic operations in the programmed-complex realization occurs near the end of WPOFA, at the expansion of the statement S = S + REAL(T*CONJG(T)) in CPOFA$_{rb}$. In WPOFA, only the real part of T*CONJG(T) is computed. In CPOFA$_{rb}$, however, a compiler might well evaluate the corresponding subexpression in complex arithmetic, computing the imaginary part and then subsequently throwing it away. This situation represents a general phenomenon accompanying the implementation of abstract data types: *instantiating a program with a particular concrete representation for an abstract data type creates possibilities for optimization*. In regard to programming language design, it is important that a language purporting to offer abstract data type facilities also offer facilities for describing optimizations thereof. In general, these optimizations must be performed *across* several abstract operators (as here). The commonly provided method of implementing operations on abstract data types, procedure or function invocation, makes such optimizations impossible; it is no doubt responsible for giving the use of abstract data types a reputation for inefficiency.

One might also note that considerable knowledge and several inferences would be required to reconstruct the abstract prototype, *i.e.*, to see that TR*TR + TI*TI is an implementation of REAL(T*CONJG(T)). This observation also illustrates a general

```
SUBROUTINE WPOFA(AR,AI,LDA,N,INFO);
PREAMBLE
    DECLARE
        REAL ZTMP10;
        INTEGER LDA,N,INFO;
        REAL AR(LDA,1),AI(LDA,1);
        REAL TR,TI;
        REAL S;
        INTEGER J,K;
        INTEGER I08,I09;
        REAL CDOT7R,CDOT7I;
    END;
ENDPREAMBLE
CHOLESKY: BLOCK;
    DO J = 1, N;
        INFO = J;
        S = 0.0E0;
        FOR K = 1, J-1;
            CDOT7R = 0.0E0;
            CDOT7I = 0.0E0;
            FOR I08 = 1, K-1;
                CDOT7R = CDOT7R + AR(I08,K)*AR(I08,J)
                                + AI(I08,K)*AI(I08,J);
                CDOT7I = CDOT7I + AR(I08,K)*AI(I08,J)
                                - AI(I08,K)*AR(I08,J);
            END;
            TR = AR(K,J) - CDOT7R;
            TI = AI(K,J) - CDOT7I;
            ZTMP10 = (TR*AR(K,K) + TI*AI(K,K))
                    /(AR(K,K)*AR(K,K) + AI(K,K)*AI(K,K));
            TI = (TI*AR(K,K) - TR*AI(K,K))
                    /(AR(K,K)*AR(K,K) + AI(K,K)*AI(K,K));
            TR = ZTMP10;
            AR(K,J) = TR;
            AI(K,J) = TI;
            S = S + TR*TR + TI*TI;
        END;
        S = AR(J,J) - S;
<-----CHOLESKY: IS EXITED IF (S .LE. 0.0E0 .OR. AI(J,J) .NE. 0.0E0);
        AR(J,J) = SQRT(S);
        AI(J,J) = 0.0E0;
    END;
    INFO = 0;
CHOLESKY: END;
RETURN;
END;
```

Figure 3. WPOFA—CPOFA$_{rb}$ Using Programmed-Complex Arithmetic.

---

phenomenon: *instantiating abstract data types and optimizing makes the resulting concrete program obscure*.

The construction of the program employing the packed representation of a symmetric matrix discussed in the preceding section provides a further example of these two phenomena. First, APR can be thought of as a concrete instantiation for the two-dimensional square array A; introducing it creates the possibility for performing extensive global optimization on the resulting program. Second, the optimized program in Figure 2 is certainly obscure.

There is yet a third point to be made about this example: *the complex data type is unnecessary in Fortran*. The realization function discussed in this section describes the semantics of the complex type in terms of other, more basic Fortran constructs. (In the absence of pointers, implementing complex functions poses a slight problem.

It can be solved by adding two arguments to receive the returned value, and converting functions to subroutines.)

Thus the complex data type could be made available outside the language simply by making available a program transformer and an appropriate set of transformations. There are definite advantages to this approach. First, it simplifies the definition of the language by removing from it the need to specify something that is really a problem-oriented (and language-independent) abstract data type. Second, it permits numerical analysts to have control over the implementation of the complex data type. Few compiler writers appreciate the numerical subtleties of implementing complex arithmetic. Thus, there are implementations in which only half the exponent range of the underlying hardware is available, because complex division and modulus needlessly overflow. The version of WPOFA displayed above also suffers this problem, but it is obvious in the Fortran code (instead of hidden in assembly code) and easily corrected in a source-code-oriented transformation rule (instead of impossible to correct in the bowels of a compiler).

2.4. Abstract Programming Transformations

The realization functions discussed above are "low-level" ones, in the sense that they construct realizations from prototype programs that can be expressed in Fortran 66. "High-level" realization functions would extend the notation available to the programmer by introducing new abstract data types and operations (including optimizations) tailored to his problem domain.

The most obviously desirable abstract data types for use in writing numerical software are matrices and vectors. When they are introduced by realization functions defined in turn by program transformations, domain-specific optimizations can also be provided. Such optimizations are based (in this case) on theorems of linear algebra. They are necessary to enable programs written using the matrix and vector data types to execute as efficiently as traditional hand-written ones (*cf*. [3]).

The benefits of writing programs at a high level of abstraction and the importance of domain-dependent optimizations can be seen by considering a subtle optimization of the program WPOFA of Figure 3. From linear algebra, one knows that the Cholesky factorization produces factors whose diagonal elements are positive real quantities. That knowledge could be used to simplify the denominator of the complex division in WPOFA to AR(K,K)*AR(K,K). To discover automatically that AI(K,K) is zero from the program of Figure 3 would be extremely difficult, for the loops on J and K and the test in the CHOLESKY: IS EXITED statement would have to be analyzed in detail. Stated in another way, the program in Figure 3 is too concrete (hence obscure) to permit this optimization. Suppose, on the other hand, that it had been generated from a program written in matrix notation. Then the act of choosing the Cholesky decomposition for the factorization would require the assumption (or guarantee) that the matrix be positive-definite. Hence the transformation that implements the decomposition could introduce an annotation that A(K,K) is real, and that annotation could trigger simplifying the denominator of the complex division.

Some preliminary work on defining matrix and vector data types and relevant domain-specific optimizations by program transformations is described in [3] and [4].

3. The Relationship between Transformations and Programming Languages

By means of the examples in the preceding section I have tried to show the close connection between program transformation on the one hand and abstract data types and program optimization on the other. Of course, the idea of abstract data type did not even exist when Fortran was designed, so Fortran offers no support for its use. Nevertheless, the examples show that modest extensions to Fortran enable one to overcome its limitations and circumvent its shortcomings well enough to apply systematic transformations to programs written in it. Evidently Fortran (so extended) is at least usable as a *base language* for program transformation.

3.1. Limitations of Fortran 66 as a Base Language for Transformation

Fortran 66 is far from an ideal base language, however.

On the one hand, it is too low-level to be an adequate language for writing proto-type programs. Unextended, it has no abstract data type capabilities. Its idiosyncrasies and *ad hoc* limitations would get in the way of smooth step-to-step transitions during transformation. (Specific examples are the inability to use fully general expressions in subscripts, loop bounds, output lists, etc,; the inability to use conditional expressions; the inability to substitute value-yielding blocks of statements for primaries in expressions; and the inability to make declarations of limited scope.) Not only must Fortran be extended along these lines (*cf.* section 2.1), but also it must be contracted to remove some of its worst horrors.

Paradoxically, Fortran is too high-level to be a completely adequate *target language* — the language into which programs are transformed. As discussed above, transformations need to be able to produce efficient programs. For the necessary optimizations to be possible, the target language must provide access to such things as pointers, addresses, and registers. A specific example of the need for them arises in the optimization phase of the **utrp** realization function, the result of which is illustrated in Figure 2. In the innermost loop, the variables I10 and I11 both increase by the same amount each time through the loop, although they start at different values. Suppose the low-level target language provided pointers and addresses. Then the transformations could express the fact that both references to the array A can be made using a single index, simply by calculating and adding appropriate offsets (which are constant in the innermost loop) in each case. Without access to these low-level features, the **utrp** transformations are forced to increment two distinct indices in order to remain within standard Fortran 66. In a similar manner, the lack of pointers makes it difficult to write transformations that extend Fortran to include "structures", and the inability to access registers makes it awkward to express the implementation of complex-valued functions (*cf.* section 2.3).

Of course, one often hears that it is "bad" for a programming language to contain such things as pointers and addresses (and labels), for using them can lead to unreliable programs. But that statement is true only with regard to their use by (human) programmers in writing prototype programs. When a transformation introduces one of these constructs, all uses of it are (in effect) introduced simultaneously. Therefore, one can easily see that all the uses will be consistent and, hence, that no further consistency checking is required. *Thus these features are completely safe when they are introduced systematically by transformations.*

3.2. What Would Be an Ideal Base Language for Program Transformation?

Answers to this question are beginning to emerge, not only from work with TAMPR, but also from Bauer's project CIP [2] and from Cheatham's work on ECL [6]. Bauer has pointed out that a *wide-spectrum* base language is needed to carry out transformation effectively. As remarked above, program transformations produce the final program in the target language by a long sequence of small steps. While this sequence of steps is being carried out, different parts of the program will be at different levels in the language spectrum simultaneously. If the transformations are forced to "funnel" the entire program simultaneously through some fixed, intermediate language-level (which is then passed to a compiler), the possibility to perform some optimizations may be lost.

The difficulties caused by funneling a program through a target language that is too high-level can be seen by considering transforming a program written at, say, the applicative level (*cf.* [1]) to one at the machine level. At the applicative level the program contains information that is very useful for its optimization: namely, that the values of certain variables remain constant within their scope in the program. If such a program is converted to the Fortran level by transformations and then handed off to a complier, this information about immutability and scope is necessarily lost. Then the complier has to do extensive analysis to determine (if it can determine at all) whether certain optimizations are possible. On the other hand, in a wide-spectrum language the transformations can carry out the optimizations while they are in possession of the relevant information, and further optimization during compilation of the target-level program is not necessary.

This discussion and that of the preceding section suggest that a base language for transformation should have the following properties: it should be a wide-spectrum language providing a sufficiently low target-language level; it should be easily extensible; and (perhaps most important of all) it should be *simple*. These three qualities are interrelated.

In my view, which is strongly influenced by that of Bauer, a wide-spectrum base language would have roughly the following appearance: Its target-level would be at about the level of PL/360 [10] for conventional machine architectures, supplemented with primitives to give access to the enhanced hardware features of vector and matrix machines. (Further supplementing might be required to permit the target language to serve highly-parallel and data-flow architectures.) At a somewhat higher level, the procedural level, it would be similar to Algol 68. What the base language must emulate is the high degree of "syntactic orthogonality" in Algol 68, which permits conditional expressions and blocks as primaries, as discussed in section 3.1. (One should note that the Ada language, in a step backward, eschews this kind of syntactic orthogonality, thereby making transformation more difficult.) At a still higher level, the applicative level, the base language would be function oriented, without assignment and side effects [1]. Above that would be parallel and non-deterministic levels, perhaps culminating in a predicate-calculus-like specification-oriented level.

Ease of extension is important in a base language for transformation because it provides a simple and flexible way to make the wide-spectrum language problem-oriented. As discussed in conjunction with programmed-complex arithmetic above, standardizing particular abstract data types makes them inflexible and makes the language complicated. The following scenario describes how a number of languages have gone down the rathole in this respect: It would be inelegant to standardize an abstract data type (*e.g.*, matrices) that is applicable in only a limited problem domain, even though it might be highly useful. Therefore, the data type should be generalized to make it more widely applicable. At first glance, a number of problem domains do seem to be using a slight generalization of the abstract data type (*e.g.*, arrays). Therefore this generalization, along with operations culled from the different problem domains, is standardized. Too late, it is discovered that the different problem domains only *appear* to be using the same abstract data type. In fact, each has slightly different requirements for its data type, and the details of these requirements irreconcilably conflict. Thus, instead of being finely tuned to, and highly useful in, one particular problem domain, the standardized abstract data type is nearly useless in all of them. Better would be a simple notation for introducing new data types and transformations to define them, their operators, and their optimizations that could be used by experts in the problem domain.

In addition, the extension mechanism itself should not be overly complicated. The elaborate mechanisms provided in Ada (especially the very complicated scope rules) make it difficult to transform. By separating the definition of abstract data types from compilation, the transformational approach should make it possible to achieve adequate security, private types, etc. with much simpler rules.

Simplicity is an important attribute of the base language because it makes extensibility possible. Much of the complexity of modern languages can be traced to including definitions of non-primitive data types as part of the language definition. As if defining such types did not add enough complexity, many languages, in a misguided attempt at "orthogonality", have defined all possible conversions among them. In any given application, most of these conversions will not be needed. When one of them is, the definition chosen by the language (for consistency) is often not the one required. (The precision rules of PL/I are an oft-cited example of this phenomenon.)

Restricting the language definition to primative (*i.e.*, hardware-oriented) data types simplifies the definition, and in addition it frees notation that otherwise would be co-opted by the non-primative types. If problem-oriented extension of the language is to be possible, notation for it must be available, and verifying that the extensions are compatible with existing language constructs must not be too difficult. A simple extensible language meets both these requirements.

We desperately need to reduce the cost and increase the reliability of writing software. Program transformation on a simple, extensible wide-spectrum base language makes this goal achievable. Such a language can be extended for high-level programming in a particular problem domain. And high-level programming need not come at the expense of efficient execution, because transformations can express domain-dependent optimizations.

Probably language extension will remain beyond the capabilities of most individual users. But it is well within the capabilities of both the developers of numerical software libraries and the developers of applications programs. Surely the language extensions and abstract data types these "domain experts" create will make programming more reliable and economical in the future.

4. References

[1] Backus, J.W., *Can Programming Be Liberated from the von Neumann Style? A Functional Style and Its Algebra of Programs*, **Comm. ACM, 21**, 8 (1978) 613-641.

[2] Bauer, F.L., *Programming as an Evolutionary Process*, **Proc. 2nd Int'l Conf. on Software Engineering**, (San Francisco, 1976) 223-234.

[3] Boyle, J. M., *Extending Reliability: Transformational Tailoring of Abstract Mathematical Software*, extended abstract, **ACM SIGNUM Newsletter, 14**, 1 (1979) 57-62.

[4] Boyle, J. M., *Towards Automatic Synthesis of Linear Algebra Programs*, in: Hennell, M.A. and Delves, L.M. (eds.), **Production and Assessment of Numerical Software**, (Academic Press, New York, 1980) 223-245.

[5] Boyle, J. M. and M. Matz, *Automating Multiple Program Realizations*, **Proc. of the MRI Symposium, XXIV: Computer Software Engineering**, (Polytechnic Press, New York, 1976) 421-456.

[6] Cheatham, T.E., Townley, J.A. and Holloway, G.H., *A System for Program Refinement*, **Proc. 4th Int'l Conf. on Software Engineering**, (IEEE, Long Beach CA, 1979) 53-62.

[7] Dongarra, J.J., Bunch, J.R., Moler, C.B. and Stewart, G.W. **LINPACK Users' Guide** (SIAM, Philadelphia, 1979).

[8] Dritz, K.W. and Boyle, J.M., *The Art and Science of Writing TAMPR Transformations*, (Argonne National Laboratory Report, Argonne IL, in preparation).

[9] Hague, S.J., *Software Tools*, in: Jacobs, D. (ed.), **Numerical Software—Needs and Availability**, (Academic Press, New York, 1978) 57-79.

[10] Wirth, N., *PL360, a Programming Language for the 360 Computers*, **J. ACM, 15**, 1 (1968) 37-74.

DISCUSSION

E. Schlechtendahl (Karlsruhe): You state that the base language should include the possibility for extensions. I disagree--there are three levels. (1) The base language should provide fundamental features, such as pointers. (2) A different set of features are needed for those working in application areas. (3) A third set of features are needed to provide tools for translation into the base language.

Boyle: I intended to advocate that people doing applications should be able to define what types of transformations are to be done; a wide-spectrum language is needed to do so.

G. Paul (IBM): I disagree with your comments on vectors, but only because we are presently unable to adequately preprocess them. Is there an implied ordering in your code transformations? Can they be applied recursively?

Boyle: Our transformation rules are applied by a generalization of Markov algorithms to tree structures. There is both bottom-up and top-down recursive application. The idea of "transformational invariant" can be used to show that transformations produce a canonical form.

S. Hammarling (NPL): Can I use TAMPR?

Boyle: We are working on a portable version. It should be available in a year or two.

A. Wilson (ICL): I disagree with your remark that the array extensions proposed for Fortran 8x do not allow the maximum performance to be leached out of the hardware. If there is any possibility for the compiler to recognize a vector operation, it can put out the best possible code for its hardware.

Boyle: I meant that there is nothing in the proposal to enable me to do the necessary optimization, such as breaking vectors into blocks of size 64 for the CRAY. In section 3.2 I discuss the problem of recognizing optimizations.

T. Lake (ICL): Avoiding an explosion of derived languages is important. Is there any sense in which what you describe is as powerful as translation from a high level language into assembly language?

Boyle: I haven't written transformations to do so, but I don't see why I could not. I believe that register allocation would be the most difficult part.

M. Wells (New Mexico State Univ.): This is in response to your negative remarks on orthogonality. If it is present, then m+n constructs, rather than m·n, need to be learned from m syntactic constructs and n semantic meanings.

Boyle: Perhaps "consistency" is what I meant.

THE RELATIONSHIP BETWEEN NUMERICAL COMPUTATION
AND PROGRAMMING LANGUAGES, J.K. Reid (editor)
North-Holland Publishing Company
© *IFIP, 1982*

PROGRAMMING PARALLEL NUMERICAL ALGORITHMS IN ADA

E.K. Blum
Mathematics Department
University of Southern California
Los Angeles, California
U.S.A.

Abstract: This is a small part of a general study of Ada's
task constructs. Within the imposed page limits, we
illustrate, by the simple example of the vector operation
scalar-times-vector, some of the problems in programming Ada
tasks to achieve some degree of parallel execution of vector
operations.

I. Parallel Vector Operations

Many numerical computations involve operations on vectors in
an n-dimensional vector space. If x and y are vectors and c is a
real scalar, the computations are usually built up from vector
addition, $x + y$, multiplication by a scalar, $c*x$ and inner-
product $\langle x,y \rangle$. If we represent x and y by arrays of components
$x(i)$ and $y(i)$ respectively, $1 \leq i \leq n$, then the array representing x
+ y has components $x(i) + y(i)$, where the latter + denotes
addition of real numbers. Similarly, $c*x$ has components $c*x(i)$,
where the latter * denotes multiplication of reals.

Several programming languages provide notation to specify
vector operations directly. However, there is no guarantee that
these operations will be implemented in a "true vector manner,"
that is, by performing the operations on all components "in
parallel." If a language does not provide "predefined" vector
operators, they can usually be introduced as defined functions.
Indeed, in Ada, the symbol "+" can be "overloaded" and used as
the name of a function subprogram, <u>function</u> "+" (x,y: VECTOR)
<u>return</u> VECTOR, which returns the vector x + y as the result.
This allows a programmer to write infix notation "x + y" in
vector expressions, presumably making the program easier to write
and read. If the function subprogram for vector + defines the
result x + y by a loop which computes sequentially the components
$x(i) + y(i)$ and stores them in components $z(i)$ of a result vector
z, then it can be argued that + is not a true vector operation
because the components are not computed in parallel. If the main
purpose in writing an algorithm in vector form is to speed up a
computation by taking advantage of available parallel processors,
then using sequential subprograms to define overloaded operators
misses the point. However, in Ada, it is possible to program the
definitions of vector operators in a way that achieves a degree
of parallelism in principle and, if an implementation provides a
multi-processing capability, in actual fact as well. This
requires the use of Ada's task construct.

297

According to the Ada reference manual (p.9-1) "tasks are
entities that may operate in parallel". However, "parallel tasks
may be implemented on multicomputers, multiprocessors, or with
interleaved execution on a single processor". Our objective is
to examine Ada's task constructs to see how well-adapted they are
to defining "parallel" vector operations. To achieve an actual
speed-up, a multi-processor implementation of tasks must be
available. Indeed, in an interleaved implementation, the
parallel-task programs for vector operations will run more slowly
than a conventional sequential loop (component-wise) program.
Ada's tasking features are usually illustrated for operating
system problems such as the producer-consumer problem. The
design of Ada's tasking constructs appears to have been motivated
more by such problems rather than by the kinds of parallelism
situations that arise in numerical applications. This may
account for some moderately surprising, circuitous programming
constructions that seem to be necessary when using tasks to
program parallel numerical computations. Because of space
limitations, we can only illustrate this here by two programs for
the "parallel" vector operation c*x. In a sequel to be
published elsewhere, we shall elaborate on the problems in
"vector programming" in high-level languages and illustrate them
by more complicated examples.

2. Ada Tasks

Tasks are defined by first declaring a task type and then
declaring one or more objects to be of that type. A task object
is a constant whose value is an instance of the declared task. A
task type is defined in two parts: a task-specification and a
task-body. The specification declares the entries of the task
that can be called (like procedure calls) by other tasks. The
body defines the statements to be executed. In particular, there
can be accept-statements for the task entries. During execution
of an Ada program, when an entry of a particular task, T, has
been called by another parallel task, S, the execution of S is
suspended until T reaches an accept statement for that entry. At
that point, the sequence of statements within the accept's do-end
delimiters is executed by T, while S remains suspended. This is
called a rendezvous. When the end of the accept is reached, both
S and T can continue execution in parallel. Rendezvous is the
main device for inter-task synchronization. It also serves as
the main means of inter-task communication, by passing parameters
in entry calls. Tasks can also communicate through shared
variables. If b is an identifier used to declare an object of a
task type having an entry declared with identifier E, then the
dot notation b.E is used to call that entry in the instance of
the task associated with b. If b is an array object of task
type, an array of tasks, all of that type, is associated with b
and the notation b(i). E denotes entry E of the ith task.
Unfortunately, there is no task attribute to determine the
identity of a task object from within the task body. The caller
must pass this information.

3. Example: Multiplication of vector x by real c.

We wish to program the n component operations c*x(i),

$1 \leq i \leq n$, in such a way as to achieve maximum overlap, that is, the program should allow a multi-processor implementation of Ada to execute the program with a maximum degree of parallelism or equivalently, in a minimum elapsed time. The absolute maximum degree of parallelism would be achieved if the elapsed time for all n multiplications is $t_k = \max \{t_i : 1 \leq i \leq n\}$, where t_i is the time to perform

$c*x(i)$. Since t_i usually depends on the particular actual

values of c and x(i), a general program cannot be based on specific t_i or t_k. Hence, the best general scheduling

strategy is to start all multiplications at the same instant. If we assume that there are n multiplier processors available in the host system, then this strategy will permit them to operate in parallel to achieve elasped time t_k in

every instance. Unfortunately, this strategy cannot be programmed. In most high-level languages, including Ada, there is no way to insist on such real time scheduling of basic operations. In principle, we would like to write a program which specifies at some point in a computation by a main task, V, when the values of all components x(i) and c become available, that n identical tasks, M(i), $1 \leq i \leq n$, begin to multiply c and x(i) respectively. Ideally, this would be done by passing the pairs (c, x(i)) to M(i), $1 \leq i \leq n$, simultaneously, doing the respective multiplies in parallel and returning the values d(i) = c*x(i) to the task V, which then proceeds with its own computation. There are two ways to pass (c,x(i)) to M(i): (1) as actual parameters in an entry call; (2) as shared variables. We illustrate this by the programs in Figures (1) and (2) respectively.

In Ada, (1) must be done sequentially, that is, successively for each M(i), in a loop within task V, because there is no Ada feature permitting parallel entry calls. Task V is suspended during its entry call to M(i). The entry call to M(i + 1) cannot be made until after M(i) completes its accept statement. If the accept included multiplication, the program would be slower than doing the n multiplications within the loop of V itself. There does not seem to be any programming device in Ada for circumventing this problem completely. The problem can be partially circumvented if the time required to pass parameters in a task entry call is small compared to the time required for execution of the corresponding accept statement; e.g. in our example, if the time needed to pass (c,x(i), d(i)) to M (i) is small compared to multiplication time for c*x(i). One technique is to design the tasks M(i) so that the accept statement merely handles the input data by storing it in local variables and ends the rendezvous as quickly as possible. Task M(i) can then begin operating on its data while task M(i + 1) is receiving its own input data during the next entry call. We have chosen to program this example using the package construct. The package VECT-OPS

```
package   VECT-OPS is
                n:  INTEGER;

    task   VM is

                entry CVOP (c: in REAL; v: in array (1..n) of REAL;
                w: out array (1..n) of REAL);

                entry FINI-MPY (z: in INTEGER);

    end VM;

private type  ACCESS-REAL is access REAL;

    task type MULT is

                entry DATA (c,x:in REAL;  j:in INTEGER;
                                          y:in ACCESS-REAL);

    end MULT;

end VECT-OPS;


package body VECT-OPS is

    M:array (1..n) of MULT;
    k:INTEGER;
    task body VM is

        d: array (1..n) of ACCESS-REAL : = new REAL;
    begin

        while TRUE loop
            accept CVOP (c:in REAL; v:in array (1..n) of REAL;
                    w: out array (1..n) of REAL) do
            for j in 1..n loop
                M(j). DATA (c, v(j), j,d(j));
            end loop;
            k: = 1;
            while k <=n loop

                accept FINI-MPY (j:in INTEGER) do
                    w(j): = d(j). all;
                end FINI-MPY;
                k: = k + 1;
            end loop;
            end CVOP;
        end loop;
    end VM;
```

Figure 1

```
task body MULT is
                TEMP1,TEMP2: REAL;  i:INTEGER;
                TEMP3: ACCESS-REAL : = new REAL;
begin
     while TRUE loop
          accept DATA (c,x: in REAL;  j: in INTEGER;
                                       y: in ACCESS-REAL)  do
                TEMP1: = c ;  TEMP2: = x;  i: = j ;
                               TEMP3 : = y ;
          end DATA;
                TEMP3. all : = TEMP1 * TEMP2;
                VM.FINI-MPY (i);
     end loop;
end MULT;
          begin
          GET( FILE:in IN-FILE; n:out INTEGER)
          --    GET is defined in generic package INTEGER  IO
end VECT  OPS;

with VECT-OPS;
procedure MAIN is

          use VECT-OPS;
          task PRIMARY;
          task body PRIMARY is
                const: REAL;
                a: array (1..n) of REAL;
                b: array (1..n) of REAL;

          begin

                    -- read or compute const and a
                VM. CVOP (const, a,b);
                --continue computation

          end PRIMARY;
begin
end MAIN;
```

Figure 1 (continued)

```
package body VECT-OPS is
     task body VM is
          TEMP2, TEMP3; array (1..n) of REAL;
          TEMP1: REAL; i:INTEGER;
               task type MULT is
                         entry DATA (j:in INTEGER);
               end MULT;
               task body MULT is
                    i: INTEGER
               begin
                    while TRUE loop
                         accept DATA (j:in INTEGER) do
                              i:=j;
                         end DATA;
                         TEMP3(i) ;= TEMP1 * TEMP2(i);
                         VM.FINI-MPY(i);
                    end loop;
               end MULT;
     M: array(1..n) of MULT;
     k: INTEGER;
begin -- execute task VM
          while TRUE loop
               accept CVOP(c:in REAL; v:in array(1..n)of REAL;
               w: out array(1..n) of REAL) do
                    TEMP1:=c;   TEMP2 :=v;
                    for j in 1..n loop

                         M(j). DATA(j);
                    end loop;
                    k: = 1;
                    while k<=n loop
                         accept FINI  MPY(j: in INTEGER) do
                              w(j) : =TEMP3(j);
                         end FINI  MPY;
                         k: = k + 1;
                         end loop;
               end CVOP;
          end loop;
end VM;
```

Figure 2

can be used as a library compilation unit, together with a "main" subprogram which includes a primary task that executes the sequential part of the computation and issues entry calls to the tasks in VECT-OPS, which then execute vector operations in an overlapped mode. The semantics of compilation, elaboration and execution of an Ada program are such that the package VECT-OPS will be compiled and elaborated before the procedure MAIN. Elaboration of the body of VECT-OPS causes all the task M(1),..., M(n) to start executing their loops. Thus, they are all ready to accept the entry call DATA. The controlling task VM is also activated and is ready to accept the entry call CVOP, which will be issued by the task PRIMARY once it begins execution.

The second method of passing the data (c, x(j)) to M(j) is through shared variables. In Figure (2), the package VECT-OPS is modified so that c and x are stored in local variables of task VM which are made "visible" to the tasks M(j) by declaring them and the type MULT in body of VM. It then suffices to pass just the index j to M(j). DATA, thereby eliminating three assignment statements as well as three parameter passes in the accept DATA statement. As before, task VM will be activated as part of the elaboration of VECT_OPS. Activation causes the body of VM to be elaborated, including the declarations of MULT and M. In this situation, the semantics of task activation stipulates that the tasks M(i) are all activated before the first statement of VM is executed. This is a slightly different activation dynamics than in the first version of VECT_OPS, in which VM could legally start executing before the M(j) tasks do. This change in dynamics of task execution does not affect the semantics of the computation of values because the M(j) are all synchronized by VM's entry calls, M(j). DATA. This second version of the body of VECT_OPS is given in Figure 2. The new specification part is as in Figure 1, but with the private part deleted.

Another difference in the task dynamics in version 2 results from the nesting of (the declaration of) M within VM. This makes the M(j) dependent tasks of VM. Consequently, VM cannot terminate until all M(j) do. Again, this does not affect the computation of values, but it raises interesting questions for the implementer, the answers to which can have a significant effect on the degree of parallelism actually achieved during execution.

REFERENCES

1. Bouknight, W.J., et. al., The Illiac IV system, Proc. I.E.E.E. 60 (April 1972), 369-388.

2. Batcher, K.E., The multidimensional access memory in STARAN, I.E.E.E. Trans. Comput. c-26 (February 1977), 147-177.

3. Enslow, P.H., Multiprocessors and Parallel Processing, (John Wiley, New York 1974).

4. Lilles, W.C., J.C. Demmel, I.S. Reed, J.D. Mallett and L.E. Brennan, Multi-domain Algorithm Evaluation, RADC-TR-78-59 (April 1978).

5. Gentleman, W., Some complexity results for matrix computation on parallel processors, JACM, (January 1978), 112-115.

6. Agerwala, T. and B. Lint, Communication in parallel algorithms for boolean matrix multiplication, Proc. 1978 Int. Conf. on Parallel Processing, I.E.E.E. Computer Society (August 1978), 146-153.

7. Dongarra, J. J. et al, LINPACK User's Guide, (SIAM, 1979).

8. Lawson, C., R. Hanson, D. Kincaid and F. Krogh, Basic linear algebra programs for Fortran usage, ACM Trans. Math. Software 5 (September 1979), 308-325.

9. Reference Manual for the Ada Programming Language, U.S. Department of Defense, July 1980 (Reprinted November 1980).

DISCUSSION

R. Brent (Australian National Univ.): When programming linear algebra in Fortran, we typically start with a formulation in matrix/vector notation, and translate it into a sequence of scalar operations. Then an optimising compiler for a parallel or pipelined machine has to take the sequence of scalar operations and figure out that we really wanted to do a matrix/vector operation. There seems to be an analogy here. In your talk you started with a simple vector operation and converted it into a highly non-transparent "package" of Ada "tasks." To implement the package with tolerable efficiency, an Ada compiler would have to figure out that all you wanted to do was a simple vector operation, and that the overhead of general task communication, etc., was avoidable. The conclusion seems to be that Ada's tasking mechanism is too general to be useful for parallel computations in numerical linear algebra (as exemplified by EISPACK and LINPACK).

Blum: I agree. In the original task proposal, we could have groups of tasks, etc.

O. Roubine (Cii Honeywell): Task families were added to preliminary Ada to solve certain problems. They have now been replaced by something more flexible--the task type. The only thing lost is an index, which enables a task to know which member of a family it is.

Blum: I found that quite critical.

Roubine: That's a hard problem. A programming solution is to pass to a task its own name.

M. Delves (Univ. of Liverpool): What's the problem with that solution? Vector multiplication doesn't seem to be a very good example, unless the individual multiplications are enormously expensive.

Blum: Yes, I have also considered elimination; it is a better example. Note that version 2 of the example shown here passes the tasks their names, in effect.

THE RELATIONSHIP BETWEEN NUMERICAL COMPUTATION
AND PROGRAMMING LANGUAGES, J.K. Reid (editor)
North-Holland Publishing Company
©IFIP, 1982

PROGRAMMING IN ALGOL 68 (as a host)

AND THE USAGE OF FORTRAN (program libraries)

C.G. van der Laan

Rekencentrum University of Groningen
Postbus 800
9700 AV Groningen, The Netherlands

A technique is described whereby a collection of FORTRAN subprograms can be made available to users of other programming
languages, notably ALGOL 68. This is illustrated with some
examples from Forsythe-Malcolm-Moler.

INTRODUCTION

The aim of this presentation is to communicate with the programming languages
community in order to make better use of FORTRAN subprograms in other languages,
especially ALGOL68, for the short term. For the long term we <u>need</u> a

　　　　real higher level programming language

which reflects (at least) the actual computing environment:
　　　. a collection of coupled (parallel) machines, with a fragmentation
　　　　of peripheral equipment and access possibilities,
　　　. various software modules organized in e.g. program libraries
　　　　written in different languages.

As long as mixed-language programming is not defined and not standardized, portability is hampered; the mixed-language program often has to be converted into
a program in one language.

1. FORMULATION OF THE ALGOL68-FORTRAN-INTERFACE

... A second novel concept is that although ALGOL, FORTRAN, and
eventually PL/1 programmers will all use the library, only one version
of most routines will exist, and it will be in the language that the
author of the routine chose. Programmers writing in the other two
languages will access the routines through interfaces ...

Gentleman & Traub (1968)

Why an interface?
Interfaces to FORTRAN reduce the need to rewrite codes for the same
algorithm into various languages as well as to rewrite the
documentation. The question why to use languages different from FORTRAN
has a long history; I feel myself comfortable by the more advanced data
types, control structures, parallelism, and parametrization of the
computing environment as well as the orthogonal and coherent design.

1.1 STATEMENT OF THE PROBLEM

The problem is that different computer languages use different internal
representations. We illustrate the situation in the right diagram below,
whereas the left diagram reflects nowadays usage.

In order to do this gracefully, it is necessary to have common data
representations, to perform subtasks in language 2 and to communicate
with the computing environment from within language 2 through language 1
as host.

In some compilers interfaces are provided; some of the limitations of
the interface in our CDC-compilers (ALGOL68, ALGOL60, SIMULA and PASCAL)
are:
• only simple types can be passed;
• arrays are stored row-wise in ALGOL-like languages and column-wise in
  FORTRAN;
• I/O from FORTRAN within an ALGOL-like host is not possible;
• a routine written in the ALGOL-like language cannot be passed as a
  parameter to a FORTRAN subprogram.

Our aim was to improve the ALGOL68-FORTRAN-interface such that numerical
work is less hampered. In particular we have realised the following
facilities:
• more complicated types can be passed;
• arrays are stored in ALGOL68 in the same way as in FORTRAN (by a
  compiler installation option);
• I/O, especially error messages, from a used FORTRAN-subprogram can be
  handled;
• ALGOL68 routines can be passed as parameter to FORTRAN subroutines and
  functions.

The restrictions are still:
. slices cannot be passed through the interface;
. ALGOL68 routines with ROW-of-TYPE as parameter must specify the dimensions
  (i.e., subscript bounds) when the routines are passed as parameter to the
  FORTRAN routine (The problem is that the information about the size of the
  array, which the ALGOL68 procedure needs, is not at hand in the FORTRAN
  layer);
. no routines are allowed as parameters of a routine which is given as a
  parameter to a FORTRAN routine;
. no correspondence exists between char and CHARACTER;
. no general correspondence exists with files.

1.2 CORRESPONDENCE OF DATA TYPES FROM A USER POINT OF VIEW
The coupling concerns some ALGOL68 protonotions resulting from MODE, and the
FORTRAN concepts INTEGER, REAL, DOUBLE PRECISION, COMPLEX, LOGICAL, DIMENSION,
SUBROUTINE and FUNCTION. In the column under FORTRAN, TYPE stands for one of
INTEGER, REAL, DOUBLE PRECISION, COMPLEX, LOGICAL; in the column under
ALGOL68, PRIMOD stands for one of int, real, long real, compl, bool.

FORTRAN	corresponds with	ALGOL68	
INTEGER		ref int	
REAL		ref real	
DOUBLE PRECISION		ref long real	
COMPLEX		ref compl	
LOGICAL		ref bool	
TYPE ( )		ref [ ]	PRIMOD
TYPE (,)		ref [,]	PRIMOD
TYPE (,,)		ref [,,]	PRIMOD
SUBROUTINE		proc void	
SUBROUTINE (...)		proc (...) void	
TYPE FUNCTION (...)		proc (...) PRIMOD	

REMARK
At the moment ref is optional.

1.3 LINKING OF A FORTRAN ROUTINE TO ALGOL 68
After the routine symbol (i.e., a colon) one must write a pragmat of the
form
       pr fortran TAG pr
followed by skip as the unit of the routine text. The TAG is the name of the
FORTRAN subprogram. The default TAG consists of at most the first 7 characters of the name of the proc.

1.4 I/O AND TRANSPUT
Within the context of the use of FORTRAN routine collections the possible
error message is passed to the file handling routines and printed as if it
handled an error from an equivalent ALGOL68 routine.

2. A METHOD OF PROGRAMMING

The most obvious way is to couple a FORTRAN routine direct to an ALGOL68
procedure. We did not follow this line of thought, because our aim was
to create virtual libraries in various languages and to treat computational as well as algorithmic technical details once in a FORTRAN layer.

2.1 ASSUMPTIONS
- The use of FORTRAN in ALGOL68 implies the knowledge of ALGOL68,
  FORTRAN and TORRIX, a matrix/vector package in ALGOL68 (VAN DER
  MEULEN, 1978).
- It is aimed as well at ALGOL68 programmers as FORTRAN programmers,
  because the former can make use of already available good software
  while the latter can benefit form the facilities of ALGOL68 and
  TORRIX.
- 'Problem-thinking' and 'method-thinking' are decoupled.
- Problems have physical meaning; singularities ought not to occur and
  are looked after in the FORTRAN layer. The accuracy of data and
  results are handled in the problem environment.
- The program set-up is done in ALGOL68 from a mathematical description
  of the problem without looking for details in the program library
  manuals.
- The brain-fag of 'reading FORTRAN' and 'writing ALGOL68'
  simultaneously is circumvented.
- One interfaces as small a number of parameters as possible.
- Work areas are introduced in ALGOL68 from bottom to top in order to be
  as problem size independent as possible, and only one per type.

REMARK
We mention TORRIX because operations on matrices and vectors are so
fundamental within a numerical context.

2.2 GLOBAL PROGRAMMING SCHEME
If the numerical problem is either not catered for in an ALGOL68 routine
collection or one considers a FORTRAN implementation more suitable, one
can use the ALGOL68-FORTRAN-interface as follows.
- translate the mathematical problem into a PROCEDURE (with identifier
  NAME, say),
- refine the routine text by a call to a procedure with FORTRAN
  manageable parameters (with identifier NAMEb, say),
- couple NAMEb with a FORTRAN subprogram called NAMEF, say,
- decide upon which FORTRAN routine collection is to be used,
- refine NAMEF and call the technical routine (example programs from the
  routine collection can be used in this stage),
- work space is introduced bottom-up as parameters in NAMEF and NAMEb;
  in the call of NAMEb this can be done by using the TORRIX routines:
  genvec or genindex, in order to have the various layers independent
  (as much as possible) of the size of the problem.

By the above method one can make a virtual ALGOL68 program library using
any FORTRAN routine collection via

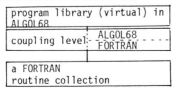

2.3 EXAMPLES OF THE USAGE OF ROUTINE COLLECTIONS

We have chosen some examples to illustrate the following coupling aspects:

a. a FUNCTION,
b. a SUBPROGRAM with at least a matrix as parameter,
c. a SUBPROGRAM with at least one ALGOL68 procedure as parameter,
d. as c) where the ALGOL68 procedure has at least a reference to ROW of INTREAL as parameter.

Problems where the above coupling aspects occur are e.g.:

a. evaluation of a special function, a random number generator,
b. solving a linear system of equations, a linear least squares problem,
c. optimization, quadrature, zerofinding,
d. solving a system of differential equations.

The FORTRAN routine collections which we have coupled - by examples - are: IMSL, NAG, CERNLIB and the routines in FORSYTHE (1977). In the following GEF stands for the routines published in FORSYTHE (1977).

A user can by the above principles couple his own FORTRAN collection and thereby create his program library in the language he choses, provided an infrastructural interface is available.

REMARK

For the modes mat,vec and the procedures genindex and genvec see VAN DER MEULEN (1978). More complicated examples such as P8.3 in FORSYTHE (1977) have been worked out, where the minimization, the solving of the ODE and the evaluation of the Besselfunction are dealt with in FORTRAN and the rest in ALGOL68.

PART OF THE VIRTUAL GEF LIBRARY IN ALGOL 68

ad a. proc random = (ref int x) real:
                 pr fortran randomf pr skip

ad b. proc linlsq = (mat a, vec b) vec:
       begin int m = 1 size a, n = 2 size a;
             proc linlsqb = (mat a, vec x, b, int m, n, vec w) void:
                          pr fortran linlsqf pr skip;
             vec x = genvec (n);
             linlsqb (a, x, b, m, n, genvec (2*(m+1)*n));
             x
       end

ad c. proc intgrl = (real a, b, proc (real) real f, real tol, ref real
                      errest) real:
                 pr fortran intgrlf pr skip

ad d. proc difsys = (real xb, xe, vec y, proc (real, vec, vec) void fxy,
                      ref real relerr, abserr) void:
       begin proc difsysb = (real xb, xe, int n, vec y, proc (real, vec,
                             vec) void fxy, ref real relerr, abserr,
                             vec w, index iw) void:
                          pr fortran difsysf pr skip;
             int n = size y;
             difsysb (xb, xe, n, y, fxy, relerr, abserr, genvec (3+6*n),
                      genindex (5))
       end

PART OF THE INTERMEDIATE LEVEL FOR GEF

ad a.
```
      REAL FUNCTION RANDOMF(IX)
C CALCULATES A "RANDOM" NUMBER DEPENDENT ON IX
      RANDOMF=URAND(IX)
      RETURN
      END
```


ad b.
```
      SUBROUTINE LINLSQF(A,C,Y,M,N,W)
C SOLUTION OF LINEAR LEAST SQUARES PROBLEM AC=Y
      DIMENSION A(M,N),C(N),Y(M),W(2*(M+1)*N)
C CALL OF GEF SINGULAR VALUE DECOMPOSITION
      CALL SVD(M,M,N,A,W,.TRUE.,W(N+1),.TRUE.,W(N*M+N+1),IERR,
     +        W(2*M*N+N+1))
      SIGMA1=0.
      DO 30 J=1,N
        IF(W(J).GT.SIGMA1)SIGMA1=W(J)
        C(J)=0.
   30 CONTINUE
C ADJUST RELERR IF DESIRED
      RELERR=1.0E-5
      TAU=RELERR*SIGMA1
      DO 60 J=1,N
        IF(W(J).GT.TAU)THEN
        W(J)=1./W(J)
        ELSE
        W(J)=0.
        END IF
   60 CONTINUE
      CALL VSUTY(W(N*M+N+1),W,W(N+1),Y,C,M,N)
      RETURN
      END

      SUBROUTINE VSUTY(V,SIGMA,U,Y,C,M,N)
C     CALCULATES V TIMES SIGMA TIMES Y TRANSPOSE TIMES Y
      DIMENSION V(N,N),SIGMA(N),U(M,N),Y(M),C(N)
      DO 60 J=1,N
        S=0.
        DO 40 I=1,M
          S=S+U(I,J)*Y(I)
   40   CONTINUE
        S=S*SIGMA(J)
        DO 50 I=1,N
          C(I)=C(I)+S*V(I,J)
   50   CONTINUE
   60 CONTINUE
      RETURN
      END
```

ad c.
```
      REAL FUNCTION INTGRLF(A,B,F,TOL,ERREST)
C CALCULATES THE INTEGRAL FROM A TO B OF THE USER DEFINED
C            FUNCTION F WITH ONE ARGUMENT
      EXTERNAL F
      CALL QUANC8(F,A,B,TOL,TOL,RESULT,ERREST,NOFUN,FLAG)
      INTGRLF=RESULT
      RETURN
      END
```

ad d.
```
      SUBROUTINE DIFSYSF(T,TFINAL,N,Y,FCN,RELERR,ABSERR,W,IW)
C CALCULATES THE VECTOR Y IN TFINAL GREATER THAN T, WHERE Y IS
C DEFINED BY THE SYSTEM OF N DIFFERENTIAL EQUATIONS:
C     DY/DT = F(T,Y)
C WITH INITIAL VALUE Y IN T.
C RELERR,ABSERR: RELATIVE AND ABSOLUTE ERROR
C TOLERANCES FOR LOCAL ERROR TEST, AT EACH STEP THE
C CODE REQUIRES THAT
C     ABS(LOCAL ERROR).LE.(RELERR+1)*ABS(Y)
C THE RHS IS EVALUATED BY THE SUBROUTINE FCN, PROVIDED
C BY THE USER AND IT SHOULD BE OF THE FOLLOWING FORM
C     SUBROUTINE FCN(T,Y,YPRIME)
C     REAL Y(N),YPRIME(N)
C        .
C        .
C FCN SHOULD EVALUATE YPRIME(1),...,YPRIME(N) GIVEN T, AND
C Y(1),...,Y(N). YPRIME(I) IS THE FIRST DERIVATIVE OF Y(I) WITH
C RESPECT TO T.
C FCN MUST APPEAR IN AN EXTERNAL STATEMENT IN THE CALLING PROGRAM AND
C T,Y(1),...Y(N) MUST NOT BE ALTERED BY FCN.
      DIMENSION Y(N),W(3+6*N),IW(5)
      EXTERNAL FCN
      IFLAG = 1
   10 CALL RKF45(FCN,N,Y,T,TFINAL,RELERR,ABSERR,IFLAG,W,IW)
      GO TO (80,20,20,20,50,60,70,80), IFLAG
   20 IF(T.GE.TFINAL)GO TO 80
      GO TO 10
   50 ABSERR = 10.0*ABSERR
      GO TO 10
   60 RELERR = 10.0*RELERR
      IFLAG = 2
      GO TO 10
   70 IFLAG = 2
      GO TO 10
   80 RETURN
      END
```

CONCLUSION

In order to create a numerical program library in another language, it is
advantageous to provide a reasonable interface between the host language and
FORTRAN; the advantages concern:
. documentation (no technical details have to be repeated)
. as well as the production of codes (duplication of codes and its
  overhead).

Most of the work is done in the intermediate level; this level is useful
for:
. FORTRAN users who like a more problem oriented parameter list,
. any other virtual program library when an appropriate interface is
  available.

I estimate that to build a NAG ALGOL68 virtual program library from the
routines in the NAG FORTRAN (mark 8) library for our compiler will take:
. three person-months to create and test the intermediate level,
. two person-months to create and test a virtual ALGOL68 library with
  documentation in comments.
Person-months must be understood with respect to a numerical analyst
familiar with NAG and with experience in software engineering.

ACKNOWLEDGEMENTS

The author is grateful to J.Ph. Kelders, H.J. van Linde and D.D. de Vries of
our computing centre and J.J.F.M. Schlichting of the Control Data
Corporation (CDC) for the discussions and the realization of the interface,
to J.P. Hollenberg and D. Velvis of our computing centre for their ideas and
work with respect to the testing and to the "Werkgroep Wiskundige
Programmatuur" (i.e. the Dutch contact group on mathematical software),
especially T.J. Dekker, for the stimulating contacts and discussions.

REFERENCES

Bos, H.J., D.T. Winter (1978):
AFLINK. Report NN17/78. Mathematisch Centrum.
(An interface in software (in contrast with a compiler adaption) is
described).

Forsythe, G.E., M.A. Malcolm, C.B. Moler (1977):
Computer methods for mathematical computations.
Prentice Hall.

Gentleman, W.M., J.F. Traub (1968):
The Bell Laboratories num. math. program library project.
Proc. 23rd nat. conf. ACM. 485-490.

Hemker, P.W., D.T. Winter (1979):
A preliminary report on numerical operators in ALGOL68. Report NW66/79.
Mathematisch Centrum.
(An ALGOL68 collection of operators is described independent of the language
of the workers).

Van der Meulen, S.G., M. Veldhorst (1978):
TORRIX, Mathematical Centre Tracts 86.

DISCUSSION

M. Hennell (Univ. of Liverpool): What about the inverse problem: calling Algol 68 from Fortran?

van der Laan: We were not willing to pay for that. We felt the host should be the richer language.

F. Ris (IBM): Professor Kahan observed in an earlier discussion that both we and the people for whom we produce software might be better off in an environment which consisted of several smaller, less general languages. What it takes to make this practical is much less a language question than it is a system implementation question.

Several people here have remarked wistfully that it used to be a lot easier to mix program modules written in separate languages. Indeed, ten and fifteen years ago I was happily mixing FORTRAN numerical routines with COBOL drivers and FORTRAN drivers with ALGOL 60 numerical routines. More recently I have had occasion to want to mix PL/I with FORTRAN, specifically to gain access to EISPACK code. I have, in other contexts, converted a substantial amount of FORTRAN to PL/I, which is fairly easy to do with an appropriate text editor which will sprinkle semicolons around and with some careful attention to declarations. However, it is exactly codes such as EISPACK that are the most dangerous to modify in this way, and so I started from the premise that it would be easier to look at the string of bits expected on the FORTRAN side of the call interface and make sure that the PL/I calling program produced exactly that calling sequence. This is not too difficult, and having to restructure two dimensional arrays (which can be done with a declaration in PL/I) seems a small price to pay.

Nevertheless, it took me the best part of two days to make the linkage work, due to a much different kind of problem. When one executes a PL/I program, in fact the compiler transparently generates linkage to a resident library routine which initializes the "PL/I execution environment" and then calls what the programmer believes to be the main routine as first level sub-routine. For those familiar with the MVS user environment and the dynamic capabilities of PL/I, this is hardly surprising. What is somewhat more surprising is that the FORTRAN execution environment is established in much the same way. That is, control is initially passed to a library routine which does FORTRAN environmental initialization and subsequently calls the FORTRAN main program. When PL/I and FORTRAN co-exist, there is a sensitive issue as to which of these otherwise invisible routines gets control, and one quickly learns that PL/I must be given the upper hand. This is easily accomplished with a straightforward loader control statement.

What is really insidious is when the program runs for a while and then quits with no diagnostic information. It turns out that the FORTRAN interface will call a main program and upon its return will assume it is returning control to the operating system. When it returns control to some other environment (e.g., PL/I execution environment) it does not cope so gracefully. In particular, it attempts to do the operating system a favour by cancelling the association between I/O streams and operating system files. When the PL/I program subsequently comes along to do its I/O, it encounters a totally unanticipated error and cannot issue a diagnostic, since the I/O stream for diagnostics has also been disconnected. It turns out that this problem can be fixed with a relatively trivial addition to the file definitions at the job control level, although understanding why that addition is necessary entails operational knowledge of the scene I have just sketched.

Now, what is it that leads to this situation? A large group develops FORTRAN; a large and independent group develops PL/I. Often these groups are separated by substantial geographical distance in addition to emotional distance. When a language gets large, the implementation group gets large, and there is enough problem getting intra-group communication to work, let alone being concerned about what, viewed from a language implementation perspective, is a bridge issue. Furthermore, language implementations are not done in parallel; they are done closer to the timetables of language standardization groups, which are not synchronized one with the other. In the press of schedule and cost, the language standard groups apply an implicit pressure for exclusive support of that language, and even the interest tends to be more of common language support across a variety of machines than a common machine support across languages. There is therefore nobody around to make the case for the community which feels that its interests are best served by the ability to mix languages in a common environment. This of course is much easier to do in smaller operating systems with smaller languages, and indeed is the natural thing to do. However, I find it unlikely that the likes of FORTRAN 8x, ADA, or ALGOL 68 would have compatible execution environments without a conscious manufacturer decision to do so. I see no source of external motivation to do that which seems so obviously sensible to us but causes the language product manager a set of headaches that he would be happy to do without.

van der Laan: The concept of simplicity is paramount.

M. Gentleman (Univ. of Waterloo):   Here is a list of hard things about mixed language programming.
1. Data structures that do not exist in the other language.
2. Passed procedures.
3. I/O abstractions that differ between the two languages, I/O implementations that require separate buffering
4. Exception handling: storage overflow, numerical overflow, etc.
5. Data structures with very different representations. (Store by row vs store by column is a trivial example of this.)
6. Labels and transfer of control. (The problem is ensuring that epilogues to clean up get executed.)
7. Incompatible storage management.
8. Debuggers, execution monitors, etc.
9. Incompatible parameter-passing semantics (e.g., SNOBOL vs BCPL; Algol 60 Jensen's device vs IBM Fortran copy-in/copy-out.)
10.Binding time differences (e.g., Pascal compile-time binding vs APL execute operator run-time interpretation.)
11.Asynchronism, if supported in either language.

van der Laan: We did not have the intention to solve the mixed language programming problem; our attainable goal was to make better use of FORTRAN program libraries in ALGOL 68 as an example how things could be combined.

T. Lake (ICL): It doesn't have to be the way Ris described if you design for different aims.   The problems are multiplied by big teams. How do you handle the row/column storage incompatibility?

van der Laan: We convinced CDC to provide a compiler option to cause Algol to store arrays columnwise.

S. Feldman (Bell Labs.): It is possible to design a system in which several languages can communicate with one another. Under UNIX, for example, we routinely mix programs written in C, Fortran 77, and Pascal, subject to certain coding rules imposed by Fortran's inflexibility.

B. Einarsson (National Defence Research Inst., Sweden):   But you wrote all the compilers yourself!

Feldman: No! C compilers were written by several people at Bell Labs.; I wrote Fortran 77 several years later at Bell Labs.; the Pascal compiler was written 3000 miles away.

J. Adams (NCAR): One large language vs a multiplicity of small ones is a serious problem.   This is not being solved, or even addressed, by X3J3.   Who is, or should be, concerned about such matters?

Ris: It has to be done by the manufacturers.

K. Roberts (Culham Lab.):   It is understandable that one reason why inter-language communication has become increasingly difficult is that a manufacturer typically has separate large teams working on each language, and these teams do not easily communicate with one another.   However, if the computer science fraternity, at meetings such as this, were to emphasize the importance of modular inter-language areas, then manufacturers might be encouraged to organize their software effort in a different way, say with one team working on I/O, another on array processing, etc. Apart from being more convenient for the user, this would also be cheaper, since it would avoid wasteful duplication. [Applause].

G. Paul (IBM): All is not lost!   There is work ongoing at the Los Angeles Scientific Center which provides for a common operating environment for the high level languages supported under S/370.

C. Lawson (JPL): Here is a good news/bad news story. UNIVAC is providing a way to mix Fortran and other language programs.   The bad news is that we can't make calls both ways between new and old Fortran routines.

THE RELATIONSHIP BETWEEN NUMERICAL COMPUTATION
AND PROGRAMMING LANGUAGES, J.K. Reid (editor)
North-Holland Publishing Company
©*IFIP, 1982*

THE FORTRAN OF THE FUTURE

Edited by F. N. Fritsch

Mathematics and Statistics Division
Lawrence Livermore National Laboratory
Livermore, California
U.S.A.

The last half of the open session was a general session on the future of Fortran as a programming language for numerical computation. We provide here edited summaries of the four formal presentations followed by an edited discussion.

Exception Handling in Fortran
Jan A. M. Snoek (University of Techn. Delft, Netherlands)

[Ed. Note: This talk was a preview of a proposal entitled "Bone Handling", to be presented by Rony Sobiesiak and Jan Snoek (EWICS/TC1) to the ANSI Fortran Standards Committee X3J3 at their meeting the week following this working conference.]

There are three basic ideas:

ERROR   EXCEPTION   EVENT

which differ only in how the programmer looks at it. Errors and exceptions are just special classes of events. EWICS/TC1 has adopted the word "BONE" as a neutral term to include all of these, but we revert to the word "event" for this presentation.

An event is a transition of the system from one state to another. This definition even includes such things as

A=B

since the execution of this statement definitely changes the system state by changing a data object.

We distinguish among:
- the event itself
- the definition of the event
- the data-item related to the event
- a handler
- a connection between the data-item and the handler

A language needs to provide the following facilities related to event handling:
1. Disconnect data-item and handler.
2. Raise the event.
3. Set the data-item.
4. Test the data-item.
5. Exchange information between the handler and the program unit that caused the event.
6. Resume.

There are (at least) five different ways to resume:
- (1)   Resume execution at the point where the program was 'interrupted' (retry).
- (2)   Transfer control to a specific label in the program.
- (3)   Terminate program (STOP).
- (4)   Terminate execution of this level of program (RETURN to the next higher subprogram).
- (5)   Return to next higher subprogram and indicate an event there.

In a real-time environment, it is necessary to add the following features to the previous list:
7. Wait for an event.
8. Raise an event in another program unit (may be another task).
9. Declare time-conditions dynamically as events.
10. Execute a piece of program undisturbed (no events will have any effect).

Conformity - Towards A Less Permissive Standard For Fortran
David T. Muxworthy (University of Edinburgh, U.K.)

> [This note describes an on-going project which has the
> objective of making the successor to the Fortran 8X
> standard less permissive, so that programs written
> using the standard should be more reliable and more
> portable.]

1. INTRODUCTION

The national standards bodies of the industrialized countries were founded before the days of computers, but their principles may be applied directly to them. Amongst the objectives of the British Standards Institution, formed in 1901, are "to eliminate the national waste of time and material involved in the production of an unnecessary variety of patterns and sizes of articles for one and the same purpose". In more recent years standards have come to be associated not only with uniformity but with quality, safety and reliablility.

A standard exists for the Fortran language, but it is possible to write a trivial, apparently standard-conforming Fortran program which is processed quite differently on each of four standard-conforming processors. The purchaser of a standard-conforming life jacket would not expect it to behave differently according to the type of water into which it was thrown.

The situation arises partly because of the permissiveness of the Fortran standard. It is understandable that permissiveness was chosen in 1962-64, when standardization followed widespread implementation. In the revision of 1970-76, no change was made in the attitude to permissiveness, although the ANSI Fortran committee was in the process of changing its role from one of post-hoc standardization to one of development. In the current revision the committee is wholly concerned with development, and there is scope for a major change in this area.

2. EXAMPLE OF DIFFERING BEHAVIOUR

Dahlstrand (1979) gives the following example of a program which behaves differently on four different processors.

```
INTEGER   K,L
K = 5000000
L = K*K
```

```
PRINT  L
END
```

This ran as follows:

System A: compile-time error because constant was too large.
System B: run-time error because of integer overflow.
System C: incorrect result without warning because of integer overflow.
System D: correct result.

Dahlstrand comments that this behaviour would not be tolerated in any other field of standardization.

3. THE CONFORMITY PROPOSAL

The Fortran working group of the British Standards Institution Programming Languages Committee is working on a proposal intended to eliminate some of these problems in Fortran 8X. On the understanding that the revised language will consist of a core language plus optional modules, it is proposed that there should be a set of facilities, tentatively known as 'conformity', which, if implemented and if called into use, would so far as possible ensure that a program was standard-conforming at compile-time and at execution-time. It is recognized that it is not practicable to make all possible checks at run-time.

Most Fortran 77 systems have taken heed of the suggestion that they should give some indication of non-conformance to the standard. In an attempt to determine what sort of tests and checks would be necessary had such an option been available for Fortran 77, the diagnostic messages from some fifteen Fortran 77 systems were collated into one list and it was estimated that about 650 of these represented violations of the standard which could reasonably be checked. Of these, about 420 could be detected at compile-time and the remainder only at run-time.

The 'reasonable' run-time errors could be grouped as follows:

> reference to undefined variables
> mathematical domain errors
> array bounds (and substring bounds) exceeded
> errors within formats (accounts for 49 items)
> errors related to input-output (accounts for 91 items)
> invalid bounds in any declarations and substring references
> use of label variables as integers and vice versa
> mismatching of argument type, number, dimension
> infinite DO iteration counts
> truncation of constants and expressions (possible warnings)
> miscellanea, e.g. misuse of variable scope after entry at ENTRY;
>   label not in assigned GO TO list.

It is possible that some of these topics may be able to be treated by the exception handling provided in Fortran 8X. What the BSI group would like to see is that, if the conformity option is in force, the user would at least be informed if any of the above standard violations occurs. It is not thought profitable to attempt to define the error messages in detail. Of more importance is whether, and how, the program should terminate after a particular error condition has occurred, but it would be too Draconian to require that this precept be applied everywhere. Nevertheless the BSI group inclines to the view, not universally accepted, that it is better to get results correct than to get them fast.

A great deal of work needs to be done. At the higher level it is not clear
whether 'conformity', if approved, should be a part of the Fortran standard,
whether it could be in an appendix or whether it should be in a separate publication. At the more detailed level a list of checks remains to be drawn up, once
the language, and particularly the event-handling, begins to take more concrete
shape. Comments, addressed to the author or to X3J3, should be most welcome.

REFERENCE

Dahlstrand, I., Language Standards and Portability, Lecture Notes for "Programming for Software Sharing", European Commission Joint Research Centre, Ispra,
Italy (September 1979).

An Alternative Approach to the Evolution of Fortran
W. S. Brown, S. I. Feldman, N. L. Schryer (Bell Laboratories, Murray Hill, N.J.)
L. D. Fosdick (University of Colorado, Boulder, Colorado)

SUMMARY

For the past three years, the Fortran standards committee (X3J3) of the American
National Standards Institute (ANSI) has been working on the design of a new language (Fortran 8x) to replace the existing standard Fortran language (Fortran
77). The goal is to extend Fortran into a powerful modern language capable of
meeting the needs of the scientific computing community in the 1990s, while at
the same time preserving practical compatibility with Fortran 66 and Fortran 77.

Although the committee's approach is ingenious, we predict that their proposal
will draw heavy fire from "conservatives," who will see it as a threat to the
stability of Fortran, and from "liberals," who will find the new language too
heavily encumbered with ugly relics of the old one. Moreover, we argue that a
standards committee is not the appropriate body to design a new language, and
that the new standard envisioned by X3J3 may fall short of the goal of adoption
by ANSI, endorsement by the U. S. government, and implementation by the major
computer manufacturers.

As an alternative, we propose that Fortran 8x should be compatible with Fortran
77, and that new features should be chosen primarily to facilitate extensions via
libraries and preprocessors. That is, we endorse Gear's suggestion (Mathematical
Software II, Purdue, 1974) that Fortran should be viewed not as a high level scientific programming language, but as a universal assembly language for the scientific computing community.

THESES AND SELECTION CRITERIA

Our point of view is summarized in the following theses:
1. A standards committee should not be a language designer.
2. Most new capabilities should be provided via libraries or preprocessors. (We
    view Fortran as a universal assembly language.)
3. Only a few fundamental capabilities need to be added to Fortran. (But they
    are crucial!)
4. X3J3 is the right body to select and specify these capabilities. (It should
    concentrate on this task.)
5. The main strengths of Fortran are its:
        (a) universality (including support),
        (b) stability,
        (c) innocence (in the sense of naivete).

Accordingly, we propose the following criteria for adding a new capability. (The answers to these questions should be Yes, Yes, No, No, and No.)
1. Is it necessary?
2. Is it ripe? (fully baked? ready to be standardized?)
3. Can it be added efficiently as a procedure?
4. Can it be added efficiently via a preprocessor?
5. If neither, does it suggest the need for a more fundamental capability?

CAPABILIIES OF PREPROCESSORS

Preprocessors are not restricted to partial languages like Ratfor [1], Sftran [2], Mortran [3], Iftran [4] and Augment [5]. They can also be used to implement complete languages like Efl [6], Altran [7] and Genstat [8]. In considering language extensions from the viewpoint of the preprocessor designer, we see a spectrum of capabilities:

1. Easy
    a. Syntactic sugar (e.g., .LE. → $<$ =).
    b. Control structures (while, case, ...).
    c. New types over Fortran types (double complex, multiple precision, interval, array, matrix, ...).
    d. Data structures.

2. Intermediate
    a. Dynamic storage allocation (static and automatic, heap).
    b. Dope vectors.

3. Help Needed:
    a. Optional arguments
    b. Exception handling
    c. Procedure-valued variables.
    d. Recursion
    e. Pointers

Help is needed for pointers if the goal is a complete list-processing facility. For many applications, an easy implementation of pointers as indexes in a workspace array is adequate. In this case, however, everything, including both automatic and static (this is, dynamic and own) storage, must be kept in the workspace. Unfortunately, this makes the initialization of global static variables exceedingly difficult.

4. Desirable Aids for Portability
    a. Environment parameters (floating point, character, input-output, tailoring).
    b. Basic functions (floating point, pack and unpack, character classes).

Environment parameters are needed because they are inherently machine dependent and cannot be provided portably in a library or preprocessor. Once the environment parameters are available, most of the basic functions can be implemented portably, but the portable versions are too inefficient for most practical purposes.

CONCLUSIONS

We summarize our approach as follows. Don't introduce a panoply of features, each unneeded (for some) and too late (for others). Instead, give us a few fundamental capabilities, so we can build new capabilities or new languages over Fortran.

1. Our approach is conservative. (No threat to existing programs, programmers, libraries, or preprocessors. No threat to Fortran's universality, stability, or innocence.)

2. Our approach is liberal. (It permits rapid evolution at the user level.)

3. The X3J3 approach is neither. It may fail:
   a. Bureaucratically (too late to arrive, competition from Ada);
   b. Technically (too ambitious, omits necessary functionality, threatening to existing programs and programmers).

ACKNOWLEDGMENTS

X3J3 has listened well. X3J3 is responding competently and vigorously (too vigorously, in our opinion). X3J3 has done everything possible to keep us (the numerical computing community) informed. We especially thank Brian Smith for his efforts on our behalf.

REFERENCES

1. Kernighan, B.W., RATFOR--A Preprocessor for a rational Fortran, Software Practice and Experience 5 (October 1975) 395-406.

2. Lawson, C.L., SFTRAN-3 Programmer's reference manual, Internal Document No. 1846-98, Rev. A, Jet Propulsion Laboratory, Pasadena, CA (April 1981).

3. Cook, A.J. and Shustek, L.J., A user's guide to MORTRAN-2, Computation Group Tech. Memo. No 165, Stanford Linear Accelerator Center, Menlo Pak, CA (February 1975;Rev. July 1975).

4. Lawson, C.L. (ed.), Proceedings of a workshop on Fortran preprocessors for Numerical software, Jet Propulsion Laboratory, Pasadena, CA (1974).

5. Crary, F.D., A versatile precompiler for nonstandard arithmetics, ACM Trans. Math. Software 5 (June 1979) 204-217.

6. Feldman, S.I., The programming language EFL, Computing Science Technical Report No. 78, Bell Laboratories, Murray Hill, NJ (June 1979).

7. Brown, W.S., Altran user's manual, 4th edition, (Bell Laboratories, Murray Hill, NJ, 1977).

8. Wilkinson, G.N. and Nelder, J.A., GENSTAT-4 user's manual, CSIRO, Adelaide, South Australia (1968). [GENSTAT is currently available from Rothemstad Experimental Station, Harpenden Harts, England.]

Notes On Fortran Evolution
E. L. Battiste (C. Abaci, Raleigh, N.C., U.S.A.)

Any innocent bystander, listening to the discussion on Data Structures, might have thought that we are attacking Fortran. The main points in that discussion were

   (1) A full, rich language was possible to envisage,
   (2) A language should be built by language experts, and
   (3) A new set of Fortran constructs was being considered for addition to the core of Fortran.

This audience should know that the important asymptotes constraining our near-term language movements are (in strength order) cost, size of the current user set, and feasibility of approach. Changes in any of these constraints changes the whole set.

This audience should know that the major difficulty we face in living up to our responsibilities to the future in computing is the collapse of timeframes in which we can attempt solutions. The collapses we have seen are small compared to those we shall see.

On the discussion points, my notes are as follows:

1. To whom does the Fortran group owe responsibility? First, to the large current user set, which is the reason for the continued existence of Fortran, and the reason for the benefits we derive from Fortran. These benefits are called "throughput". Who are these users? Not us. They are the engineering groups in a million shops, and they demand current throughput.

Second, responsibility is owed to us because we can aid in honing Fortran into a passable scientific language, and because our aid is needed to protect users. The current tendency of users to incorporate black boxes into their problem solving tools will only increase, and our best aid is in the area of protection.

2. Can we continue to allow Fortran to stand in the way of good language development, and still meet our obligations to the future? We have absolutely no control in this case. We can only shape Fortran while attempting circumvention.

3. What is possible?

    A. Those of us who work in numerical software can make clear statements of need, presented on time, to language standards groups.

    B. We might attempt to obtain agreement from the Fortran standards group that Fortran should be a scientific language with no responsibility to special groups. The matrix enhancements described do not belong in the core of Fortran. If Fortran is allowed to grow to serve all people, I feel that none will be served.

    C. The language standards group could agree that numerical mathematicians (numerical software groups) are not special interest groups. That is the case.

    D. The language standards group should agree that the collapse of timeframes must be understood and considered in their evolutions. This can only lead to delimitation of core growth.

We fear that our few requests will be delayed so long that ability to produce adequate robust code for users will be delayed too long. We can and should keep the core of Fortran small. That aim will impact micro-computer users, and these will soon be the largest set of users.

Specific to points (1) and (2) above: Full-rich seems to imply unused, and that poor fate will affect new language developments for some time. We might consider that we have time to develop audience oriented languages which are full and rich in their context, and that an attendant responsibility is to establish standards for interfaces between languages. In this mode of operation, where time collapse is not a large factor, we can allow language experts to perform well.

DISCUSSION

W. Kahan (Univ. of Calif.) [to Snoek]: I think we have come to opposite conclusions. IEEE events are exceptions--departures from what a reasonable user would expect. They are quite different from events in a real-time situation. Your event requires an asynchronous action by the system. Ours requires only a good arithmetic engine.

Snoek: You may feel that way, but we are trying to provide a general language facility.

E. Schlechtendahl (Karlsruhe): How do you deal with events happening in the event handler? For example, a divide by zero in the handler for the divide by zero event.

Snoek: Event handlers can be nested.

T. Hull (Univ. of Toronto): You two seem to be dealing with different kinds of "bones". Kahan is working with bones at the hardware level, and has shown how to handle them, whereas Snoek is working at the language level, and is proposing a very flexible way of handling the bones that arise at this level. I feel that it is important to work together to see how these two approaches can be made to complement one another.

Kahan: Isn't this another example of "spurious generality"? Floating point exceptions are much less general and do not require the elaborate event handlers needed for real time systems.

Snoek: I am not talking only about real time systems.

M. Hennell (Univ. of Liverpool) [to Muxworthy]: It appears there are a few systematic errors here. What about errors the compiler writers have not found?

Muxworthy: With 15 systems, that's unlikely.

S. Feldman (Bell Labs.): How do you get around the problem: Integer I; Real R; Equivalence (I,R). If we set I=3, R becomes undefined. How can you check for uses of R? Do any compilers warn about this?

Muxworthy: That is too expensive to check.

Kahan: I think your first two types of errors are different from the rest. A "mathematical domain error" should not stop the machine. (I may be trying to test the domain of my function.) A "reference to an undefined variable" should not stop it, either. Statistical analysts frequently want to use undefined variables merely to determine whether data has been provided.

Snoek: We want the item checked, but will not necessarily stop. We may have an event handler to deal with it.

Muxworthy: We have a compiler similar to WATFIV which stops execution on such exceptions. At first it is a pain, but you soon find this makes you design your program more carefully.

J. Rice (Purdue Univ.): I feel you have overlooked the fact that lots of programs are not written by people. "IF (3.NE.2) ..." shouldn't stop execution when generated by a preprocessor. Similar constructs generated by preprocessors are arrays declared but not used, successive assignments of different values to a variable. While I fully support the use of the checking you describe for program development, it should be a compiler option, so automatically generated code does not lead to lots of spurious warnings or, worse yet, spurious program stops.

L. Meissner (LBL): That was only to give a warning.

F. Ris (IBM): When that has come from a macro which has been expanded hundreds of times, you tend to lose significant diagnostics in the forest of warnings.

B. Einarsson (National Defense Research Inst.): Isn't there a FIPS (Federal Information Processing Standard) for Fortran 77 that requires validation to be included with every compiler?

Meissner: I understand that the validator is required to detect any compile-time errors.

Einarsson: Is it now in effect?

Meissner: I believe the FIPS was written by the National Bureau of Standards, and has been published, to become effective some time in 1982.

Hennell [to Brown]: What do you propose to do about Muxworthy's conformity? Strip it out and make everybody use a preprocessor?

Feldman: Yes, the preprocessor generates perfect Fortran. Only if the compiler has bugs should you get incorrect code.

M. Delves (Univ. of Liverpool): It is not true, as you claimed, that bugs in Fortran programs have only localized effect. Consider a statement changing the value of a common variable, or of an equivalenced variable; or possible changes to a procedure parameter called with a constant value.

J. Adams (NCAR): Fortran 8X is a standard; it isn't a general language. We hope to write the first draft by the end of next summer. The remaining steps may take three years. Funding has forced us to cut down to four meetings per year, which will slow down the process by 1 to 1-1/2 years. We have a very dedicated committee doing a difficult job; there are both technical and human aspects of the work. We have tried to be very open to input. We also have certain constraints from X3 in the program of work that X3J3 is charged to do.

G. Paul (IBM): I am very sympathetic with Brown's view. Most of Vectran is preprocessable to Fortran. However, the remaining features, such as identify, require vector extensions.

Kahan: There are problems with comparisons that I will not dwell on. I observe that at least half the items on Brown's "wish list" (e.g., environmental inquiry) would fall off that list if his own criteria were applied. Even so, I think his criteria make sense.

Delves: I would sum up Brown's talk as a plea to X3J3 to put only primary new features, such as pointers, dynamic storage allocation, and (I think) structures, into Fortran, leaving the provision of extra "syntactic sugar" and features which can either be provided easily as procedures, or can be defined wholly in terms of existing "basic" new features, to the writers of preprocessors. To this I would make two comments. (1) I think the implied sense of priorities is very appropriate. (2) However, reliance on preprocessors loses portability. It is quite possible to wish to use two facilities (such as subroutines) simultaneously which have been written for different and incompatible preprocessors, and it may not be possible to get over this by preprocessing the two separately, since the features

used may be needed by the calling routine. Also, who will standardize preprocessors?

Brown: The time has not yet come to standardize preprocessors. We don't have
enough experience. When fully baked, we'll give that job to another committee
(not X3J3).

L. Fosdick (Univ. of Colorado): It is not nearly so important to standardize
preprocessors.

J. Boyle (ANL): The marketplace will standardize them.

Kahan: Doesn't a preprocessor convert Fortran to a universal portable "assembly
language"?

Fosdick: Yes, and we should encourage the scientist to build his own preprocessor for his own application. It may be incompatible with his colleague's, but
who cares? They can still share at the Fortran level.

Delves: It is important to standardize preprocessors. Suppose a well-written
NAG routine uses a structure as one of its arguments?

Boyle: Give him the standard NAG preprocessor.

Delves: But this means everybody has to drag around the preprocessor.

Boyle: It would come with the NAG Library.

Delves: What if you have two libraries?

A. Wilson (ICL): The new array processing facilities have been considered by and
gotten a positive reaction from numerous scientific users of Fortran who are routinely solving large problems. [He displayed a sample of remarks from a recent
Gordon Conference on numerical solution of PDE's.]

Battiste: But why do the array facilities need to be in the core?

B. Ford (NAG): They are needed by big programs running on the CRAY's, etc.

C. Lawson (JPL): I look at preprocessors and carefully structured libraries as
one approach to providing extensible languages. There are advantages and disadvantages. It is relatively easy to provide a new extension and there is less
need for standardization. There are problems with incompatible prologues in
Algol 68.

Meissner: For the moment you've got to use what you have. In a few years we
will be able to make a separation between standardizing core Fortran on the one
hand and standardizing extension facilities on the other. Unfortunately, with
the current state of the art, X3J3 must put the extension facilities into the
"Fortran Language". We have to get the core right before it goes to press. We
won't be able to come back two years later with improvements.

Boyle: Maybe we could effect a trade: we do some of this and you put in
pointers.

J. Wagener (Amoco Production Research): I don't see a great deal of difference
between Brown's proposal and what X3J3 is trying to accomplish. We have always
envisioned that many application modules would be preprocessors, although this
has not been mentioned in any of our documents. If there is one thing for us to

take back to X3J3 from this conference, it is that preprocessors are important. Another dramatic oversight, evidently, is the inclusion of pointers. The other items on Brown's list have by-in-large already been adopted.

Boyle: I agree with Wagener on pointers.

Meissner: I believe we could standardize on a pointer facility, but it would either be too simple (and unsafe) or too complex to satisfy you.

Anonymous: Is a simple LOC function (to provide the machine address of a variable) useful for anything?

Brown: If pointers and recursion are not available in the base language, the person who decides to implement them via a preprocessor will need a few machine-dependent primitive functions. In this context, a LOC function may be very useful.

Rice: I wish to express general support for Brown's and Battiste's comments. In a universal assembly language, however, there has to be some way to get at important hardware features. This implies the need for some vector capability in the core. One point bothers me greatly: next summer seems too soon to resolve all your differences. I fear I may retire before Fortran 8X has any impact. If there is any misstep by X3J3, we may be stuck with Fortran 77 forever.

P. Fox (Bell Labs.): I'm in a dirty back shop trying to get good Cray code. The preprocessor doesn't give users code that will vectorize. We need something better than we have now for parallel and vector machines.

Wilson: In suggesting that arrays and array operations can easily be added to Fortran via a preprocessor, have you considered the question of efficiency on computers that provide vector or array operations in hardware?

Brown: If a preprocessor expands array operatrions into Fortran DO loops, then I agree that it would be unreasonable to expect the Fortran compiler to reverse the process and reconstruct the intended array operations. As an alternative, I suggest that the preprocessor produce array-level function calls. On scalar machines, an implementation of the required functions as portable library procedures would be quite satisfactory, but on vector and array processors they would be re-implemented as assembly language primitives. Since the overhead of a function call may be unacceptable in this context, and even the existence of a function call may prevent essential optimizations, we need the ability to define primitives that can be expanded into in-line assembly code; that is, we need user-defined intrinsics in Fortran.

G. Paul: It would be worthwhile to explore that idea in detail, but I fear it would still be too difficult for the Fortran optimizer to produce satisfactory object code.

Brown: You may be right, and in any case I am open-minded about adding array processing to the Fortran language. Our point is that it isn't feasible to solve every problem by adding features to Fortran, and therefore we think X3J3 should concentrate on a small number of fundamental capabilities. If a compelling case can be made for array processing, then so be it.

Kahan: A plea to X3J3: I request that any "optimization" that could change the result of a calculation from what would have followed from a slavish adherence to the guidance of parentheses, including changes to side-effects, should be reflected in a comparable change to the listing of the source-code, so that the "optimization" can be debugged at the source level.

Meissner: Even if I decide to compute A/B as (1/B)*A?

Kahan: Yes, you must tell me! Otherwise, your rearrangement could cause an overflow that no analysis of the source-code will explain.

Feldman: I wish the core could be a small compact language, but I don't think that will be possible in the presence of extension modules and the obsolete features module. Further, some features of the present proposals, such as dependent compilation, will require extensive operating system changes. I wish to dissent slightly from the view of Fortran as a universal intermediate language. If we're going to have an intermediate language, let's get a <u>good</u> one!

T. Lake (ICL): I must point out that an input preprocessor is not a complete language system. We need satisfactory diagnostic and post-mortem facility, and a well-defined exception handling environment. Hooks for run-time errors are difficult to provide. The "four musketeers" must extend their proposal if they are serious about preprocessors. I believe user-satisfactory preprocessor technology, even at the input level, to be a relatively recent achievement.

Lawson: We have been routinely using preprocessors for many years.

G. Paul: We [X3J3] appreciate your comments about array/vector facilities. Other areas of the language are also being changed, and we want comments on those, too.

Brown: John Rice remarked that many of us may be stuck for the rest of our careers with Fortran 77. This viewpoint is in fact one of the primary motivations for working with (and on) preprocessors. If we really want a better language (or just a new feature), we can't afford to wait for X3J3 (and then the computer manufacturers) to produce it, so we have to do the best we can for ourselves.

Adams: Let me conclude with a quick review of X3J3's milestones.

Next week: – The core proposal will be presented.

                – The S6 document, containing all Fortran 8X proposals approved to date, has been prepared and is being refined by X3J3.

Next summer: A "talking document" will be produced as the basis for committee discussion.

SESSION 8
SOURCE TO SOURCE TRANSFORMATIONS AND LANGUAGE COMPLEXITY

Chair
J.M. Boyle

Discussant
Th.J. Dekker

THE RELATIONSHIP BETWEEN NUMERICAL COMPUTATION
AND PROGRAMMING LANGUAGES, J.K. Reid (editor)
North-Holland Publishing Company
©IFIP, 1982

SPECIFICATION LANGUAGES and PROGRAM TRANSFORMATION

P. Pepper

Department of Computer Science, Stanford University, Stanford, Ca 94305
and
Institut für Informatik, Technische Universität München, Postfach 202420, D-8000 München 2

Specification languages provide tools for describing *what* a program is to do (as opposed to *how* it shall do it). These tools are usually taken from mathematics — predicate calculus, set theory, graph theory, functional equations, algebraic structures etc. — and they are put into notations that are compatible with those of classical programming languages. In order to encourage programmers to actually use these tools the initial specifications must become an integrated part of the overall development process. This can be achieved by using the method of program transformation. Here one starts from an initial specification and proceeds through a series of intermediate versions until a final, efficient program is achieved. The transitions between subsequent versions are done according to formal rules, called transformation rules. This formalization of the transitions makes it even possible to support the development process by an automated system.

1. Introduction

Classical programming languages are a compromise between two objectives: On the one side, they should allow one to formulate programs in a problem-oriented way. The catchwords for this goal are "expressive power", "high level" languages and — to top it — "very high level" languages. On the other side, an automatic code generation is still required and an efficient code generation at least desired. For this reason the majority of today's programming languages belongs to the so-called *procedural* (or *imperative*) *style*: Their main features are variables, assignments, loops, gotos and, on the side of data structures, arrays and pointers. By contrast to this, the so-called *applicative* (or *functional*) *style* is more oriented towards the traditional tools of mathematics, namely functions, expressions, predicate calculus and set theory. On the side of data structures this list has in the last years been extended by "recursive type declarations" and "abstract data types" (see section 5). The most prominent examples of applicative languages are LISP and — more recently — the language described in [2].

Applicative programming is claimed to be safer than the traditional procedural style mainly for two reasons (cf. [2]): Firstly, any program part can be understood on its own account, for an expression is completely characterized by its textual representation. By contrast, whenever a variable occurs in a piece of program text one generally needs some knowledge about its current value. In other words, the meaning of a part of a procedural program usually depends on the history of computations preceding the execution of this part. Secondly, since the applicative style provides so many tools that have proved good in mathematics, more tasks can be formulated in a more problem-oriented way than would be possible in the procedural style where these tools are not available.

What kept applicative languages from being more widely accepted is their inefficiency: They do not directly match the architecture of today's computers. An attempt to combine the increased reliability of applicative programming with the efficiency of the procedural style is

program transformation. Here one starts from an initial applicative program and proceeds through a series of intermediate versions until a final, efficient program is achieved. The transitions between subsequent versions are done according to formal rules, called "transformation rules". This formalization of the transitions makes it even possible to support the development process by an automated system.

It is a characterisitic of the transformational approach that the first versions of a program — though already obeying the syntactic requirements of a programming language — are usually not intended for an immediate code generation by a compiler. As a consequence, efficiency considerations do not yet play a major role in these first versions. It is even possible to employ "non-executable" (also called "non-operational") constructs in the programs, i.e. constructs for which an automatic code generation is in general not possible. Examples are the aforementioned concepts of predicate calculus and set theory as well as abstract data types. Since these constructs allow one to specify *what* a program is to do (rather than *how* it is done), they are often subsumed under the name **specification constructs**. It is the task of a transformational program development to replace these specification constructs by equivalent operational and efficient formulations.

A transition between two successive stages in such a development process usually does not change the whole program but only a (very small) part of it. As a consequence, programs will in general contain specification constructs, recursive functions, loops and assignments and maybe even gotos side by side. Therefore, languages supporting this methodology have to provide a common syntactic frame for all those levels of formulation.

At the Technical University Munich such a "wide spectrum language" — named CIP-L — has been developed in the course of the projet CIP ("Computer-aided, Intuition-guided Programming"). This project — carried out under the joint guidance of Prof. F. L. Bauer and the late Prof. K. Samelson — deals with the design of a system for program development by formal transformation. Actually, the language CIP-L comes in several "dialects", i.e. external notations that are oriented towards those of classical programming languages such as Algol, Pascal or Lisp. Throughout this paper we will use the Pascal dialect.

To illustrate the transformational approach we will use as a running example the concept of "enumeration". This paradigm is frequently employed in programs containing loops and arrays. So it seems interesting to formalize the standard considerations taking place here in terms of transformation rules. Enumeration is based on (linearly ordered) sets for which an explicit "successor"-function exists. To shorten the writing-down, we will use throughout this paper a particular representant for such sets, viz. the finite interval **type** $index = [j .. k]$ with integers j and k.

2. A Short Example

To give a first impression of the language we consider a very simple example, viz. the computation of a lower bound of a given set of elements $y_0, \ldots, y_n$ — under the assumption that there is a "primitive" operation min such that $min(a, b) \leq a \wedge min(a, b) \leq b$. This task is formally specified by the function (where m denotes the type of the elements y_i, nat stands for the nonnegative integers and the symbol •, which is used to avoid the extensive overloading of the symbol ":", may be read as "such that" or "holds")

(1)
$$\textbf{function } lob\,(n{:}nat){:}m;$$
$$lob := \textbf{some } x{:}m \bullet \forall\, i{:}[0 .. n] \bullet \ x \leq y_i$$

The meaning of the universal quantifier is the usual one known from mathematics with the addition that the clause $(\forall\, x{:}m \bullet p(x))$ is undefined if $p(x_0)$ is undefined for at least one x_0. The **some**-construct introduces nondeterminism. Its definedness can be explained in terms of the existential quantifier: (some $x{:}m \bullet p(x)$) is defined iff ($\exists\, x{:}m \bullet p(x)$) yields true. In this case (some $x{:}m \bullet p(x)$) may yield any value x_0 for which $p(x_0)$ is true.

Note that (1) is nothing but a notational formalization of the classical verbal definition "a lower bound of a set is an object that is not greater than any element of the set". Although it is not possible to directly generate executable code for this specification we may use it as the starting point of a program development process. In the sequel we will present such a process by showing a number of intermediate program versions. The transformation rules according to which the various transitions take place will be given in the following sections.

After introducing the case distinction $n = 0$, $n \neq 0$ (rule T1) we can derive the version (by applying rule T3)

(2)
$$\begin{aligned}
&\textbf{function } lob\,(n{:}nat){:}m; \\
&\quad \textbf{if } n = 0 \textbf{ then } lob := \textbf{some } x{:}m \bullet (\forall\, i{:}[0\,..\,0] \bullet x \leq y_i) \\
&\quad\quad\quad \textbf{else } \quad lob := \big(\textbf{some } x{:}m \bullet (\forall\, i{:}[0\,..\,n-1] \bullet x \leq y_i) \wedge (x \leq y_n)\big) \textbf{ endif}
\end{aligned}$$

With the help of the primitive operation *min* mentioned above this becomes (rule T4)

(3)
$$\begin{aligned}
&\textbf{function } lob\,(n{:}nat){:}m; \\
&\quad \textbf{if } n = 0 \textbf{ then } lob := y_0 \\
&\quad\quad\quad \textbf{else } \quad lob := min\big((\textbf{some } x{:}m \bullet \forall\, i{:}[0\,..\,n-1] \bullet x \leq y_i),\, y_n\big) \quad \textbf{endif}
\end{aligned}$$

The remaining some-expression is an instance of the original definition of *lob*. Hence, it can be replaced by a — now recursive — call of *lob* (rule *folding*):

(4)
$$\begin{aligned}
&\textbf{function } lob\,(n{:}nat){:}m; \\
&\quad \textbf{if } n = 0 \textbf{ then } lob := y_0 \\
&\quad\quad\quad \textbf{else } \quad lob := min\big(lob(n-1),\, y_n\big) \quad \textbf{endif}
\end{aligned}$$

The development so far has led from a nonoperational specification to a recursive function (which is — though inefficiently — executable). The next step is the transition to a procedural program. We give here two slightly different variants resulting from applying different transformation rules (T5 and T6, respectively).

(5)
```
function lob (n:nat):m;              function lob (n:nat):m;
    begin                                begin
    var i:nat := n;                      var i:nat := 0;
    var z:m := yₙ;                       var z:m := y₀;
    while i ≠ 0 do z := min(yᵢ, z);      while i ≠ n do i := i + 1;
                  i := i - 1                           z := min(z, yᵢ)
    endloop;                             endloop;
    lob := min(y₀, z)                    lob := z
    end                                  end
```

The subtle differences between these two programs — in particular the proper treatment of the termination cases — show the particular benefit of program transformation: It takes care of details and allows the programmer to concentrate on the essentials of his development.

3. Transformation Rules

(This section is a compendium of the more detailed exposition given in [8].) A **program transformation** is the derivation of a new program from a given one. Therefore, a **transformation rule** is a mapping from programs to programs. Actually, it is a partial mapping, since it is in general only *applicable* to a certain class of programs.

There are essentially two ways of representing transformation rules. Either they are given in the form of **algorithms** that take programs as input and produce programs as output. Or they are given as ordered pairs of **program schemes** together with an **applicability condition**. We will concentrate here on the second variant, which has turned out to provide greater flexibility.

Our transformation rules are denoted in the form

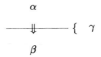

where α and β are called input- and output-scheme, respectively, and γ is the applicability condition (To ease the readability we will suppress throughout this paper those applicability conditions that have a purely technical purpose such as avoiding name clashes between newly introduced and existing identifiers etc.). The domain of such a rule, i.e. the set of programs to which it is applicable, is determined by the form of the input scheme ("success of the matching process") and by the applicability condition.

Examples: The following selection of elementary rules is guided by the *lob*-example of section 2. The first rule allows one to introduce case distinctions into a program. The "free" occurrence of p in the output-scheme means that a suitable predicate p has to be provided by the user upon application of the rule.

$(T1)$
$$e$$
$$\Downarrow$$
$$\textbf{if } p \textbf{ then } e \textbf{ else } e \textbf{ endif}$$

Many basic transformation techniques are applicable to a number of language constructs. Consider for example "distributivity over conditionals"; this idea is realized by the rules

$(T2)$
$$f(\textbf{if } p \textbf{ then } e_1 \textbf{ else } e_2 \textbf{ endif})$$
$$\Downarrow$$
$$\textbf{if } p \textbf{ then } f(e_1) \textbf{ else } f(e_2) \textbf{ endif}$$

$$x := \textbf{if } p \textbf{ then } e_1 \textbf{ else } e_2 \textbf{ endif}$$
$$\Downarrow$$
$$\textbf{if } p \textbf{ then } x := e_1 \textbf{ else } x := e_2 \textbf{ endif}$$

Just as the validity of the above rules follows from the semantics of the if-construct, the semantics of the universal quantifier directly leads to the following rule:

$(T3)$
$$\forall i{:}[j \mathbin{..} k] \bullet p(i)$$
$$\Downarrow \qquad \{k \geq j$$
$$\left(\forall i{:}[j \mathbin{..} k - 1] \bullet p(i)\right) \wedge p(k)$$

The next rule is more complex in the sense that the choice of a suitable function h usually means a major design decision — as can be seen in the transition from version (2) to version (3) of the *lob*-example. The effect of this transformation is a simplification of the predicate in the **some**-expression. Such simplifications often are the prerequisite for deriving recursive programs from specifications.

$(T4)$

$$\frac{\text{some } x{:}m \bullet p(x) \wedge q(x)}{h(\text{some } x{:}m \bullet p(x))} \Big\Downarrow \quad \begin{cases} \forall\, x \bullet p(x) \Rightarrow p(h(x)) \\ \forall\, x \bullet p(x) \Rightarrow q(h(x)) \end{cases}$$

There are two very simple rules that have proved valuable in many applications. They have become known under the names *fold* and *unfold* (cf. [13]). *Unfolding* ("macro expansion") means to replace an applied occurrence of an identifier by the right-hand side of its declaration. For instance, the call of a function is substituted by its body and a variable is substituted by the (dynamically) preceding assignment. *Folding* is the converse operation, i.e. an instance of the right-hand side of a declaration is replaced by the corresponding identifier. For example, unfolding of the call (where $high(y)$ gives the highest index k of the set $y_0, \ldots, y_k$)

$lob(high(y))$

yields

if $high(y) = 0$ **then** y_0 **else** $min\big(lob(high(y) - 1), y_{high(y)}\big)$ **endif**.

We conclude this section with two remarks on the theoretical background of program transformation.

The claim that program transformation provides a rigorous, formal methodology for program development depends crucially on the existence of a precise notion of the "correctness" of transformations and transformation rules. Naturally, such a notion has to be based on suitable relations between programs. Given such a relation ρ the transition from a program P to a program P' is said to be **correct** if $P\,\rho\,P'$ holds. The study of the semantics of programming languages offers various suitable relations ρ. For example, P and P' may be required to be "functionally equivalent" or even "operationally equivalent". (The latter requirement is in general too restrictive for providing interesting transformations.) For nondeterministic programs it may even suffice that P' is "included" in P, i.e. that the possible results of the new program form a subset of the possible results of the original program (cf. [11]). Whichever relation ρ is chosen here, it has to be *reflexive* (the identity should be a valid transformation) and *transitive* (the composition of two transformations should be a valid transformation). Furthermore, ρ has to be *monotonic* with respect to all language constructs, for the transformation of a small part of a program should also be correct in the context of the overall program.

But there is an even closer relationship between transformation rules and semantics: When a programming language is designed for the purpose of assisting the transformational development of programs then the transformation rules should serve as guidelines for the definition of the semantics of the language. In other words, one requires the validity of certain rules and defines the semantics such that these requirements are met. (For details on this subject we refer to the literature, e.g. [10], [12] and [16].) This so-called **transformational semantics** leads to a natural modularization of the definition of a programming language and therefore allows one to cope with the inherent complexity of large languages. Its feasibility and usefulness is demonstrated by the wide spectrum language CIP-L, which is completely specified in this technique (cf. [6]). In addition, it has also been shown that transformational semantics can be used to explain concurrency in terms of nondeterministic sequential statements (cf [7]).

4. Compact Transformation Rules and Transformation Algorithms

From a practical point of view the number of rules to be applied during a development process must not be unduly long — which would be the case if we restricted ourselves to elementary rules such as T1 – T4. This may be achieved in two ways: Firstly, we may use "compact" transformation rules, which are adapted to particular, frequently occurring situations. Naturally, there is no upper bound for the set of such compact rules; collections may even vary from task to task.

Examples:
The following two rules achieve "recursion removal", i.e. they convert certain recursive functions into iterations. (A more detailed presentation of such rules can be found in [4] and [5].) The first rule depends on certain algebraic properties of the primitive operations occurring in the body of the function. (The underlying idea of this rule goes back to [14].)

$(T5)$ **function** $f\,(x{:}m){:}r$;
   if $p(x)$ **then** $f := a(x)$
        else $f := b(c(x), f(d(x)))$ **endif** $\begin{cases} \forall\, x, y, z \bullet b(x, b(y, z)) \\ \qquad\qquad = b(b(x, y), z) \\ \forall\, x \bullet b(x, x) = x \end{cases}$
   ――――――――――――――――――――⇓
   function $f\,(x_0{:}m){:}r$;
      begin var $x{:}m := x_0$; **var** $z{:}r := c(x_0)$;
         while $\neg p(x)$ **do** $z := b(z, c(x))$;
                 $x := d(x)$
         endloop;
         $f := b(z, a(x))$
      end

This rule uses the associativity of b to change the order in which intermediate results are computed. (For a proof cf. [14] or [5].) The property $b(x, x) = x$ is one of several ways for avoiding a special treatment of the case where $f(x_0)$ terminates immediately. Another solution is to put a conditional around the **while**-loop.

The second rule uses the idea of simple enumeration.

$(T6)$ **function** $f\,(i{:}nat \bullet j \le i \le k){:}r$; $f := e$
   ――――――――――――――――――⇓ $\begin{cases} \forall\, i : j \le i \le k - 1 \bullet \\ \text{DEFINED}[f(i + 1)] \Rightarrow \text{DEFINED}[f(i)] \end{cases}$
   function $f\,(i_0{:}nat \bullet j \le i_0 \le k){:}r$;
      begin var $i{:}nat := j$;
      var $z{:}r := e$;
      while $i < i_0$ **do** $i := i + 1$;
                 assert $z = f(i - 1)$;
                 $z := e$
      endloop;
      $f := z$
      end

This transformation is correct in any case. Whether it is useful or not depends on the form of the expression e: If e contains recursive calls of the form $f(i - 1)$, then the assertion generated by the transformation allows us to substitute these calls by z.

Remark: Although assertions provide a convenient and easily understandable tool here, we need not introduce them into our language in order to formulate transformation rules such as (T6). In the place of the above assertion we could as well give a re-declaration of f:

$\therefore$ **do function** $f\,(j{:}nat){:}r;$
$$\textbf{if } j = i - 1 \textbf{ then } f := z \textbf{ else } f := e[j/i] \textbf{ endif}$$
$z := e$
endloop $\ldots$

Unfolding of the calls $f(i-1)$ occurring in the assignment $z := e$ will then lead to the expression **if** $i-1 = i-1$ **then** z **else** $e[j-1/j]$ **endif**, which can be simplified to z. Still another possibility would be to modify the main assignment in the output scheme into $z := e[z/f(i-1)]$. However, this kind of substitution operation complicates both the correctness considerations and the technical realization of transformations. (*End of remark*)

The second method of decreasing the necessary interactions of the user is by way of "transformation algorithms", i.e. by the automatic combination of or selection from a number of transformation rules. The verification of such algorithms simply comes from the transitivity of our notion of correctness. But in addition one has to guarantee that the algorithm terminates. The above example already shows a transformation algorithm: After the application of the main rule (T6) we have to apply substitutions according to the given assertion. Similarly, the application of compact rules often leads to terms such as $e + 0$, **if** *true* **then** $\ldots$ etc., which allow an immediate simplification. Therefore the application of compact rules should be combined with the subsequent application of such simplifying rules.

A more complex problem is to prepare the application of compact rules: Consider for instance the rule (T5) above; it requires that the recursive call occurs in the **else**-branch. If this rule is to be applied to a function having the recursive call in the **then**-branch we have to prepare this application by the elementary rule

(T7) $\qquad$ **if** p **then** e_1 **else** e_2 **endif**

$$\rule{6cm}{0.4pt}$$
$$\Downarrow$$

$\qquad$ **if** $\neg p$ **then** e_2 **else** e_1 **endif**

The distinction between compact rules and transformation algorithms is a completely pragmatic one. Consider again the derivation process for our *lob*-example. We can either make the sequence of applied transformations, viz. T1, T3, T4 and T6 together with some simplifying rules, into a transformation algorithm or we can "freeze" them into a new rule. Thus we get by abstracting from the particularities of the *lob*-example the new transformation rule

(T8) $\qquad$ **some** $x{:}m \bullet \forall\, i{:}[j \mathinner{\ldotp\ldotp} k] \bullet p(x, i)$

$$\rule{8cm}{0.4pt} \qquad \begin{cases} k \geq j \\ \forall\, x, i, i' \bullet p(x, i) \Rightarrow p(h(x, i'), i) \end{cases}$$
$$\Downarrow$$

$\qquad$ **begin**
$\qquad$ **var** $x{:}m := (\textbf{some } x{:}m \bullet p(x, j)\,)$;
$\qquad$ **for** $i := j + 1$ **to** k **do** $x := h(x, i)$ **endfor**;
$\qquad$ x
$\qquad$ **end**

5. Applicative and Procedural Data Structures

Classical programming languages are quite poor when it comes to data structures. Arrays, records and pointers still are the essential features put at the programmers disposal. More powerful tools are included only in a few languages such as SETL (finite sets). But even there all structures are oriented towards the procedural style. Yet, recent developments allow us to establish for the area of data structures the very same conceptual partitioning that we have made for control structures. To illustrate this point, we will use the concept of arrays as a running example.

On the highest level, viz. on the level of nonoperational applicative specifications, arrays are described algebraically by way of their characterisitic properties. This technique has become known under the name "abstract data types". Consider the example (the notation of which is explained below)

type   ARRAY (type m) $array, vac, alter, value$;
        based on: INDEX

    sorts: $array$

    opns: $vac :$ $\rightarrow$ $array$
                   $alter :$ $array \times index \times m$ $\rightarrow$ $array$
                   $value :$ $array \times index$ $\rightarrow$ m

    axioms: $\forall\, a{:}array;\ i,j{:}index;\ x,y{:}m \bullet$
                   $value(alter(a,i,x),i) = x$
                   $i \neq j \Rightarrow value(alter(a,i,x),j) = value(a,j)$
                   $alter(alter(a,i,x),i,y) = alter(a,i,y)$
                   $i \neq j \Rightarrow alter(alter(a,i,x),j,y) = alter(alter(a,j,y),i,x)$

    restrns: $\forall\, i{:}index \bullet$
                   $value(vac,i) = error$
end of type

As usual, the header line of this declaration specifies a name, the kind of the parameters (arrays may be defined for elements of any type m) and the "result", i.e. the list of identifiers that are made available to the surrounding program. The body of the type consists of a syntactic and a semantic part. The former gives an identifier — viz. $array$ — for the set of objects defined by the type and the functionalities of all operations introduced along with this new set. The latter specifies the behavior of all operations (and thus of the newly introduced objects) by way of conditional equations. The element $error$ is introduced as a particular object in order to make all operations of the type "total". (Alternatively, one may employ partial functions. For more defails see [17] where our approach and notation are explained in more detail and a list of further references can be found.) The declaration of this type allows us to write programs like

        var $a,b{:}array$ **of** $char$; $a := vac$; $b := alter(a,i,c)$

It is a relatively simple exercise to design abstract data types for all the classical structures of computer science such as sequences, queues, stacks, trees, graphs etc. But in addition, this technique also allows us to specify restrictions and extensions of these basic structures in a precise and straightforward manner, for example bounded queues, balanced trees, graphs with "built-in" traversal operations and the like. Thus, abstract data types are a most powerful tool for working with data structures in an applicative style (cf. [9]).

While abstract data types allow us to *specify* applicative data structures, recursive type declarations which are based on finite enumeration, direct product and direct sum, provide us with *operational* data structures on the applicative level (i.e. an automatic, though inefficient code generation is possible). A simple example is given by

> **type** *tree* = **atomic** (*empty*) | *cons*:**record** *left*:*tree*; *node*:*nat*; *right*:*tree* **endrecord**

where **atomic**$(c_1, \ldots, c_n)$ introduces n (identifiers for) new objects, "|" stands for direct sum and "**record** ...**endrecord**" for direct product. The identifiers *cons*, *left* etc. denote operations which may be used to write expressions such as

> **const** *t*:*tree* = *cons*(*empty*, 5, *cons*(*empty*, 6, *empty*));
> **const** *x*:*nat* = *node*(*right*(*t*)).

Also, we may test to which variant t belongs: the expression (t **is** *empty*) yields false, whereas (t **is** *cons*) yields true. Obviously, there is a simple abstract data type specifying the meaning of these new operations in a straightforward way.

Remark: There is also an operational applicative version of the concept of arrays, viz. functions over finite domains:

> **function** *vac* :**function** (*index*)*char*;
>   *vac* := *error*;
> **function** *alter* (**function** *a*:(*index*)*char*; *i*:*index*; *c*:*char*):**function** (*index*)*char*;
>   *alter* := ((j:*index*):*char*; **if** $j = i$ **then** c **else** $a(j)$ **endif**)
> **function** *value* (**function** *a*:(*index*)*char*; *i*:*index*):*char*;
>   *value* := $a(i)$

It is easily verified that these functions meet the specifications of the type ARRAY and thus provide an (operational) implementation. Unfortunately, we have to work here with "functionals", i.e. functions having functions as arguments and results. However, there are many cases where the notation is not so awkward. For example the declaration

> **const** a':**function** (*index*)*char* = *alter*(*a*, *i*, *c*)

is equivalent to

> **function** a'(j:*index*):*char*; **if** $j = i$ **then** $a' := c$ **else** $a' := a(j)$ **endif**

Similar correspondences result for the other functions. (*End of remark*)

As in the case of control structures we need the possibility of passing from the applicative to the generally more efficient procedural level in a simple and straightforward way. Again, the first step is the introduction of a corresponding procedure for each function. Following [15] this is achieved by the declaration

> **module** ⟨*chararray*, *init*, *update*, *get*⟩ **isoftype** ARRAY(*char*)

For each operation having a result of sort *array* we thus get a procedure having a variable parameter of sort *array*. In other words, the above module declares the procedures and functions

> **procedure** *init*(**var** *a*:*chararray*);   *a* := *vac*
> **procedure** *update*(**var** *a*:*chararray*; *i*:*index*; *c*:*char*);   *a* := *alter*(*a*, *i*, *c*)
> **function** *get*(*a*:*chararray*; *i*:*index*):*char*;   *get* := *value*(*a*, *i*).

These examples also show that the meaning of the new procedures is explained in terms of the operations of the corresponding abstract data type. Again, modules may have various implementations. For instance, the one given here describes the behavior of direct access files and may therefore be taken for developing programs using such files. But of course, the most straightforward implementation is given by the classical arrays. In this case we get the following correspondences (which we will use to simplify notations in the rest of the paper):

$$
\begin{array}{lcl}
\textbf{var } a{:}char\,array & \triangleq & a{:}\textbf{ array } [j \mathinner{..} k] \textbf{ of } char \\
update(a, i, c) & \triangleq & a[i] := c \\
get(a, i) & \triangleq & a[i]
\end{array}
$$

The converse direction is more complex, since in multiple assignments $\langle a[i], a[j] \rangle := \langle e_1, e_2 \rangle$ the problem of aliasing, i.e. the case $i = j$, has to be taken care of.

6. Transformations and Data Structures

There are two obvious connections between abstract data types and transformation rules. First, the axioms of a type may be used to verify the applicability condition of a transformation rule. Consider for example the rule (T8); if we interpret

$$
\begin{array}{lll}
\text{the predicate } p(a, i) & \text{by} & value(a, i) = f(i) \quad \text{and} \\
\text{the function } h(a, j) & \text{by} & alter(a, j, f(j))
\end{array}
$$

then the first two laws of the type ARRAY immediately show the validity of the applicability condition, which with these interpretations reads:

$$
\forall\, a, i, j \bullet value(a, i) = f(i) \Rightarrow value(alter(a, j, f(j)), i) = f(i).
$$

Secondly, the (conditional) equations of a type may be used as transformation rules themselves. For example, the axiom $i \neq j \Rightarrow value(alter(a, j, x), i) = value(a, i)$ gives rise to the transformation rule

$$(T9) \qquad value(alter(a, j, x), i)$$
$$\overline{\qquad\qquad \text{\textparallel} \qquad\qquad} \quad \{i \neq j$$
$$value(a, i)$$

One has to be aware of a technical problem here: The free variables a, i, j and x of the conditional equation become scheme variables standing for expressions. Therefore we have to ensure that only "good" expressions are instantiated for these variables, i.e. defined, determinate and side-effect free expressions.

Besides these general connections between types and transformations there are, of course, also rules connected to particular data types. Actually, these are the more interesting rules, since they often reflect classical programming techniques.

Example:
The following transformation rules realize classical programming techniques (called "dynamic programming" in [1]) that are used to compute certain recursive functions efficiently by storing intermediate results in arrays. Consider the function *tabulate* with the specification

$$
\textbf{function } tabulate\,(j, k{:}nat;\ \textbf{function } f{:}(nat)m\,){:}array;
$$
$$
tabulate := \textbf{some } a{:}array \bullet \left(\forall\, i{:}nat \bullet j \leq i \leq k \Rightarrow a[i] = f(i)\right)
$$

i.e. in the interval $[j \mathinner{..} k]$ the elements of the array a are the values of f. Using this function, we may now introduce the declaration $\textbf{const } a{:}array = tabulate(j, k, g)$ into a given program according to the rule

$(T10)$ **function** $g\,(i{:}nat){:}m;$ $g := e$

$$\overline{\hspace{4cm}\Big\Downarrow\hspace{4cm}}\quad \{\forall\,i \bullet j \le i \le k \Rightarrow \mathrm{DEFINED}[g(i)]\}$$

     const $a{:}array = tabulate(j,k,g);$
     function $g\,(i{:}nat){:}m;$
         if $j \le i \le k$ **then** $g := a[i]$
                   else $g := e$ **endif**

The additional if-clause in the body of g guarantees that a reference to a is made wherever possible in the place of (recursive) calls of g. If j and k have been suitably chosen subsequent unfoldings of the calls of g in the surrounding program will only leave references to the array a. This transformation effects recursion removal "by tabulation", which is particularly suited for multiple recursion such as in the simple example

     function $fib\,(n{:}nat){:}nat;$
         if $n > 1$ **then** $fib := fib(n-1) + fib(n-2)$
             else $fib := 1$ **endif**

It remains to implement the function *tabulate*. This leads to the standard and well-known problems of programming with arrays, and we will show in the rest of this section, how the tools we have presented so far may be used in this task. As pointed out in the remark after (T6) there are a number of technical realizations of these ideas. We choose here to work with transformation algorithms in order to keep the rules and their applicability conditions simpler. This simplicity also justifies e.g. the temporary introduction of auxiliary variables, which facilitate a main transformation and can be eliminated at the end of the algorithm.

As has been seen at the beginning of this section, the rule (T8) is applicable here, which leads to a simple generation algorithm for the above specification. But now the kind of replacement enabled by (T10) for the surrounding program is also possible during the generation process itself, provided that the recursive calls of $g(i)$ have arguments between j and i.

$(T11)$ **const** $a{:}array = tabulate(j,k,g)$

$$\overline{\hspace{5cm}\Big\Downarrow\hspace{5cm}}$$

     $a{:}$ **array** $[j\,..\,k];$
     for $i := j$ **to** k **do assert** $j \le i_0 < i \Rightarrow a[i_0] = g(i_0);$
                          $a[i] := g(i)$
     endfor

Again, it depends on the good or bad choice of the values j and k, whether the assertion allows us to eliminate the calls of g: After unfolding $g(i)$ in the assignment $a[i] := g(i)$ the recursive calls of g must have arguments between j and $i-1$. We get, of course, a trivial variant of the rule (T11) by counting downwards instead of upwards, and it depends on the arguments of the recursive calls of g, which variant is the useful one.

Another variant is particularly suited for "overwriting" arrays. Assume that in the following rule the function g_a depends on values of the array a. Then we temporarily introduce an auxiliary array b by the transformation (where g_b is derived from g_a by substituting all occurrences of a by b)

$(T12)$ $a := tabulate(j,k,g_a)$

$$\overline{\hspace{5cm}\Big\Downarrow\hspace{5cm}}$$

     $b{:}$ **array** $[j\,..\,k] := a;$
     for $i := j$ **to** k **do assert** $i_0 \ge i \Rightarrow b[i_0] = a[i_0];$
                          $a[i] := g_b(i)$
     endfor

Application of this rule again starts a transformation algorithm: If — again — the values j and k are suitably chosen it should be possible to replace not only all recursive calls of g but also all references to the auxiliary array b. If so, we may eliminate b.

The transformational approach is, of course, not restricted to the derivation of code for sequential machines. If we replace in the above transformation algorithm the central rule (T12) by the variant (T13) below then the assignment is transformed into a form that is suited for execution on array or vector machines. (The notation $\llbracket \ldots \| \ldots \| \ldots \rrbracket$ denotes concurrent processes).

$(T13)$ $\qquad a := tabulate(j, k, g_a)$

$$\Downarrow$$

b: **array** $[j .. k] := a;$
$\llbracket$ **assert** $b[j] = a[j]; a[j] := g_b(j) \| \ldots \|$ **assert** $b[k] = a[k]; a[k] := g_b(k) \rrbracket$

7. Another Simple Example

The example of section 2 illustrated the derivation of an iterative program from a non-operational specification given in terms of the quantifiers **some** and $\forall$. Now we consider an example, where the initial specification is given by recursive equations and where the implementation uses data structures.

The interpolation algorithm of Aitken and Neville is based on recursive functions (equations) of the form

$\qquad$ **function** $p\,(i, k{:}nat \bullet 1 \le k \le i \le n){:}real;$
$\qquad\qquad$ **if** $k = 1$ **then** $p := y_i$
$\qquad\qquad\qquad$ **else** $\quad p := e_{i,k}\big(p(i, k - 1), p(i - 1, k - 1)\big)$ **endif**

where $e_{i,k}$ is some arithmetic expression in x and the points x_i, x_k. Once these recursive equations have been developed (based on principles of mathematics rather than of computer science), the remaining task of efficiently implementing them can be solved with the transformations we have developed so far.

Since in both recursive calls the term $k - 1$ occurs, it seems promising to apply the rule (T6). However, the other parameter does not fit into the pattern. Following the well-known mathematical isomorphism $[A \times B] \to C \simeq A \to [B \to C]$ we therefore define a new function q such that $q(k)(i) = p(i, k)$ ("currying")

$\qquad$ **function** $q\,(k{:}nat \bullet 1 \le k \le n){:}$**function** $(nat)real; \quad q := \big((i{:}nat \bullet k \le i \le n){:}real; \, p(i, k)\big)$

As usual, unfolding of the call of p, some minor simplifications and a final folding lead to

$\qquad$ **function** $p\,(i, k{:}nat \bullet 1 \le k \le i \le n){:}real; \quad p := q(k)(i)$

$\qquad$ **function** $q\,(k{:}nat \bullet 1 \le k \le n){:}$**function** $(nat)real;$
$\qquad\qquad$ **if** $k = 1$ **then** $q := \big((i{:}nat \bullet 1 \le i \le n){:}real; \, y_i\big)$
$\qquad\qquad\qquad$ **else** $\quad q := \big((i{:}nat \bullet k \le i \le n){:}real;$
$\qquad\qquad\qquad\qquad\qquad e_{i,k}(q(k - 1)(i), q(k - 1)(i - 1))\big)$ **endif**

As indicated by the transformation (T10), q may be transformed into a function that returns an array instead of another function. Again, we start by defining a new function

$\qquad$ **function** $q'\,(k{:}nat \bullet 1 \le k \le n){:}array$ **of** $real;$
$\qquad\qquad q' := tabulate(k, n, q(k))$

and the same kind of unfold/fold steps as before — but now applied to the call $q(k)$ — lead to the version

function $p\,(i, k{:}nat \bullet 1 \le k \le i \le n){:}real; \quad p := q'(k)[i]$

function $q'\,(k{:}nat \bullet 1 \le k \le n){:}array$ **of** $real;$
   if $k = 1$ **then** $q' := tabulate\big(1, n, (i{:}nat \bullet 1 \le i \le n){:}real;\; y_i\big)$
    else $\quad q' := tabulate\big(k, n, (i{:}nat \bullet k \le i \le n){:}real;$
            $e_{i,k}(q(k-1)[i], q(k-1)[i-1])\big)$
   endif

This version fits into the pattern of (T6) and we get — together with an unfolding of the remaining call of q' within p

function $p\,(i_0, k_0{:}nat \bullet 1 \le k_0 \le i_0 \le n){:}real;$
   begin
   var $k{:}nat := 1;$
   var $a{:}array$ **of** $real := tabulate\big(1, n, (i{:}nat \bullet 1 \le i \le n){:}real;\; y_i\big);$
   while $k < k_0$ **do**
    $k := k + 1;$
    $a := tabulate\big(k, n, (i{:}nat \bullet k \le i \le n){:}real;\; e_{i,k}(a[i], a[i-1])\big)$
   endloop;
   $p := a[i_0]$
   end

The application of the transformation rule (T12) and some "cleaning up" yield the final program for computing $p(n, n)$

$a{:}$ **array** $[1 .. n]$ **of** $real;$
for $i := 1$ **to** n **do** $a[i] := y_i$ **endfor**;
for $k := 2$ **to** n **do**
  for $i := n$ **downto** k **do** $a[i] := e_{i,k}(a[i], a[i-1])$ **endfor**
endfor
assert $a[n] = p(n, n)$

We will conclude this section by sketching very briefly how to produce a variant of the Aitken-Neville algorithm that is suited for execution on vector or array machines. Unfortunately, in the last version of p above the assignment $a := tabulate(\ldots)$ contains two references to a with different indices. Hence, the assertions generated by (T13) will not suffice to replace all occurrences of the auxiliary array b. To overcome this deficiency we define in the place of the function q two functions r and s, one for the odd and the other for the even arguments, such that $r(k)(j) = p(2 \times j - 1, k)$ and $s(k)(j) = p(2 \times j, k)$:

function $r\,(k{:}nat \bullet 1 \le k \le n){:}$**function** $(nat)real;$
   $r := \big((j{:}nat \bullet \frac{k+1}{2} \le j \le \frac{n+1}{2}){:}real;\; p(2 \times j - 1, k)\big)$
function $s\,(k{:}nat \bullet 1 \le k \le n){:}$**function** $(nat)real;$
   $s := \big((j{:}nat \bullet \frac{k}{2} \le j \le \frac{n}{2}){:}real;\; p(2 \times j, k)\big)$

After the same unfold/fold steps as above the transformation (T6) — applied to both functions simultaneously — will lead to a version where the assignment within the loop reads

$\langle a, b \rangle := \langle tabulate(\ldots a[j], b[j-1] \ldots), tabulate(\ldots b[j], a[j] \ldots) \rangle.$

The sequentialization of this multiple assignment needs an auxiliary variable (array) c. Then (T13) is applicable to each of the resulting assignments and we finally get a version of our program where the inner loop is replaced by three concurrent operations that are suited for execution on array or vector machines. We leave the details of this modified derivation to the reader, in particular the justification of the highly optimized expressions for determining the lower bounds of the arrays.

> **var** $\langle amin, bmin \rangle{:} nat := \langle 1, 1 \rangle$;
> **var** $\langle amax, bmax \rangle{:} nat :=$ **if** $even(n)$ **then** $\langle \frac{n}{2}, \frac{n}{2} \rangle$ **else** $\langle \frac{n+1}{2}, \frac{n-1}{2} \rangle$ **endif**;
> a: **array** $[1 .. amax]$ **of** $real$; b: **array** $[1 .. bmax]$ **of** $real$; c: **array** $[1 .. amax]$ **of** $real$;
> $[\![\, a[1] := y_1 \,]\!] \dots [\![\, a[amax] := y_{2 \times amax-1} \,]\!]$; $[\![\, b[1] := y_2 \,]\!] \dots [\![\, b[bmax] := y_{2 \times bmax} \,]\!]$;
> **for** $k := 2$ **to** n **do**
>   **if** $amin = bmin$ **then** $amin := amin + 1$ **else** $bmin := amin$ **endif**;
> $[\![\, c[amin] := a[amin]$ $\|\dots\| c[amax] := a[amax]$ $]\!]$;
> $[\![\, a[amin] := e_{\dots}(a[amin], b[amin-1]) \,\|\dots\| a[amax] := e_{\dots}(a[amax], b[amax-1]) \,]\!]$;
> $[\![\, b[bmin] := e_{\dots}(b[bmin], c[bmin]) \quad \|\dots\| b[bmax] := e_{\dots}(b[bmax], c[bmax]) \quad]\!]$
> **endfor**

Note that the definitions of r and s lead to the assignment

$$\langle amin, bmin \rangle := \textbf{if } odd(k) \textbf{ then } \langle \tfrac{k+1}{2}, \tfrac{k+1}{2} \rangle \textbf{ else } \langle \tfrac{k}{2} + 1, \tfrac{k}{2} \rangle \textbf{ endif}$$

for the lower array indizes. Transformation rules that are known in the field of optimizing compilers as "strength reduction" then yield the more efficient version of the above program.

8. Concluding Remarks

It is *not* the main purpose of program transformation to produce new and ingenious programming techniques. It shall rather provide a means for formalizing known techniques such that they can be used in a safe and convenient way. The area we have chosen in this paper should indicate this particularly well: The proper treatment of arrays with overwriting is not new, but the same intrinsic difficulties occur in every program anew — causing the same errors and the same efforts to avoid them again and again. This effort can be invested once into the derivation of appropriate transformation rules (samples of which we have demonstrated here). From then on the development of programs belonging to this class is simply done by applying these previously verified rules.

When the transformations are convenient to use — e.g. through support by a system — then the programmer should be encouraged to start his program development on a higher (and thus safer) level. For instance, whenever an iterative process can be described as producing matrices $A^{(i+1)}$ from A^i, it suffices to give the recursive equation $a(i+1, j, k) = e(a(i, \dots), \dots)$. This pertains to the algorithm of Gauss as well as to the Warshall algorithm for computing the transitive closure of a graph (cf. [8]). Analogously, when transformations allow one to derive algorithms from specifications by applying the classical rules for quantifiers and the like in a convenient way, they might encourage the programmer to use these formulations — which considerably increases the transparency and reliability of the production of programs.

A further important point has not been mentioned explicitly so far, although it occurred several times in our examples: By including nondeterminism into specifications and programs we may delay design decisions until more information about the particular implementation circumstances is gained. Furthermore, we may also employ nondeterminism to develop families of related algorithms instead of single, specific programs.

Acknowledgement

This paper reflects ideas that have been developed in the CIP-group in Munich. John Reid has considerably influenced its present form by a number of valuable and constructive suggestions. I am also indebted to Manfred Broy and Jim Boyle for reading a draft version.

References

[1] A. V. AHO, J. E. HOPCROFT, J. D. ULLMAN: The Design and Analysis of Computer Algorithms. Reading, Mass.: Addison-Wesley, 1974.

[2] J. BACKUS: Can Programming be Liberated from the von Neumann Style? A Functional Style and its Algebra of Programs. *Comm. ACM* **21**, 613–641 (1978).

[3] F. L. BAUER, M. BROY (EDS.): Program Construction. Lecture Notes of the International Summer School on Programm Construction, Marktoberdorf 1978, *Lecture Notes in Computer Science* **69**, Berlin-Heidelberg-New York: Springer 1979.

[4] F. L. BAUER, H. PARTSCH, P. PEPPER, H. WÖSSNER: Techniques for Program Development. In: Software Engineering Techniques. *INFOTECH State of the Art Report* **34**, 1977.

[5] F. L. BAUER, H. WÖSSNER: Algorithmic Language and Program Development. Berlin-Heidelberg-New York: Springer (to appear).

[6] F. L. BAUER ET AL.: Report on the Wide Spectrum Language CIP-L. Technische Universität München, Institut für Informatik (1981).

[7] M. BROY: Transformational Semantics for Concurrent Programs. *Information Processing Letters* **11**:2, 87-91 (1980)

[8] M. BROY, P. PEPPER: Program Development as a Formal Activity. *IEEE Transactions on Software Engineering*, *SE-***7**:1, 14-22 (1980).

[9] M. BROY, P. PEPPER: Combining Algebraic and Algorithmic Reasoning: An Approach to the Schorr-Waite Algorithm. *ACM TOPLAS* (to appear)

[10] M. BROY, M. WIRSING: Programming languages as abstract data types. 5eme Coll. Les Arbres en Algebre et Programmation, Lille 1980, Univ. de Lille, 160-177

[11] M. BROY, P. PEPPER, M. WIRSING: On Relations between Programs. In: B. ROBINET (ed): 4th Int. Symposium on Programming. Paris, April 1980, *Lecture Notes in Computer Science* **83**, 59–78, Berlin-Heidelberg-New York: Springer (1980)

[12] M. BROY, P. PEPPER, M. WIRSING: On the Algebraic Specification of Programming Languages. (to appear)

[13] R. M. BURSTALL, J. DARLINGTON: Some Transformations for Developing Recursive Programs. *J.ACM* **24**:1, 44–67 (1977).

[14] D. C. COOPER: The Equivalence of Certain Computations. *Comp. J.* **9**, 48–52 (1966)

[15] A. LAUT: Safe Procedural Implementations of Algebraic Types. *Information Processing Letters* **11**:4, 5, 147-151 (1980).

[16] P. PEPPER: A Study on Transformational Semantics. Dissertation Munich 1979. (Also in [3])

[17] M. WIRSING, P. PEPPER, H. PARTSCH, W. DOSCH, M. BROY: On Hierarchies of Abstract Data Types. Technische Universität München, Institut für Informatik, TUM-I8007 (1980).

DISCUSSION

Summary by discussant (Dekker)
The speaker described how correctness preserving transformation rules can be used to transform a formal problem specification into an (efficiently) executable program. A language supporting this methodology is CIP–L, a "wide spectrum language" developed at the Technical University Munich. The method was explained with various examples, among others a sequence of transformations concerning the Aitken-Neville scheme.

van der Laan. I am very much in favour of efforts like the PLANKALKUL, CIP-L and TAMPR, the more because from different starting points similar tools will emerge. I would like to know, however, your answers to the following questions.
1. Assuming that these transformations do preserve correctness, do they also preserve stability?
2. The transformations produce programs starting from scratch; how about using already available software such as numerical program libraries?

Pepper. 1. The programmer has to be aware of the shuffling of the operations. The notions of correctness (i.e., of equivalence of programs) have to be chosen very carefully in connection with floating-point arithmetic, etc. Ultimately, this is reflected in the choice of axioms for the corresponding abstract data types.
2. We can stop the transformations at any required level (e.g., according to what optimization is available in the compiler). One such level is certainly defined by available library routines and one may therefore direct a transformation process into that direction. However, we have not yet worked this out in detail.

M.Paul. You pointed out that the "some" operator reflects a kind of non-determinism, leaving unspecified which of the elements of the extension of a predicate is to be chosen at a particular instant. I want to add that the bounded universal quantifier also reflects a kind of non-determinism, leaving unspecified the order (sequential and/or concurrent) in which the elements of the bound variable are to be treated. This kind of non-determinism is intimately related to concurrency as it occurs in (without being restricted to) array programming.

Pepper. I agree. The two notions of non-determinism are different in a subtle way, but basically lead to the same programming problems and techniques.

Wilson. Can the formal methods you have described be used to prove the formal equivalence of serial and parallel algorithms? For example, consider the recursive definition of the partial sums of a given vector; can it be shown that the parallel algorithm (recursive doubling) is equivalent to the conventional serial algorithm (a DO–loop generating the partial sums)?

Pepper. Yes. Starting from the formal recursive definition of the partial sums, the system will allow one to follow two different development paths leading to either the sequential or the parallel version; this shows their equivalence. It is very difficult, however, to go from the sequential to the parallel program directly.

THE RELATIONSHIP BETWEEN NUMERICAL COMPUTATION
AND PROGRAMMING LANGUAGES, J.K. Reid (editor)
North-Holland Publishing Company
©IFIP, 1982

Tools for Numerical Programming[†]

Webb Miller

Department of Computer Science
The University of Arizona
Tucson, Arizona   85721

Software tools ameliorate certain of the problems characteristic of the production of *numerical* software. The need within the mathematical software community for such tools has ramifications for the design of numerical programming languages. Also, the practicality of these and related tools suggest simplifications for the processes by which those languages are standardized and made available.

1. Introduction

A *software tool* is a computer program that assists in the production of other programs. Generally it reads, as its data, a program and produces, as its output, a related program and/or information about the original program. Here we are interested in software tools of particular value for production of *numerical* programs, i.e., programs that call for floating-point operations.

The paper begins with a sampling of capabilities that are realized by existing tools whose usage is both convenient and affordable. We discuss capabilities for rounding error analysis (during the algorithm design process), for structuring (during progam maintenance) and for automatic generation of programs that evaluate Jacobians or that utilize nonstandard arithmetic.

Our sampling of tool capabilities concludes with two examples of "program prototypes." Here the desire is to prepare a master version of a program in an expressive language. From the master program, versions tailored to specific machines and numerical precisions are generated automatically. This capability, as illustrated here, is realized in the M4 macro processor and the EFL compiler from the UNIX[TM] operating system.

As the sampling of capabilities indicates, a common use of software tools in the production of numerical software is to generate Fortran programs from higher-level specifications. As Gear [1] suggested some years ago, Fortran has become a "universal machine language" for numerical programming.

Not represented in the sampling are tools to assist program testing. While such tools can be profitably applied to numerical software [2], they generally make no use of

---

[†]This work was carried out as part of the Toolpack project and was supported by NSF Grant No. MCS79-26441.

the special characteristics of this class of programs.

Following our sampling of tool capabilities we turn briefly to the question, "What do we need in a programming language?" To appreciate the tool writer's needs involves understanding that he deals with three levels of programming languages, so this is where our discussion begins.

We conclude the paper by making some recommendations to the Fortran committee, X3J3. Our hope is that Fortran will strengthen its position as a language for numerical programming, and not fall into the trap [3] of trying to be ideal for every programming task (e.g., for tool building). The language standardization effort is simplified by the realization that certain language extensions can be left to preprocessors and that tools can assist in reclaiming pre-standard programs.

2. Automatic roundoff analysis

A developer of matrix software may have the need for automatic error analysis. From a specification of the floating-point operations in a matrix algorithm (i.e., omitting the many additional details that must be determined to turn the algorithm into an acceptable piece of software) a software tool [4,5] can decide if the algorithm is numerically stable. In addition, the tool should have the capability to indicate where an unstable algorithm goes wrong.

For example, the tool user might prepare the following input to the tool.

```
    # Solution of the linear least-squares problem.
        test m = 4, n = 3
        dimension a(m,n), t(n,n), l(n,n), b(m), s(n), x(n), y(n)
        input a, b
#
# Step 1. Form the normal equations, Tx = s.
        do i = 1 to n {
            do j = 1 to i
                t(i,j) = sum(a(k,i)*a(k,j), k = 1 to m)
            s(i) = sum(a(k,i)*b(k), k = 1 to m)
            }
#
# Step 2. Solve the normal equations.
        do i = 1 to n {
            do j = 1 to i-1
                l(i,j) = (t(i,j) - sum(l(i,k)*l(j,k), k = 1 to j-1))/l(j,j)
            l(i,i) = sqrt(t(i,i) - sum(l(i,k)*l(i,k), k = 1 to i-1))
            }
        do i = 1 to n
            y(i) =(s(i) - sum(l(i,j)*y(j), j = 1 to i-1))/l(i,i)
        do i = n downto 1
            x(i) = (y(i) - sum(l(j,i)*x(j), j = i+1 to n))/l(i,i)
        output x
```

The computer can determine that not only is this algorithm unstable, it is *inherently* unstable. That is, it will remain unstable even if other methods are used to form and to solve the normal equations.

For this capability it is desirable to employ a very restricted "user language" (i.e., the language in which the user specifies a matrix algorithm). Doing so helps the tool writer keep the user from applying the tool to algorithms for which the tool is not appropriate. A less important benefit to the tool writer is that of making the tool easier to write.

3. Automatic structuring

One of the first steps when reading a program written by someone else (or by one-self in the distant past) is to get an understanding of the program's control flow. With Fortran programs one must decipher GO TO statements.

For instance, consider the following Fortran program.

```
            SUBROUTINE TQL1(N,D,E,IERR)
            INTEGER N,IERR,L,J,M
            REAL D(N),E(N),MACHEP,F,B,ABS
            MACHEP = 9.54E-7
            IERR = 0
            IF (N .EQ. 1) GO TO 1001
            CALL INIT(N,E,F,B)
            DO 290 L = 1,N
            J = 0
            H = MACHEP*(ABS(D(L)) + ABS(E(L)))
            IF (B .LT. H) B = H
            DO 110 M = L,N
            IF (ABS(E(M)) .LT. B) GO TO 120      ←
 110        CONTINUE
 120        IF (M .EQ.L) GO TO 210
 130        IF (J .EQ. 30) GO TO 1000
            J = J + 1
            CALL SHIFT(N,D,E,L,F)
            CALL QLTRAN(N,D,E,L,M)
            IF (ABS(E(L)) .GT. B) GO TO 130      ←
 210        CALL ORDER(N,D,F,L)
 290        CONTINUE
            GO TO 1001
1000        IERR = L
1001        RETURN
            END
```

Understanding control flow in this program involves, among other things, seeing that the "IF (· · ·) GO TO 120" statement merely breaks out of a loop, and that the "IF (· · ·) GO TO 130" statement repeats a loop until the condition fails.

A tool [6] can be employed to ease the difficulty of understanding control flow in a Fortran program by making explicit the nature of many of its GO TO statements. For instance, from the above program a tool might generate the following program (in the Ratfor language).

```
subroutine tqll(n,d,e,ierr)
integer n,ierr,l,j,m
real d(n),e(n),machep,f,b,abs
ierr = 0
if (n!=1) {
      call init(n,e,f,b)
      do l = 1,n {
           j = 0
           h = machep*(abs(d(l))+abs(e(l)))
           if (b < h)
                b = h
           do m = l,n
                if (abs(e(m)) < b)
                     break
           if (m != l)
                repeat {
                     if (j == 30)
                          go to 10
                     j = j + 1
                     call shift(n,d,e,l,f)
                     call qltran(n,d,e,l,m)
                     }
                until (abs(e(l)) <= b)
           call order(n,d,f,l)
           }
      return
      10 ierr = 1
      }
return
end
```

4. Automatic generation of Jacobians

Many numerical programs involve a procedure to evaluate the partial derivatives of a given (vector-valued) function. Differentiating expressions by hand is extremely error prone, so a software tool should be used.

For instance, the following input defines a function $f_1(x_1, x_2, x_3)$ whose evaluation includes invocation of the intrinsic arc tangent function *atan* and of the user-supplied function h. The second line specifies that $h'(u)$ is the product of the cosine of u and $g(u)$, where g is another user-supplied function.

```
common /blk/ param
h'(u) = cos(u)*g(u)
temp = atan(x1 + 0.1*(x2 + x3))
redund = x3 + x2
fcnval = h(param*redund)
f1 = temp/fcnval
```

From this input, one tool [7] generates the following Fortran procedure to compute the Jacobian matrix $D(i,j) = \partial f_i / \partial x_j$.

```
SUBROUTINE JACOBN(X,F,D,M,N)
DIMENSION X(N), F(M), D(M,N)
COMMON /BLK/ PARAM
T1 = X(3) + X(2)
T2 = X(1) + 0.1*T1
T3 = PARAM*T1
T4 = H(T3)
T5 = ATAN(T2)/T4
T6 = 1.0 + T2*T2
T7 = T6/T4
T8 = PARAM*COS(T3)*G(T3)*(-T5/T4) + 0.1*T7
F(1) = T5
D(1,1) = T7
D(1,2) = T8
D(1,3) = T8
RETURN
END
```

This capability to generate Fortran subroutines that compute partial derivatives is available from some general-purpose symbolic manipulation systems [8]. However, a special-purpose tool can often generate faster-running subroutines and may itself run several orders of magnitude faster than the symblic manipulation systems on even modest-sized problems.

The software tool used in the above example employs a very restricted user language. As with the error analysis tool of Section 2, the reasons are to lessen the danger of inappropriate use of the tool and, secondarily, to simplify the tool building process.

5. Preprocessors for nonstandard arithmetic

On certain occasions, one may want to write an otherwise standard program that makes use of nonstandard arithmetic, perhaps multiple precision arithmetic [9] or interval arithmetic [10]. For instance, a nonstandard Fortran program might include the following declarations and assignment statements, which set multiple precision at 30 decimal digits, declare a variable *discrm* to be multiple precision, and then use multiple precision operations in arithmetic expression.

```
SET PRECISION TO 30 DIGITS
MULTIPLE discrm
double precision a, b, c, root1
        .
        .
        .
discrm = sqrt(EXTEND(b)*EXTEND(b) - 4.0d0*EXTEND(a)*EXTEND(c))
root1 = (-b - discrm)/(2.0d0*a)
        .
        .
        .
```

A preprocessor like AUGMENT [11] can be employed to generate a pure Fortran program like the following, which uses MPCDM and MPCMD to convert between double and multiple precisions and invokes other procedures to perform multiple precision operations using an integer temporary array.

```
INTEGER INTTMP(32,7), DISCRM(32)
DOUBLE PRECISION A, B, C, ROOT1
        .
        .
        .
CALL MPCDM(B, INTTMP(1,1))
CALL MPMUL(INTTMP(1,1), INTTMP(1,1), INTTMP(1,2))
CALL MPCDM(4.0D0, INTTMP(1,3))
CALL MPCDM(A, INTTMP(1,4))
CALL MPMUL(INTTMP(1,3), INTTMP(1,4), INTTMP(1,5))
CALL MPCDM(C, INTTMP(1,6))
CALL MPMUL(INTTMP(1,5), INTTMP(1,6),INTTMP(1,7))
CALL MPMUL(INTTMP(1,2), INTTMP(1,7), DISCRM)
CALL MPCDM(-B, INTTMP(1,1))
CALL MPSUB(INTTMP(1,1), DISCRM, INTTMP(1,2))
CALL MPCDM(2.0D0*A, INTTMP(1,3))
CALL MPDIV(INTTMP(1,2), INTTMP(1,3), INTTMP(1,4))
CALL MPCMD(INTTMP(1,4), ROOT1)
        .
        .
        .
```

6. The first program prototype: matrix multiplication

The following prototype program multiplies square matrices of declared order lda and actual order n. The second and fourth parameters of the inner product macro give the spacing between successive memory references.

```
procedure matmlt(a, b, c, lda, n)
TYPE a(lda,n), b(lda,n), c(lda,n)
integer lda, n, i, j
do i = 1, n
    do j = 1, n
        INNER_PRODUCT([a(i,_)], lda, [b(_,j)], 1, [c(i,j)], 1, n)
return
end
```

If the computer has been instructed that for IBM computers a single-precision inner product should be accumulated in double precision, then the automatically generated version of *matmlt* for that case might be the following.

```
          SUBROUTINE MATMLT(A, B, C, LDA, N)
          INTEGER LDA, N
          REAL A(LDA,N), B(LDA,N), C(LDA,N)
          INTEGER I, J, INTTMP
          DOUBLE PRECISION DBLE, TEMP
          DO 3 I = 1, N
              DO 2 J = 1, N
                  TEMP = 0.0D0
                  DO 1 INTTMP = 1, N
                      TEMP = TEMP + DBLE(A(I,INTTMP))*DBLE(B(INTTMP,J))
1                 CONTINUE
                  C(I,J) = TEMP
2             CONTINUE
3         CONTINUE
          RETURN
          END
```

On the other hand, a request for the double precision IBM version could generate this program.

```
          SUBROUTINE MATMLT(A, B, C, LDA, N)
          INTEGER LDA, N
          DOUBLE PRECISION A(LDA,N), B(LDA,N), C(LDA,N)
          INTEGER I, J, INTTMP
          DO 3 I = 1, N
              DO 2 J = 1, N
                  C(I,J) = 0.0D0
                  DO 1 INTTMP = 1, N
                      C(I,J) = C(I,J) + A(I,INTTMP)*B(INTTMP,J)
1                 CONTINUE
2             CONTINUE
3         CONTINUE
          RETURN
          END
```

Naturally, it should be possible to alter the definition of the inner product macro to check that the lower limit does not exceed the upper limit, to "unroll" the loop, to invoke a function, etc. Thus one can enjoy many of the benefits provided to the Linpack project by the TAMPR system [12]. For instance, a request for the single-precision variant of *matmlt* in which the inner product is evaluated by a function invocation might produce the following.

```
          SUBROUTINE MATMLT(A, B, C, LDA, N)
          INTEGER LDA, N
          REAL A(LDA,N), B(LDA,N), C(LDA,N)
          DO 2 I = 1, N
             DO 1 J = 1, N
                C(I,J) = SDOT(N, A(I,1), LDA, B(1,J), 1)
   1            CONTINUE
   2         CONTINUE
          RETURN
          END
```

7. The second program prototype: the hyperbolic sine function

$Sinh(x)$ can be safely evaluated as $\frac{1}{2}(e^x - e^{-x})$, provided (1) x is large enough that serious cancellation in the subtraction is impossible and (2) neither e^x nor e^{-x} can suffer overflow. This second condition can be rephrased as $|x| \leq \ln(EPHUGE)$, where the environment parameter [13] $EPHUGE$ is the machine's largest floating-point number. If $|x|$ is so small that e^x is close to e^{-x}, then a series expansion can be used, where the appropriate number of terms depends upon the number of digits carried in the computation. For instance, the following program prototype (based on a NAG Library routine [14]) uses 7, 5 or 4 terms depending upon whether the computer carries at least 14, 11-13, or no more than 10 decimal digits, respectively.

```
define(TERMS, ifelse(eval(DIGITS>=14), 1, 7, eval(DIGITS>=11), 1, 5, 4))

real function sinh(x)
real bj, bjp1, bjp2, c(TERMS), expx, x, x2
initial c(1) = TRIM(2.1730421940472)
initial c(2) = TRIM(8.75942219228)e-2
initial c(3) = TRIM(1.0794777746)e-3
initial c(4) = 6.374849e-6
ifelse(eval(TERMS>4), 1,
      initial c(5) = 2.20237e-8
      )
ifelse(TERMS, 7,
      initial c(6) = 4.99e-11
      initial c(7) = 1.0e-13
      )
integer j
if (abs(x) > log(EPHUGE)) {
      write(,' sinh:', x, ' is too big')
      stop
      }
if (abs(x) > 1.0) {
      expx = exp(x)
      return(0.5*(expx - 1.0/expx))
      }
x2 = 2.0*(2.0*x*x -1.0)
bjp2 = 0.0
bjp1 = c(TERMS)
j = eval(TERMS-1)
repeat {
      bj = x2*bjp1 - bjp2 + c(j)
      if (j == 1)
            return(x*0.5*(bj - bjp2))
      bjp2 = bjp1
      bjp1 = bj
      --j
      }
end
```

If the version-generating tool "knows" that IBM computers carry 7 digits in single-precision, overflow just above 7.24E75 and write to unit 6, then it can automatically produce the following IBM realization of sinh.

```
      REAL FUNCTION SINH(X)
      REAL X
      INTEGER J
      REAL ABS, BJ, BJP1, BJP2, C(4), EXP, EXPX, X2
      DATA C(1)/2.173042/
      DATA C(2)/8.759422E-2/
      DATA C(3)/1.079477E-3/
      DATA C(4)/6.374849E-6/
      IF (ABS(X) .LE. ALOG(7.24E75)) GOTO 2
         WRITE(6, 1) X
1        FORMAT(6H SINH:, 1PE15.7, 11H IS TOO BIG)
         STOP
2     IF (ABS(X) .LE. 1.0) GOTO 3
         EXPX = EXP(X)
         SINH = 0.5*(EXPX - 1.0/EXPX)
         RETURN
3     X2 = 2.0*(2.0*X*X - 1.0)
      BJP2 = 0.0
      BJP1 = C(4)
      J = 3
4        BJ = X2*BJP1 - BJP2 + C(J)
         IF (J .NE. 1) GOTO 5
            SINH = 0.5*X*(BJ - BJP2)
            RETURN
5        BJP2 = BJP1
         BJP1 = BJ
         J = J - 1
         GOTO 4
      END
```

8. Languages for tool writers

The writer of software tools must deal with three classes of programming languages: (1) the *implementation* languages in which the tools are written, (2) the *user* languages in which the users of his tools prepare their programs and, in the case of program generators, (3) the *target* languages of the generated programs. For the tools discussed above, the situation is as follows. (The structurer has two implementations, one in C and one in EFL.)

tool	implemention	user	target
roundoff analyzer	Fortran 66	special	-
structurer	C, EFL	Fortran	Ratfor
Jacobian generator	Fortran 66	special	Fortran 66
AUGMENT	Fortran 66	extended Fortran	Fortran
M4 macro processor	C	arbitrary	arbitrary
EFL compiler	C	EFL	Fortran 66

In terms of simplifying the job of tool building, a good implementation language has much in common with a good language for compiler construction. Nontrivial numerical capabilities are rarely useful, even when building tools for numerical programming. (The roundoff analyzer is an exception to this rule. It solves generalized singular value problems and uses a numerical optimization procedure.)

On the other hand, a good target language is very permissive. For instance, pointers to objects of arbitrary type and procedure-valued variables would be useful [12,15]. (Unfortunately, we do not see how such dangerous constructs can be kept from the user if they are made available to program generators.) In addition, sophisticated numerical capabilities are useful in a target language for automatically generated numerical programs.

Employing Fortran as an implementation and/or target language is often quite costly in terms of toolsmith time. When a more appropriate language approaches Fortran's ubiquity, so that its use will not extract a serious penalty in user acceptance, we will change our current practice of using Fortran (and languages that can be preprocessed into Fortran).

9. Recommendations to the Fortran Committee

1. Do not worry about language features to help the tool writer. His needs are often orthogonal to, or even in conflict with, those of the vast majority of numerical programmers.

2. One of your most difficult challenges is that of providing adequate language support for recent hardware advances, like processors conforming to the proposed IEEE floating point standard [16] and non-serial architectures [17,18]. Support must be provided in a compiled language; preprocessors alone are not adequate.

3. Leave out of the standard any untested features that can be provided by a (practical) preprocessor.

4. The need for "upward compatability" can be satisfied without requiring that Fortran 77 be a subset of Fortran 8X (with the "obsolete features module"). For most purposes, it is satisfactory to have the capability to automatically translate Fortran 77 into Fortran 8X. (For old programs existing only in object form, the new compilers should obey the same conventions about module linkage as do the old compilers for the same machine.) In fact, such a translation may even improve the program, e.g., by automatic structuring.

Acknowledgments

We would like to thank Stan Brown and Wayne Cowell for their comments on a preliminary version of this paper, and Peter Downey for comments on the current manuscript.

References

[1]   Gear, C.W., "What do we need in programming languages?," *Informal Proceedings of Mathematical Software II,* Purdue University (May 1974), 19-24.

[2]   Howden, W.E., "Applicability of software validation techniques to scientific pro-
      grams," *ACM Trans. on Programming Languages and Systems 2* (1980), 307-320.

[3]   Hoare, C.A.R., "The emperor's old clothes," *CACM 24* (1981), 75-83.

[4]   Miller, W., and Wrathall, C., *Software for Roundoff Analysis of Matrix Algo-
      rithms* (Academic Press, New York, 1980).

[5]   Miller, W., and Spooner, D., "Algorithm 532: Software for roundoff analysis,"
      ACM Trans. Math. Software 4 (1978), 388-390.

[6]   Baker, B.S., "An algorithm for structuring flowfraphs," *JACM 24* (1977), 98-120.

[7]   Miller, W., "A program to generate Fortran programs that compute Jacobians,"
      submitted for publication.

[8]   Fateman, R.J., "Symbolic manipulation languages and numerical computation:
      trends," in this volume.

[9]   Hull, T.E., "The use of controlled precision," in this volume.

[10] Reinsch, C.H., "A synopsis of interval arithmetic for the designer of programing
      languages," in this volume.

[11] Brent, R.P., Hooper, J.A., and Yohe, J.M., "An AUGMENT interface to Brent's
      multiple precision arithmetic package," *ACM Trans. Math. Software 6* (1980),
      146-149.

[12] Boyle, J.M., "Practical applications or progam transformations," in this volume.

[13] Cody, W.J., "Floating-point parameters, models and standards," in this volume.

[14] Schonfelder, J.L., "Special functions in the NAG library," in: Cowell, W. (ed.), *Por-
      tability of Numerical Software* (Springer-Verlag, Berlin, 1977).

[15] Brown, W.S., Feldman, S.I., Fosdick, L., and Schryer, N., "An alternative
      approach to the evolution of Fortran," in this volume.

[16] Feldman, S.I., "Language features to support the IEEE arithmetic unit," in this
      volume.

[17] Lake, T., "Exception handling in array languages," in this volume.

[18] Kuck, D., Padua, D., Sameh, A., and Wolfe, N., "Languages and high-performance
      computations," in this volume.

Summary by discussant (Dekker). The speaker described the strengths and weaknesses of
various software tools in a Fortran programming environment and paid attention especially to numerical software aspects. Moreover, he presented some recommendations for
Fortran 8X to the X3J3 committee.

Lake. User preferences have not been well represented here. I would like to point to C.
Wetherell's survey of large laboratories ("Design considerations for array processing
languages," *Software — Practice and Experience 10* (1980), 265-271), in which element-
wise array operations and dynamic space allocations were most popular. The people who
propose making the array functions external to Fortran must accept extensions to parameter passing that make this easy.

Miller. As proof of user support for the proposed array features in Fortran 8X we were shown two handwritten evaluations by specialists in reservoir modelling. Even the published article that you have pointed out does not represent what I consider to be an adequate "human factors" study upon which to base such an important decision.

Wilson. Preprocessing Fortran is not a satisfactory approach to accessing the full power of the large vector and array machines. Even on the very much smaller array processors such as the Floating-Point Systems FP120B, the preprocessor method can release no more than 50% of the 12 megaflop power that is potentially available.

Miller. I thank you for the figures supporting my belief that one of X3J3's biggest challenges is the one posed by non-serial computer architectures.

Smith. It is not clear how X3J3 can reconcile your recommendation that we "work hard on new numeric capabilities to support recent hardware advances" with the other recommendation that we omit "untested features that can be provided by a practical preprocessor." The language facilities needed to support the complete IEEE standard appear more specialized and more complex to the X3J3 committee than anything that they have considered to date. Following the example of the facilities implemented on the IBM 704 at Toronto, which included a forerunner of the IEEE standard, it is conceivable that a preprocessor could be used to implement and evaluate the language features needed to support the IEEE facilities. But, since this has not been done, the IEEE facilities are clearly "untested." From X3J3's point of view, your recommended criteria immediately disqualify any language features for the IEEE standard for consideration by X3J3. You should supply specific recommendations that would provide X3J3 with enough information to make a proper judgment of the situation.

Lawson. Do I understand you correctly? Did you say, in effect, that you have the language *C* and therefore do not need any more special features in Fortran?

Miller. *C* is one of the programming languages that I use. My production tools are implemented in Fortran 66 (i.e., only Fortran goes out on the tape). My computer files may contain *EFL* programs, *M4* macro definitions, editor scripts, *C* programs, etc., to generate those Fortran programs. For numerical computation (e.g., for the numerical maximizer and the matrix codes embedded in my software for roundoff analysis), I wish that Fortran could give me easy access to exception handling mechanisms like those in the proposed IEEE floating-point standard.

Ris. You cited as an example your view that unrestricted pointers need not be included in Fortran to accomodate tool builders. Is this because you are prepared to do without pointers in Fortran, or because you are prepared to do without Fortran when you need pointers?

Miller. I am prepared to pay the cost of building tools in Fortran without pointers, for I believe the cost to be no more than a factor of two in programming time, which is itself a small fraction of the total effort.

Ris. That is a matter of taste on which we disagree.

THE RELATIONSHIP BETWEEN NUMERICAL COMPUTATION
AND PROGRAMMING LANGUAGES, J.K. Reid (editor)
North-Holland Publishing Company
©IFIP, 1982

LANGUAGELESS PROGRAMMING

A. van Wijngaarden

Mathematical Centre,
Amsterdam

ABSTRACT

This paper demonstrates that it is readily possible to express an algorithm through a two-level grammar, with an ease comparable to but under avoidance of an intervening programming language.

1. INTRODUCTION

If a programmer wants to specify a process to be performed on data then (s)he writes a particular program in some programming language, e.g., ALGOL 68. The particular program might be:

$(INT\ n;\ read(n);\ print(("p(", whole(n,0), ")\underline{\ } - \underline{\ }25\underline{\ } = \underline{\ }", whole(prime(n)-25,0))))$

where the marks are representations of the ALGOL 68 terminal symbols.

This particular program does not completely specify the process to be performed for two reasons. First, RR68, i.e., the definition of ALGOL 68 [2] does not provide a declaration for *prime*. This declaration might however be given in a particular prelude. Let us assume that this is the case and that the call *prime(n)* delivers what the wording suggests, viz., the n-th prime number. The second reason is that the call *read(n)* might assign a nonpositive value to n. Thus we must assume that *prime* can cope with this situation in some way or another, or, otherwise, the programmer should correct the particular program, e.g., by writing *read(n); n:=ABSn+1* instead of *read(n)*. We shall assume the first possibility.

This particular program may be elaborated by a computer. Let us assume that *read(n)* has the same effect as $n:=125$ would have had. Since the 125-rd prime number is, as well known, 691 the result would be

$$p\ (\ 1\ 2\ 5\ )\quad -\quad 2\ 5\quad =\quad 6\ 6\ 6$$

where the marks are from a totally different vocabulary, only partially defined by RR68.

The definition of the programming language, here ALGOL 68, obviously also belongs to the specification of the process because, otherwise, the meaning of the program would be undefined. This definition is given in RR68 partly in the English language and partly by means of a specific two-level grammar with production rules like

program : strong void new closed clause.

where the marks are again from a totally different vocabulary.

The definition of the concept "two-level grammar" obviously also belongs to the specification of the process because, otherwise, the meaning of the specific two-level grammar of the language would be undefined. This definition was first given in [1] in a loose manner and later more elaborately in [2]. A formal operational definition was given in [3] and is included here in section 2.

There are therefore, apart from the definition in the English language, at least four levels of definition involved in the specification, viz., the definition of the two-level-grammar concept by the tool maker; the specific two-level grammar of a language by the language designer; the particular program in that language by the programmer; the output of that particular program by the computer.

This is a considerable hierarchy and, moreover, a voluminous one: RR68 contains more than two hundred pages. The reward is only that ALGOL 68 programs can be elaborated by actual present-day computers but still specification is poor; in effect the (truth) value of $0<0$ is not better defined by RR68 than by the loose remark 'usual mathematical meaning' and the (truth) value of $0-0=0$ is not defined at all by RR68.

The reason for this deplorable situation is that we have on one hand the extreme definitional power of two-level grammars and on the other hand a modest process that has to be defined. The insertion between these two of a large programming language not specifically adapted to that process is — apart from that elaboration by an actual present-day computer — like the use of an elephant to carry groceries home from the supermarket; the elephant may eat part of them.

In this paper it is shown how the language can be eliminated on the hand of a specific example. In fact, a set of metarules and a set of hyperrules is given whose power exceeds by far this example and only the choice of a specific hyperrule for the startword restricts it to this example. In some sense these sets of metarules and hyperrules may be regarded as a language but there is an essential difference: the "grammarian", i.e., the person who specifies a process by means of a two-level grammar, never writes down a representation of a terminal letter; on the contrary, these representations are equivalent with the marks that are written down by the computer introduced above. The language is thereby completely eliminated.

The critical remarks on ALGOL 68 should not be misinterpreted; remarks on other languages might very well have been much more critical indeed.

2. TWO-LEVEL GRAMMARS

A "vocabulary" is a set; its elements are termed "letters". A "word" over a vocabulary V is a mapping $[1:n] \mapsto V$, for any $n \in \mathbb{N}_0$, and is thus a set of n ordered pairs (i, v_i), for $i = 1, \cdots, n, v_i \in V$. Therefore, v_i is termed the "i-th letter" and n the "length" of the word. If $n = 0$, then the word is the empty set, also termed the "empty word". For each vocabulary V, V^* denotes the set of words over V, and V^+ the set of nonempty words over V. A "sentence" over a vocabulary V is a word over the vocabulary whose letters are the words over V; hence V^{**} is the set of sentences over V.

A "rule" is an ordered pair (v, w) where v and w are words over certain vocabularies.

A "two-level grammar" VWG is an ordered sextuple $(V_m, V_o, V_t, R_m, R_h, w_s)$, where V_m, V_o, V_t are finite vocabularies, whose letters are termed "metaletters", "ortholetters" and "terminal letters" respectively, R_m and R_h are finite sets of rules, termed "metarules" and "hyperrules" respectively, and w_s is some word over V_o, termed the "startword". Let $V_h := V_m \cup V_o$. It is required that $V_m \cap V_o = \{\}$, $V_t \subset V_o^+$, $R_m \subset V_m \times V_h^*$, $R_h \subset V_h^+ \times V_h^{**}$, $w_s \in V_o^+$.

The grammar VWG "produces" a sentence set L defined as follows:

Let $R_{mo} := V_m \times V_o^*$, $R_{oo} := V_o^+ \times V_o^{**}$ and $R_{st} := \{w_s\} \times V_t^*$. A set R_m' identical with R_m and a set R_h' identical with R_h are introduced and then extended by applying, arbitrarily often, if possible, the following extension, where at each application, out of the three alternatives separated by ',' and enclosed by '( )', consistently either the first, or the second, or the third must be chosen:

Extension: To (R_m', R_h', R_h') a rule is added, obtained by replacing, in a copy of some rule $(v, w) \in (R_m', R_h', R_h')$ and for some rule $(v', w') \in (R_m' \cap R_{mo}, R_m' \cap R_{mo}, R_h' \cap R_{oo})$, (some, each, some) occurrence of v' in $(w, v$ and $w, w)$ by w'.

Then, $L := \{w \mid (w_s, w) \in R_h' \cap R_{st}\}$.

The set of "terminal metaproductions" of a metaletter M is the set $\{w \mid (M, w) \in R_m' \cap R_{mo}\}$.

In this paper a metaletter is a conventional 'capital letter' possibly followed by one or more times '″' (apostrophe); an ortholetter is a conventional 'lower case letter', 'decimal digit', 'opening parenthesis' or 'closing parenthesis'. Four other letters play a role, viz., ':' (colon), '.' (point), ',' (comma) and ';' (semicolon).

Letters are "written" one after the other in such a way that the order in which they have been written is clear, in this paper conventionally to the 'right of' the letter lastly written before, or 'on the next line' or 'on the next page', whatever this may mean.

A word is written when its writing starts by writing its first letter, if it exists, and, when its i-th letter has been written by writing its $(i + 1)$-th letter, if it exists.

A metarule (v,w) is written by writing v, then writing twice a colon, then writing w and then writing a point.

A hyperrule (v,w) is written by writing v, then writing a colon, then writing w and then writing a point. Writing w, however, poses the problem that a sentence over V_o might later on be misread. In natural languages this problem is overcome by separating the words of the sentence by blanks. Here, traditionally, one separates the words by writing a comma after a word has been written and before the next word of the sentence is going to be written, which leaves open the use of blanks for display purposes inside the words.

The grammar mechanism as defined so far is, however, not yet complete. The terminal letters are words over V_o but this internal structure is of no relevance to the user. Therefore, V_t is mapped onto another vocabulary W_t, the set of "representations" of the terminal letters, which may be marks chosen by the user at his convenience as long as they differ from all letters mentioned above.

A terminal production of a VWG, i.e., an element of L, is represented by replacing each terminal letter in a copy of that element by its representation and taking out the comma that follows it, if any.

A useful shorthand notation for rules is the following one: If two rules have the same left-hand side up to and including the colon or double colon, then they may be combined into one rule consisting of the first rule in which the point has been replaced by a semicolon, followed by the right-hand side of the second rule.

Thus **H :: 9 ; 8 ; 7.** stands for **H :: 9. H :: 8. H :: 7.**.

Another useful convention stems from the fact that one frequently needs metarules differing only in the left-hand side, because one wants to circumvent the effect of the, utterly necessary, word 'each' in the Extension. Therefore, by convention, it holds:

Let M stand for any element of V_m. Then any occurrence of M' in a VWG tacitly implies that $M' :: M.$ is an element of R_m.

Thus, the occurrence of **V″** implies the metarule **V″ :: V′.**, which implies the metarule **V′ :: V.**, so that **V**, **V′** and **V″** have the same terminal productions over V_o.

3. AN EXAMPLE OF A TWO-LEVEL GRAMMAR FOR LANGUAGELESS PROGRAMMING

The feasibility of languageless programming is shown by the following two-level grammar.

The metaletters are in first instance **H, J, K, L, M, N, P, Q, R, V, W** and, moreover, if M stands for any metaletter a metaletter M' may be added.

The ortholetters are **n, p, s, 9, 8, 7, 6, 5, 4, 3, 2, 1, 0, (,)**.

The metarules are in first instance

M1 **H :: 9 ; 8 ; 7 ; 6 ; 5 ; 4 ; 3 ; 2 ; 1 ; 0.**
M2 **J :: HJ ; .**
M3 **K :: n ; p.**
M4 **N :: nN ; .**
M5 **P :: pP ; .**
M6 **M :: nN ; pP ; .**
M7 **Q :: HQ ; KQ ; .**
M8 **V :: sM.**
M9 **W :: sQ ; (R).**
M10 **L :: VL ; .**

M11 **R :: WR ; .**

The terminal letters are terminal metaproductions of **V**.

The startword is **s0**.

The hyperrules are

H1   **L(R′)R : LR′R.**

H2   **s0 : sp25sn8spPs1ns10sn9sn2sn41sn2sp25s1ns10sn2sn36sn2spPs50sp25s41s1ns10sps64ps2.**

H3   **Ls00R : .**

H4   **LsKHJR : Lss0KHJR.**

H5   **LVs0KHJR : LVsppppppppppps42s00KH0123456789s40s0KJR.**

H6   **LVs0KR : LVR.**

H7   **LsM00KHH′JR : LsKM00KHJR.**

H8   **LsM00KHHJR : LsMR.**

H9   **LVs1MR : LVss24s4sn41sMVs03ps9s01s01R.**

H10 **Ls01PsP′L′s01NR : LsP′sp10s34s4(sP′sp10s41s9(s01pP)L′s01N)**
                (sPss22s4(s01sPsP′L′s01nN)(sP′L′snN))R.

H11 **LVs2R : V , LR.**

H12 **LsKMs3M′R : LsMsKs4s3nM′s3pM′R.**

H13 **Lss3M′R : LsM′R.**

H14 **LVs03KR : LVVs4( )s3sKs4( )s3R.**

H15 **LsPs4WW′R : LWR.**

H16 **LsNs4WW′R : LW′R.**

H17 **LVs5(WW′R′)R : LVss32s4W(Vsps41s5(W′R′))R.**

H18 **LVs5(W)R : LWR.**

H19 **Ls6pPVR : LVs6PR.**

H20 **Ls6pPs6HMWR : LR.**

H21 **LVV′s6nNR : LVs6NV′R.**

H22 **Vs6nNLs6HMWR : VLR.**

H23 **Ls7WW′R : LWs4(W′s7WW′)( )R.**

H24 **LVV′V″s8WW′R : LV′ss36s4sp(VV″V′s4s32s34)VV′V″s8pWW′R.**

H25 **LsMVV′V″s8KWW′R : LsMs4(sKs4(WVV′V″s8n)(W′VV′s40V′V″s8)WW′)sKR.**

H26 **LVs9WR : LWVR.**

H27 **LVs10R : LR.**

H28 **LVs11R : LVVR.**

H29 **LsMs12WR : LWMR.**

H30 **LVs13R : LVs4spsnR.**

H31 **LVV′s20R : LVV′s21s11s3s21R.**

H32 **LVV′s21R : LVV′s24s4VV′R.**

H33 **LVV′s22R : LVV′s41s022R.**

H34 **LspPs022R : LspPR.**

H35 **LsNs022R : LsnNR.**

H36 **LVV′s23R : LVR.**

H37 **LVV′s24R : LV′Vs22R.**

H38 **LVV′s25R : LV′R.**

H39 **LVV′s26R : LVV′s22VV′s24s27R.**

H40 **LVV′s27R : LVV′s22s4VV′R.**

H41 **LVV′s3HR : LVV′s2Hs3R.**

H42 **LsMsM′s40R : LsMM′R.**

H43 **LsnNpPR : LsNPR.**

H44 **LspPnNR : LsPNR.**

H45 **LVV′s41R : LVV′s3s40R.**

H46 **LVV′s42R : LV′ss36s4s(V′s4(VV′sps41s42Vs40)(VV′s3s42s3))R.**

H47 **LVV′s43R : LsVs03pV′s03ps043Vs4(V′s4( )(sns64s3))(V′s4(sns64ps3)s3)R.**

H48 **LsP″sPspP′s043R : LsPspP′s34s4(spP″sPspP′s41spP′s043)(sP″sP)R.**

H49 **LVsPs44R : LsPss36s4sp(VsPsps41s44Vs42)R.**

H50 **LVV'spPs45R : LVV'V's40s03pspPs22s4(V's13s40)( )R.**

H51 **LVV's40pPR : LVV's40R.**

H52 **LVV's41pPR : LVV's41R.**

H53 **LVV's42pPR : LVV's42spPs60s43spPs60s45R.**

H54 **LVV's43pPR : LVspPs60s42V's43spPs60s45R.**

H55 **LVV's44pPR : LV'ss36s4(spPs60)(V's4(VV'sps41s44pPVs42pP) (spPs60VV's3s44pPs43pP))R.**

H56 **LspPs50R : LspPspspps050R.**

H57 **LspPsppP'sppP''s050R : LsppP'sppP''s43sns64s10ss22 s4(spPsppP'spppP''s050)**
         (sppP'sppP''s22s4(spPsppP'spps050)(spPsps22s4(sPspppP'spps050)sppP'))R.

H58 **LsM's6HMWR : sM'ss22s4(s6M'L)(Ls6M')s6HMWR.**

H59 **LVs6L's60R : LVL'VR.**

H60 **LVs6L'V's61R : LV'L'R.**

H61 **Ls6L'V's62R : LV'L'R.**

H62 **Ls6L's63KR : LsKs4L'LL'R.**

H63 **LVs6L's64MWR : LVWsMss36s4(L')(s6ML's64MW)R.**

4. INTERPRETATIONS

A two-level grammar defines a formal play with letters without the need of an interpretation. We might, therefore, leave the reader with the lists of metarules and hyperrules without comment. However, he may wonder what they are good for and he may find it hard to understand their effect. Therefore, we shall help him with some "interpretations", in this section, and "explanations", in section 5, but he should keep in mind that these interpretations and explanations have no defining power, they are just comment.

The terminal productions of **W** are termed "syllables". A syllable is either a "value" or an "operator". Values are the terminal productions of **V**. They are therefore

     ... , **snnn** , **snn** , **sn** , **s** , **sp** , **spp** , **sppp** , ...

and operators are therefore, e.g.,

     s23 , **s3n** , **sp25** , **spppnn** , **(sns64s10)** .

Each, possibly empty, sequence of syllables is a terminal production of **R** (for right) and, more specifically, each, possibly empty, sequence of values is a terminal production of **L** (for left). The idea is that a "word" typically consists of a, possibly empty, sequence of values, followed by an operator, the "first operator", followed by a, possibly empty, sequence of syllables. Hyperrules the define how the word produces another word due to the first operator in its context. The location of the first operator is specified by the "name" F; the location of any value to the left of the first operator is specified by a name which "refers to" the value at that location. These names are, counting from the left to the right, Sp , Spp , $Sppp$, ... , F and also, counting from the right to the left, ... , $Snnn$, Snn , Sn , S , F . Each value to the left of the first operator is therefore referred to ·by two names, a "positive" one and a "nonpositive" one. A pair formed by a name and the value to which that name refers is a "variable". A variable with a positive name is a "declared" variable, i.e., a variable whose name does not depend on the location specified by F; a variable with a nonpositive name is a "stack" variable, i.e., a variable whose name does depend on that location.

Values may be "interpreted", e.g., in the sense of ALGOL 68, i.e., integers, real numbers, truth values, names, etc. The interpretation of the above sequence of values as integers is

     ... , $-3, -2, -1, 0, 1, 2, 3, ...$

The interpretation as real numbers is

     ... , $-3/B, -2/B, -1/B, 0/B, 1/B, 2/B, 3/B, ...$

where B, the "base", stands for some integer greater than one.

The interpretation as truth values in a many-valued logic is

... , *fff, ff, f, u, t, tt, ttt,* ...

where *f* suggests FALSE, *t* TRUE and *u* UNDEFINED.

For a given word the interpretation as names is

... , *Snnn* , *Snn* , *Sn* , *S* , *Sp* , *Spp* , *Sppp* , ...

names which have been defined above.

Which of these interpretations is in a given case the most appropriate one depends on the first operator and its context.

The terminal letters of the grammar are certain values if they stand on their own, i.e., not as syllables of a larger word. Strictly speaking, an interpretation of the terminal letters is unnecessary since the grammarian never writes down the representation of the terminal letters, in contrast to the situation that a programmer writes a program in some programming language where he writes only representations of the terminal letters of the grammar defining that language. It is, however, more helpful to provide a list of representations in order to show an interpretation of the terminal letters. In the list we write for the terminal letter consisting of **s** followed by, e.g., 25 times **p** simply **sp25** and for **s** followed by, e.g., 8 times **n** simply **sn8**. This is not so sloppy as it might seem at first sight since, according to the hyperrules H4-H8, the operator **sp25** indeed produces **s** followed by 25 times **p** and the operator **sn8** indeed produces **s** followed by 8 times **n**. The list might then run as follows:

sn42	×		**sn41**	−	**sn40**	+	**sn36**	=	**sn34** ≥
sn32	⩽	⌐	**sn29**	}	**sn28**	{	**sn26**	≠	**sn24** <
sn22	>		**sn19**	]	**sn18**	[	**sn17**	\	**sn16** |
sn15	/		**sn13**	←	**sn12**	→	**sn11**	↑	**sn10** ↓
sn9	)		**sn8**	(	**sn7**	:	**sn6**	,	**sn5** .
sn4	(newpage)				**sn3**	(newline)			
snn	(blank space)				**sn**	(invisible)			

s	*0*	**sp**	*1*	**spp**	*2*	**sp3**	*3*	**sp4**	*4*
sp5	*5*	**sp6**	*6*	**sp7**	*7*	**sp8**	*8*	**sp9**	*9*
sp10	*a*	**sp11**	*b*	**sp12**	*c*	**sp13**	*d*	**sp14**	*e*
sp15	*f*	**sp16**	*g*	**sp17**	*h*	**sp18**	*i*	**sp19**	*j*
sp20	*k*	**sp21**	*l*	**sp22**	*m*	**sp23**	*n*	**sp24**	*o*
sp25	*p*	**sp26**	*q*	**sp27**	*r*	**sp28**	*s*	**sp29**	*t*
sp30	*u*	**sp31**	*v*	**sp32**	*w*	**sp33**	*x*	**sp34**	*y*
sp35	*z*	**sp36**	*A*	**sp37**	*B*	**sp38**	*C*	**sp39**	*D*
sp40	*E*	**sp41**	*F*	**sp42**	*G*	**sp43**	*H*	**sp44**	*I*
sp45	*J*	**sp46**	*K*	**sp47**	*L*	**sp48**	*M*	**sp49**	*N*
sp50	*O*	**sp51**	*P*	**sp52**	*Q*	**sp53**	*R*	**sp54**	*S*
sp55	*T*	**sp56**	*U*	**sp57**	*V*	**sp58**	*W*	**sp59**	*X*
sp60	*Y*	**sp61**	*Z*						

The choice of the representations of **s** up to **sp61** is obvious. The representations of the remaining terminal letters are partially chosen for mnemonic reasons. Since, e.g., **s40** is the operator that performs the addition of two integers, + is chosen as representation of **sn40**.

5. EXPLANATIONS

We now start to explain the effect of the hyperrules for the operators in some detail. We shall not follow the order in which they are given in Section 3 but shall deal with them in such an order that on the right side only operators occur that have been dealt with already or that are under discussion. That this is possible proves at the same time that the grammar does not contain viciously circular definitions.

It often occurs that in a hyperrule **W** occurs whereas in a certain application of that hyperrule we actually like to replace that **W** not by one syllable but by a sequence of syllables, e.g.,

sns64s10. We then replace it instead by a "compound operator", in this case **(sns64s10)**, which is a terminal production of **W**, i.e., a syllable, but not of **V**, i.e., not a value and hence an operator. If this operator becomes the first operator then the general rule

H1   **L(R')R : LR'R.**

decomposes it into a sequence of its constituent syllables, after which the hyperrule for the new first operator takes over control.

In the context **LVs4WW'R** the operator **s4** "chooses" between **W** and **W'** roughly as *IF* **V** *THEN* **W** *ELSE* **W'** *FI* does:

H15 **LsPs4WW'R : LWR.**
H16 **LsNs4WW'R : LW'R.**

Here the obvious interpretation of **V** is a truth value. Since our logic is many-valued, **V** may be **s**, interpreted as UNDEFINED, and then both hyperrules are applicable, leaving it undefined whether **W** or **W'** is chosen.

In the context **LVs3R** the operator **s3** "negates" **V**, i.e., it transforms **s** followed by a number of times **p** resp. **n** into **s** followed by the same number of times **n** resp. **p**. If **V** is interpreted as an integer or a real number then this negation means a change of the sign of **V**, whereas if **V** is interpreted as a truth value then it means logical negation. Its definition is an application of the operator **s4**:

H12 **LsKMs3M'R : LsMsKs4s3nM's3pM'R.**
H13 **Lss3M'R : LsM'R.**

The operator **s3** might also have been defined without using the operator **s4**. This would cost four hyperrules instead of two, and our strategy is always to minimize the number of hyperrules.

In the context **LVs03KR** the operator **s03p** delivers *ABS* **V** and the operator **s03n** delivers $-ABS$ **V**:

H14 **LVs03KR : LVVs4( )s3sKs4( )s3R.**

In the context **LVs9WR** the operator **s9** "swaps" **V** and **W**:

H26 **LVs9WR : LWVR.**

In the context **LVs10R** the operator **s10** "voids" **V**:

H27 **LVs10R : LR.**

In the context **LVs11R** the operator **s11** "duplicates" **V**:

H28 **LVs11R : LVVR.**

In the context **LsMs12WR** the operator **s12** "adorns" **W**:

H29 **LsMs12WR : LWMR.**

In the context **LVs13R** the operator **s13** "projects" **V** onto **sK**:

H30 **LVs13R : LVs4spsnR.**

An application of this projection is the transformation of many-valued logic into Boolean logic, but see also H50.

We now turn to the simplest arithmetic operations on integers.

In the context **LVV's40R** the operator **s40** delivers **V + V'**:

H42 **LsMsM's40R : LsMM'R.**
H43 **LsnNpPR : LsNPR.**
H44 **LspPnNR : LsPNR.**

In the context **LVV's41R** the operator **s41** delivers **V − V'**:

H45 **LVV's41R : LVV's3s40R.**

Now that we have subtraction of integers at our disposal we turn to operators that deal with relations between values that can be interpreted as truth values, integers or real numbers at will. For some of them the interpretation as truth values and for other ones the interpretation as integer or real numbers is more obvious. Anyhow, we start from the sixteen Boolean operators and drop then the restriction that the two arguments and the result be two-valued. We follow their natural order and number them from **s20** up to **s27** and from **s30** up to **s37**. We shall treat them in a different order so that the reader does not need to look ahead. The context is always **LVV's2HR** resp. **LVV's3HR**.

The operator **s22** delivers $V > V'$:

H33 **LVV's22R : LVV's41s022R.**

H34 **LspPs022R : LspPR.**

H35 **LsNs022R : LsnNR.**

The operator **s24** delivers $V < V'$:

H37 **LVV's24R : LV'Vs22R.**

The operator **s21** delivers V *AND* V' or min(V,V'):

H32 **LVV's21R : LVV's24s4VV'R.**

The operator **s27** delivers V *OR* V' or max(V,V'):

H40 **LVV's27R : LVV's22s4VV'R.**

The operator **s20** delivers (V *AND* V') *AND NOT* (V *AND* V'); i.e., FALSE of some size:

H31 **LVV's20R : LVV's21s11s3s21R.**

The operator **s23** delivers **V**, its first operand:

H36 **LVV's23R : LVR.**

The operator **s25** delivers **V'**, its second operand:

H38 **LVV's25R : LV'R.**

The operator **s26** delivers $V \neq V'$:

H39 **LVV's26R : LVV's22VV's24s27R.**

The operator **s3H** delivers *NOT* (**VV's2H**):

H41 **LVV's3HR : LVV's2Hs3R.**

The operators **s20**, **s21**, **s23**, **s25**, **s27**, **s30**, **s31**, **s35** and **s37** can deliver s, i.e., UNDEFINED, whereas **s22**, **s24**, **s26**, **s32**, **s34** and **s36** cannot.

We now consider operators which have to do with names. In the context **LsM's6HMWR**, where the syllable **sM'** can best be interpreted as a name which refers to *SM'*, the operator **s6HM** generates an operator **s6M'** and takes out **sM'** but not itself; an operator **s6pP** is inserted to the left of *Sp*, an operator **s6N** is inserted to the right of *S*:

H58 **LsM's6HMWR : sM'ss22s4(s6M'L)(Ls6M')s6HMWR.**

An inserted operator **s6pP** steps over the value on its right losing one **p** at that step and an inserted operator **s6nN** steps over the value on its left losing one **n** at that step. In doing so, an inserted operator **s6M'** either comes to rest as **s6** at the right of *SM'* or, if it threatens to step out of **L**, it disappears together with **s6HMW**:

H19 **Ls6pPVR : LVs6PR.**

H20 **Ls6pPs6HMWR : LR.**

H21 **LVV's6nNR : LVs6NV'R.**

H22 **Vs6nNLs6HMWR : VLR.**

What happens after the operator **s6** has come to rest, to the right of *SM'*, depends on **H** and **M**.

In the context **LsM's60R** the operator **s60** "dereferences" **sM'**:

H59 **LVs6L's60R : LVL'VR.**

In the context **LV'sM's61R** the operator **s61** "assigns" **V'** to **sM'**:

H60 **LVs6L'V's61R : LV'L'R.**

In the context **LV'sM's62R** the operator **s62** "inserts" **V'** to the right of *SM'*:

H61 **Ls6L'V's62R : LV'L'R.**

In the context **LsM's63KR** the operator **s63p** duplicates the part of **L** to the right of *SM'* and the operator **s63n** duplicates the part of **L** that ends with *SM'*:

H62 **Ls6L's63KR : LsKs4L'LL'R.**

In the context **LsM's64MWR** the operator **s64** makes the operator **W** operate on *SM'* whereas the operators **s64p** resp. **s64n** make **W** operate on *SM'* and all syllables of **L** to the right resp. to the left of it:

H63 **LVs6L's64MWR : LVWsMss36s4(L')(s6ML's64MW)R.**

Now that we have relations and operations concerning names at our disposal we can continue defining operations on integers.

In the context **LVV's42R** the operator **s42** delivers $V \times V'$:

H46 **LVV's42R : LV'ss36s4s(V's4(VV'sps41s42Vs40)(VV's3s42s3))R.**

In the context **LVV's43R** the operator **s43** delivers the quotient $V \div V'$ followed by the remainder $V \div \times V'$:

H47 **LVV's43R : LsVs03pV's03ps043Vs4(V's4( )(sns64s3))(V's4(sns64ps3)s3)R.**

H48 **LsP''sPspP's043R : LsPspP's34s4(spP''sPspP's41spP''s043)(sP''sP)R.**

In the context **LVsPs44R** the operator **s44** delivers $V{\uparrow}sP$:

H49 **LVsPs44R : LsPss36s4sp(VsPsps41s44Vs42)R.**

In the context **LVV'spPs45R** the operator **s45** "rounds" V if $2 \times ABS\,V' > spP$:

H50 **LVV'spPs45R : LVV'V's40s03pspPs22s4(V's13s40)( )R.**

Next we consider operations on real numbers. The mathematical interpretation of a value as a real number is its interpretation as an integer divided by the "base", i.e., some positive integer greater than one, e.g., $10{\uparrow}18$. The operator **s4HpP** interprets syllable SpP as that base.

In the context **LVV's40pPR** the operator **s40pP** delivers $V + V'$:

H51 **LVV's40pPR : LVV's40R.**

In the context **LVV's41pPR** the operator **s41pP** delivers $V - V'$:

H52 **LVV's41pPR : LVV's41R.**

In the context **LVV's42pPR** the operator **s42pP** delivers $V \times V'$:

H53 **LVV's42pPR : LVV's42spPs60s43spPs60s45R.**

In the context **LVV's43pPR** the operator **s43pP** delivers V/V':

H54 **LVV's43pPR : LVspPs60s42V's43spPs60s45R.**

In the context **LVV's44pPR** the operator **s44pP** delivers $V{\uparrow}V'$, where V' is interpreted as an integer:

H55 **LVV's44pPR : LV'ss36s4(spPs60)(V's4(VV'sps41s44pPVs42pP) (spPs60VV's3s44pPs43pP))R.**

So far, we have discussed only operators of general and widely applicable character like those in the standard-prelude of ALGOL 68. We shall now give one example of a very specific operator like one that one might expect in a particular-prelude of ALGOL 68.

In the context **LspPs50R** the operator **s50** delivers the **spP**-th prime number:

H56 **LspPs50R : LspPspspsps050R.**

H57 **LspPsppP'sppP''s050R : LsppP'spppP''s43sns64s10ss22 s4(spPsppP'spppP''s050)**
        (sppP'sppP''s22s4(spPspppP'spps050)(spPsps22s4(sPspppP'spps050)sppP'))R.

Rule H56 introduces two new values both set to two, the first of which is a candidate for being a prime number and the second of which is a trial divisor to investigate whether the candidate is actually a prime number, and a new operator **s050**. Rule H57 determines the remainder of the candidate upon division by the trial divisor; if this remainder is positive then the trial divisor is not a divisor of the candidate and the next trial divisor is tried; if the remainder is zero then the rule determines whether the candidate is greater than the trial divisor; if this is so then the candidate is a composite number and the next candidate is investigated starting again with a trial divisor two; otherwise, the candidate is a prime number and the rule investigates the number of prime numbers that were still to be found; if this number is greater than one then that number is diminished by one and the next candidate is investigated; otherwise, the candidate is the sought prime number.

Next we define some ALGOL 68-like constructs.

In the context **LVs5(WR')R** the operator **s5** "selects" a syllable from the sequence **WR'** roughly as *CASE V IN W, R' ESAC* does:

H17 **LVs5(WW'R')R : LVss32s4W(Vsps41s5(W'R'))R.**

H18 LVs5(W)R : LWR.

Here the obvious interpretation of **V** is an integer. The syllables of **WR'** are counted from 0 up to k, say; if **V**$\leq$0 then **W** is selected, if **V**$\geq k$ then syllable k and, otherwise, syllable **V** is selected.

In the context **Ls7WW'R** the operator **s7** "repeats" **W'** subject to the condition **W** like *WHILE **W** DO **W'** OD* does:

H23 Ls7WW'R : LWs4(W's7WW')( )R.

In the context **LVV'V"s8WW'R** the operator **s8** repeats **W'** subject to a bound test and the condition **W** like *FROM **V** BY **V'** TO **V"** WHILE **W** DO **W'** OD* does:

H24 LVV'V"s8WW'R : LV'ss36s4sp(VV"V's4s32s34)VV'V"s8pWW'R.

H25 LsMVV'V"s8KWW'R : LsMs4(sKs4(WVV'V"s8n)(W'VV's40V'V"s8)WW')sKR.

In contrast to the ALGOL 68 clause a non-void result is delivered, viz., **sp** when the test on **V** and **sn** when the test on **W** failed.

Now we consider operators that perform the transition from conventional decimal notation to our unary notation and vice versa.

In the context **LsKHJR** the operators **spHJ** resp. **snHJ** deliver **s** followed by **HJ** times **p** resp. **n**:

H4 LsKHJR : Lss0KHJR.

H5 LVs0KHJR : LVspppppppppps42s00KH0123456789s40s0KJR.

H6 LVs0KR : LVR.

H7 LsM00KHH'JR : LsKM00KHJR.

H8 LsM00KHHJR : LsMR.

Rule H4 introduces **s**, i.e., zero as preliminary result, and the new operator **s0KHJ**. Rule H5 multiplies the preliminary result by ten, applies the operator **s00KH0123456789** to add the number "suggested" by the leading decimal digit **H** with the sign as given by **K**, and then deals with the next digit of **J**, if any, by a repeated application of rule H5. The rules H7 and H8 deal with this new operator. Notice that this operator in H5 has zero letters **p** or **n** between **s** and **00** and that **H** is facing the digit **0**. Application of rule H7 inserts a letter **p** or **n** after **s**, so that there is now one letter **p** or **n** between **s** and **00** and the digit **0** that **H** was facing is taken out so that **H** is now facing the digit **1**. In general, after a number of applications of rule H7 the number of letters **p** or **n** between **s** and **00** equals the number suggested by the digit **H'** that **H** is facing. Repeated application of the rule H7 eventually leads to a blind alley, the only one in our grammar, because **H** is no longer facing a digit **H'**. However, in this process **H** will necessarily face itself once and then rule H8 delivers the desired result, i.e., the value suggested by **H**. When all digits of **HJ** have been dealt with in this way rule H6 delivers the final result.

The operator **sKHJ** enables the grammarian to enter constants in his grammar like, in decimal notation, *125* resp. -256 by writing **sp125** resp. **sn256**.

In the context **LVs1MR** the operator **s1M** translates **V**, in our unary notation, into a sequence of values, viz., the sign of **V**, the successive decimal digits of **V** and an indication of the number of digits obtained:

H9 LVs1MR : LVss24s4sn41sMVs03ps9s01s01R.

H10 Ls01PsP'L's01NR : LsP'sp10s34s4(sP'sp10s41s9(s01pP)L's01N)
$\qquad$**(sPss22s4(s01sPsP'L's01nN)(sP'L'snN))R.**

If **V** is negative then the sign is **sn41**; if **V** is nonnegative then the sign is **sM**. This enables the grammarian to choose, e.g., between the plus mark **sn40**, the blank-space mark **snn** or the invisible mark **sn** by writing **sn40s12s1**, **s1nn** or **s1n** respectively. Rule H9 first determines the sign and then swaps the absolute value of **V** between the operators **s01** and **s01**. Rule H10 then decomposes **V** gradually into a sequence of decimal digits, in unary notation, of course, thereby using the first **s01** as portmanteau for intermediate values of the quotient upon division by ten and the second **s01** as portmanteau for intermediate values of the indication of the number of decimal digits obtained. The final indication is then the negation of that number. The reason for not building up that number itself will soon become clear. Anyhow, if the grammarian does not want the indication he can void it immediately by means of **s10**.

In the context **LVs2R** the operator **s2** "outputs" **V**:

H11 **LVs2R : V , LR.**

It is our only hyperrule in which the comma occurs. Since **V** is a member of the terminal vocabulary V_t it plays no further role in the production process and it can be replaced by its representation. This is our equivalent of output in a programming language.

Since **LVs1MR** produces a sign followed by a number of decimal digits followed by the negation of that number, **LVs1Ms64ps2R** outputs **V** in decimal notation; hence, **Lsp25s2R** outputs *p* and **Lsp25s1ns64ps2R** outputs *25*.

In the context **Ls00R** the operator **s00** "finishes" the production process:

H3 **Ls00R : .**

In the context **s0**, at last, the operator **s0**, i.e., the startword, produces the specific problem that the grammarian has in mind. A very specific hyperrule is, by way of example:

H2 **s0 : sp25sn8spPs1ns10sn9sn2sn41sn2sp25s1ns10sn2sn36sn2spPs50sp25s41s1ns10sps64ps2.**

First we observe that on the right-hand side the metaletter **P** occurs which does not occur on the left; hence its terminal metaproduction can be freely chosen. This is our equivalent of 'input' in a programming language. Suppose that we choose a sequence of one hundred and twenty four times the letter **p** for that terminal production. Then the operators on the right-hand side are harmless ones like **sp25** that delivers **s** followed by twenty five times **p**, and **sn8** that delivers **s** followed by eight times **n**. The first interesting combination is **spPs1ns10**, which produces **snspsppsppppp**. The second interesting combination is **spPs50** which produces **s** followed by six hundred and ninety one times **p** because the 125-rd prime number is 691, which follows from rule H56. Hence the combination **spPs50sp25s41s1ns10** produces **sppppppspppppppspppppp** because 691 − 25 = 666. At last, the combination **sps64ps2** selects the left-most syllable and outputs it together with all its right-hand side syllables. This then produces

$$p(125) \; - \; 25 \; = \; 666$$

a result which is the same as that of the ALGOL 68 particular program in the introduction but, of course, now rigorously defined.

6. REFERENCES

[1]   WIJNGAARDEN, A. VAN, *Orthogonal design and description of a formal language*, Mathematical Centre, Amsterdam, MR76 (1965).

[2]   WIJNGAARDEN, A. VAN et al., eds., *Revised Report on the algorithmic language ALGOL 68*, Acta Informatica **5** (1975) 1-234.

[3]   WIJNGAARDEN, A. VAN, *Thinking on two levels*, in Proc. Bicentennial Congress of the Wiskundig Genootschap, part 2, P.C. Baayen, D. van Dulst & J. Oosterhoff, eds., Mathematical Centre, Amsterdam, Mathematical Centre Tracts, Vol. 101 (1979) 417-428.

LIST OF PARTICIPANTS

J.C. ADAMS,
National Center for Atmospheric Research,
Boulder, Colorado 80307,
U.S.A.

E.L. BATTISTE, C. ABACI,
P.O. Box 5715,
Raleigh, North Carolina 27650,
U.S.A.

R. BRENT,
Department of Computer Science,
Australian National University,
Box 4, Post Office,
Canberra, ACT 2600,
Australia.

E.K. BLUM,
Mathematics Department,
University of South California,
Los Angeles, California 90009,
U.S.A.

J.M. BOYLE,
Applied Mathematics Division,
Argonne National Laboratory,
9700 South Cass Avenue,
Argonne, Illinois 60439,
U.S.A.

W.S. BROWN,
Computing Science Research Center,
Bell Laboratories,
Murray Hill,
New Jersey 07974,
U.S.A.

F. CHATELIN,
IMAG,
University of Grenoble,
BP 53,
F-38041 Grenoble,
France.

W.J.CODY, Jnr.,
Applied Mathematics Division,
Argonne National Laboratory,
9700 South Cass Avenue,
Argonne, Illinois 60439,
U.S.A.

W. COWELL,
Applied Mathematics Division,
Argonne National Laboratory,
9700 South Cass Avenue,
Argonne, Illinois 60439,
U.S.A.

Th.J. DEKKER,
Department of Mathematics,
University of Amsterdam,
Roetersstraat 15,
Amsterdam NL-1018 WB,
The Netherlands.

L.M. DELVES,
Computer Science Department,
University of Liverpool,
Liverpool L69 3BX,
U.K.

B. EINARSSON,
Forsvarets Forskningsanstalt,
National Defense Research Institute,
Department 4,
S-901 82 Umeå,
Sweden.

S.C. EISENSTAT,
Computer Science Department,
Yale University,
New Haven, Connecticut 06520,
U.S.A.

R.J. FATEMAN,
Department of Electrical Engineering
and Computer Science,
University of California,
Berkeley, California 94720,
U.S.A.

S.I. FELDMAN,
Bell Laboratories,
600 Mountain Avenue,
Murray Hill, New Jersey 07974,
U.S.A.

L.D. FOSDICK,
Department of Computer Science,   ·
Campus Box 430,
University of Colorado,
Boulder, Colorado 80309,
U.S.A.

P.A. FOX,
Bell Laboratories,
600 Mountain Avenue,
Murray Hill, New Jersey 07974,
U.S.A.

B. FORD,
NAG Central Office,
7 Banbury Road,
Oxford OX2 6NN,
U.K.

F.N. FRITSCH,
Mathematics and Statistics Section,
Lawrence Livermore Laboratory,
P.O. Box 808,
Livermore, California 94550,
U.S.A.

J. GARY,
Department of Computer Science,
Campus Box 430,
University of Colorado,
Boulder, Colorado 80309,
U.S.A.

W.M. GENTLEMAN,
Department of Computer Science,
University of Waterloo,
Waterloo, Ontario N2L 3Gl,
Canada.

S.J. HAMMARLING,
Division of Numerical Analysis
and Computing,
National Physical Laboratory,
Teddington,
Middlesex TW11 OLW,
U.K.

E.C. HEHNER,
Computing Science Department,
University of Toronto,
Toronto, Ontario,
Canada M5S 1A7.

P.W. HEMKER,
Mathematical Center,
Kruislaan 413,
1098 SJ Amsterdam,
The Netherlands.

M.A. HENNELL,
Computer Science Department,
University of Liverpool,
Liverpool L69 3BX,
U.K.

G.S. HODGSON,
NAG Central Office,
7 Banbury Road,
Oxford OX2 6NN,
U.K.

T.E. HULL,
Department of Computer Science,
University of Toronto,
Toronto, Ontario,
Canada M5S 1A7.

W. KAHAN,
Computer Science Division,
University of California,
Berkeley, California 94720,
U.S.A.

D. KAHANER,
Computer Science Division,
National Bureau of Standards,
Washington, DC 20234,
U.S.A.

P. KEMP,
Computing Laboratory,
University of Newcastle-upon-Tyne,
Claremont Tower,
Claremont Road,
Newcastle-upon-Tyne NE1 7RU,
U.K.

T.W. LAKE,
I.C.L.,
Lovelace Road,
Southern Industrial Estate,
Bracknell, Berkshire,
U.K.

C.L. LAWSON,
Jet Propulsion Laboratory,
Mail Stop 125/128,
4800 Oak Grove Drive,
Pasadena, California 91109,
U.S.A.

L.P. MEISSNER,
Lawrence Berkeley Laboratory,
Berkeley, California 94720,
U.S.A.

W. MILLER,
Department of Computer Science,
The University of Arizona,
Tucson, Arizona 85721,
U.S.A.

C.B. MOLER,
Department of Computer Science,
University of New Mexico,
Albuquerque, New Mexico 87131,
U.S.A.

D.T. MUXWORTHY,
Program Library Unit,
University of Edinburgh,
King's Buildings,
Mayfield Road,
Edinburgh EH9 3JZ,
U.K.

J.A. NELDER,
Rothamstead Experimental Station,
Harpenden,
Hertfordshire AL5 2JU,
U.K.

L.J. OSTERWEIL,
Department of Computer Science,
Campus Box 430,
University of Colorado,
Boulder, Colorado 80309,
U.S.A.

G. PAUL,
Department 441,
IBM T.J. Watson Research Center,
P.O. Box 218,
Yorktown Heights,
New York 10598,
U.S.A.

M. PAUL,
Technische Universität München,
Institut für Informatik,
Arcisstrasse 21,
D-8000 München 2,
Germany.

P. PEPPER,
Department of Computer Science,
Stanford University,
Stanford, California 94305,
U.S.A.
and
Institut für Informatik,
Technische Universität München,
Postfach 202420,
D-8000 München 2,
Germany.

J.K. REID,
Computer Science and Systems Division,
Atomic Energy Research Establishment,
Harwell, Oxfordshire OX11 ORA,
U.K.

C. REINSCH,
Department of Mathematics,
Technical University,
21 Arcisstrasse,
8000 München 2,
Germany.

J.R. RICE,
Division of Mathematical Sciences,
Purdue University,
West Lafayette,
Indiana 47907,
U.S.A.

F. RIS,
IBM T.J. Watson Research Center,
P.O. Box 218,
Yorktown Heights,
New York 10598,
U.S.A.

K.V. ROBERTS,
Culham Laboratory,
United Kingdom Atomic Energy Authority,
Abingdon, Oxfordshire OX14 3DB,
U.K.

O. ROUBINE,
CII Honeywell Bull,
68 Route de Versailles,
78430 Louveciennes,
France.

A.H. SAMEH,
Department of Computer Science,
University of Illinois at Urbana-Campaign,
Urbana, Illinois 61801,
U.S.A.

E.G. SCHLECHTENDAHL,
Institut für Reaktor-Entwicklung,
Kernforschungszentrum Karlsruhe,
7500 Karlsruhe,
Germany.

N.L. SCHRYER,
Bell Laboratories,
600 Mountain Avenue,
Murray Hill, New Jersey 07974,
U.S.A.

A.E. SEDGWICK,
Department of Computer Science,
University of Toronto,
Toronto, Ontario,
Canada M5S 1A7.

T. SMEDSAAS,
Department of Computer Science,
University of Uppsala,
Sturegatan 4B,
S752 23 Uppsala,
Sweden.

B.T. SMITH,
Applied Mathematics Division,
Argonne National Laboratory,
9700 South Cass Avenue,
Argonne, Illinois 60439,
U.S.A.

J.A.M. SNOEK,
University of Techn.
Delft,
The Netherlands.

H.J. STETTER,
Institut für Numerische Mathematik,
Technische Universität Wien,
A-1040 Wien,
Gusshausstrasse 27-29,
Austria.

P.N. SWARTZTRAUBER,
National Center for Atmospheric Research,
P.O. Box 3000,
Boulder, Colorado 80307.
U.S.A.

D.B. TAYLOR,
Regional Computing Center,
Edinburgh University,
King's Buildings,
Mayfield Road,
Edinburgh EH9 3JZ,
U.K.

M.J. TIENARI,
Department of Computer Science,
University of Helsinki,
Tukholmankatu 21,
Helsinki 25,
Finland.

C.G. van der LAAN,
Rekencentrum University of Groningen,
Postbus 800,
9700 AV Groningen,
The Netherlands.

A. van WIJNGAARDEN,
Mathematical Center,
Kruislaan 413,
1098 SJ Amsterdam,
The Netherlands.

J.L. WAGENER,
Amoco Production Company,
4502 East 41st Street,
P.O. Box 591,
Tulsa, Oklahoma 74102,
U.S.A.

M.B. WELLS,
Computer Science Department,
New Mexico State University,
Las Cruces, New Mexico 88003,
U.S.A.

T. WIBERG,
Department of Information Processing,
University of Umea,
Umea,
Sweden.

A. WILSON,
I.C.L.,
ICL House Annexe,
Brewhouse Street,
Putney,
London, SW15 1SW,
U.K.